Technology, Transgenics and a Practical Moral Code

The International Library of Ethics, Law and Technology

VOLUME 4

For further volumes:
http://www.springer.com/series/7761

Dennis R. Cooley

Technology, Transgenics and a Practical Moral Code

Springer

Dennis R. Cooley
P.O.Box 6050
Fargo ND 58108-6050
18 Putnam Hall
USA
dennis.cooley@ndsu.edu

ISSN 1875-0044 e-ISSN 1875-0036
ISBN 978-94-007-3059-5 e-ISBN 978-90-481-3021-4
DOI 10.1007/978-90-481-3021-4
Springer Dordrecht Heidelberg London New York

Printed on acid-free paper

Springer is part of Springer Science+Business Media (www.springer.com)

Contents

Introduction

Most philosophers still like to feel that they have a special subject matter, well insulated from anything that the social scientists, and scientists in general, have to tell them. That is not healthy for philosophy; and it is all too likely to lead to an ethics that continues, as of old, to plead for its ultimates-the fact that one is totally ineffectual being decently concealed by an impressive terminology. (Stevenson 1963, pp. 114–5)

Many so-called moral theories do not even attempt to explain or justify common morality but are used to generate guides to conduct intended to replace common morality. These proposed moral guides, those generated by all of the standard consequentialist, contractarian, and deontological theories, are far simpler than the common moral system and sometimes yield totally unacceptable answers to moral problems. Since these philosophers who put forward these theories have usually dismissed common morality as confused, they are completely unaware of the complexity involved in making moral decisions and judgments. It is not surprising that many who take morality seriously and try to apply it to real problems faced by actual people are so critical of moral theory. (Bernard Gert 1998, p. 6)

As both Stevenson and Gert note, ethics requires social and other sciences for by its very nature, ethics is a practical enterprise. In addition, the study of morality was intended to explain ethics to all thoughtful people so that each had the best tools to make decisions affecting not only themselves but the societies in which they live.

To ethics' detriment, those who study the field have sometimes attempted to sever the intimate connection between theory and application. Although morality's main goal has always been to produce useful results for society and its members, some academic ethics have become more and more limited to university preserves than being something all citizens can apply in their everyday lives.[1] The only common feature in the range of simple to complex, multi-tiered theories appears to be their implausibility to anyone other than their inventors and adherents to employ them in decision processes and defend the results.

The trend from usefulness to impractical abstraction has caused ethics as a discipline a degree of schizophrenia. Pure theorists tend to worry more about developing a clear and consistent theory, a set of normative principles, and a value theory than they do about whether any person in society will be able to apply their work. Their primary goal is to create a theory unassailable from all criticism, regardless of

[1] I am not claiming that no philosopher today is interested in the practical, only that many of the most influential ones seem to have misplaced the purpose of ethics.

whether the attack is reasonable. Applied ethical issues, such as stem cell research, bribery in the Developing World, and genetically modified or transgenic organisms, are considered to be philosophically unimportant and uninteresting. Since applied ethics' nature entails that purely rational arguments are impossible, many times it is dismissed with contempt (Callicott 1999, p. 28).[2] At best, it is considered to be a "soft" alternative to real philosophy. Making matters worse are a number of applied ethicists who have helped foster this opinion by appearing to know only the simplest versions of ethical theory and principles, e.g. equating Mill's nuanced normative principle with simplistic standard act-utilitarianism. Classes and research in applied philosophy are tolerated in many departments because the university demands them, but if some department members had their way, they would be eliminated for more sections of pure metaphysics and epistemology.[3]

So how did this sorry state of affairs come about? Over the last 2,500 years, the change from practical to abstract was slow and subtle. Although sexist and elitist – only wealthy men could rule in his view – Aristotle still understood that all citizens in every society need practical ethics in their lives. States are creations of nature, and since states are comprised of human persons, all persons are political animals (Aristotle 1941d, 1253a). Furthermore, each state is a community whose end is the highest good (Aristotle 1941d, 1252a). The best state is one ruled by the ethical principle of justice for "justice is the bond of men in states for the administration of justice, which is the determination of what is just, is the principle of order in political society" (Aristotle 1941d, 1253a). Without justice, there cannot be order. Without order, the best state is impossible. Hence, in order to achieve the highest good, citizens have to know how to be ethical. Ethics is not only practical; it *must* be practical so that the state and its citizens can survive.

Moving ahead to the 18th century, David Hume defends practical morality by rejecting attempts to create an ethics based upon pure reason alone. If moral principles could only be used correctly by abstract theorists, then few others would be able to make decisions with the actual principles. At best, they would be lucky guessers or have to make do with a simpler procedure method based upon the actual principle(s), which generally classifies correctly but is not guaranteed always to do so. The result is that no one other than pure reasoners could be certain about their duties (Hume 1948a, p. 177).

For Hume, however, true ethics is obvious and practical to most people. He states that the mental qualities we should pursue and foster are those that are useful to us and others. Anyone using the general sentiment all normal persons have can evaluate alternative actions and select one that is correct, while also knowing which things are good or bad and why they have that status (Hume 1948a, p. 251). Not only are all persons possessing the requisite emotional and reasoning capacities competent

---

[2] Social and political philosophy often faces the same problem.

[3] Some years ago at a university that will have to remain nameless, an interviewer was incredulous at my assertion that "real" philosophers could be interested in business ethics research. For those of us in applied ethics, this is an all too common occurrence.

of moral reasoning, ethics itself is not as difficult as many would have us believe. Where we go astray is by doing too much theory and too little social science. Hume states that "it seems a reasonable presumption that systems and hypothesis have perverted our natural understanding when a theory so simple and obvious could so long have escaped the most elaborate examination"(Hume 1948a, p. 249). In other words, the obfuscation problem lies with well meaning academicians who have made the practical impractical.

The Father of Modern Philosophy, Immanuel Kant, bears part of the blame for the severing of the practical from the theoretical. Unlike Hume, Kant rejects natural sentiments or desires as being part of morality's basis (Kant 1956, pp. 68–71; Callicott 1999, p. 102). According to Kant, right action in a particular set of circumstances is always what a purely rational person would do in that situation. In fact,

> All moral concepts have their seat and origin in reason completely a priori... In this purity of their origin is to be found their very worthiness to serve as supreme practical principles, and everything empirical added to them is just so much taken away from their genuine influence and from the absolute value of the corresponding actions. (Kant 1956, p. 79)

In other words, if morality's concepts were even partly the result of real world experiences, they would not be as good or as useful to those making difficult moral decisions. Moreover, the only way they can be fully understood is a priori, that is, through pure reason alone. Hence, those who can be truly moral are limited to the few individuals in society who have the resources, such as time and proper education, and an intellect capable of this type of reasoning.

Although Kant claims his theory to be practical, it is not. Kant's ethical framework creates a situation in which it is impossible to decide what to do because, as Hume correctly pointed out, reason alone cannot give any moral agent the power to make decisions. Suppose, for instance, a person is faced with an unlikely choice between the destruction of the world and the pricking of her finger. Reason tells us that many people would die if the first alternative is chosen, while little is lost if the second one occurs. If applying cost/benefit analysis to the situation, then the latter would be preferred over the latter. After all, a little pain for one person is much less a loss than the destruction of all things on the earth. However, if desire is not included in the decision process, the person trying to figure out what she is supposed to do will never be able to choose one thing over another. Even though she knows the outcomes of the two distinct alternatives, without being able to care about either one, she neither understands nor appreciates the differences. Much like Buridan's ass, she will be unable to do anything at all to her own peril. This absurd result shows that in order to choose, that agent must be motivated to choose, and motives by definition incorporate emotions or desires. Hence, emotions and desire are essential to ethics, which Kant's theory, because it so heavily depends on the theoretical, abstract reasoning of the purely rational person fails to understand.

Even John Stuart Mill, an advocate of women's suffrage, the elimination of slavery and other useful democratic ideas, produced an extremely impractical theory. Mill's consequentialism would seem initially to be more useful because it relies on observable results and is merely a formulation of the intuitively appealing prescription to do the best one can in all one's actions. According to Mill, "actions are right in proportion as they tend to promote happiness, wrong as they tend to produce the reverse to happiness. By happiness is intended pleasure, and the absence of pain; by unhappiness, pain, and the privation of pleasure" (Mill 1988b, p. 7). Since everyone knows what pleasure and pain is, and can do basic cost/benefit analysis, Mill's theory appears useful on its surface.

But relying on the unknowable consequences of actions destroys any theory's practicality. We should begin to worry about how useful Mill's theory is when he argues there are types of pleasure and pain that can be adequately evaluated only by those who have experienced both; thereby eliminating the input of those who have not experienced intellectual pleasures (Mill 1988b, pp. 10–11). Mill further obscures the principle by claiming that even those who have experienced both will not come to unanimous agreement on their value rankings. In fact, he writes the best result we can get is pleasure and pain types' values will be recognized by the majority. The majority will select high quality, intellectual pleasures over low quality physical pleasures (Mill 1988b, p. 12). What should concern us is that already, there is an elitism problem that precludes those lacking a higher education from being able to recognize true value, which might omit them from full membership in society's moral community.

Of even greater concern is Mill's claim that utilitarianism is not action guiding. In order to figure out what to do, we rely on rules-of-thumb, which generally maximize utility, but need not do so in all particular circumstances. Average citizens who are not philosophers are obliged to use these general principles as their best tools, while philosophers apply them only until they find better (Mill 1988b, p. 25). In other words, the majority of people use the rules to make their ethical decisions, while enlightened philosophers are able to understand and apply the actual moral code as it should be. Since most people will not be able to study ethics to the degree required to enable them to use the theory correctly, they must make do with second best rules of conduct. Once again, morality is beyond the grasp of the majority of thoughtful people.

The 20th century philosopher, R. M. Hare, continued the tradition of making ethics too abstract and theoretical. To deal with moral conflicts, Hare proposed that two levels of ethical reasoning exist. In the lower level, conflicts occur because people are applying principles too general and broad to guide their actions rather than reasoning at the consistent higher level. The higher second level excludes moral conflicts because there is always at least one clearly right action to those with enough cognitive capacity to grasp it. Hare characterizes the two thinker types as Archangels and Proles. An Archangel makes his decisions based upon pure critical reasoning incorporating all the moral elements of any situation confronting him. On the other hand, Proles "like most of us, have to rely on intuitions and sound prima facie principles and good dispositions for most of the time; he is totally incapable of critical thinking" (Hare 1992, pp. 44–5). In other words, most people using their rules-of-thumb can get common day morality right much of the time, but they do not actually grasp the true nature of morality as the Archangel does. Under Hare's theory, the difficulty for the Proles is they have to utilize clunky, ill fitting rules different from those of the Archangels. If moral principles were tools, then Proles use the side of a crescent wrench to hammer a nail into the wall, while the Archangels use hammers. Hence, in this theory, real morality is impossible for the very people it is intended to help.

The unreal abstraction of ethics' absurdity is characterized by a passage from Richard Taylor's memorial minutes. "[Taylor] marveled at how some philosophers could discuss seriously whether earthworms have souls but scoff at an examination of love and marriage" (Holmes 2004, p. 170). In the honorable pursuit of discovering reality's nature, many have lost sight of it, including why morality is needed, i.e., to help us lead good lives in the actual world. Abstract theory addressing non-existent worlds has replaced the common moral sense of what ethics is and why it is needed in our lives.

The impracticality of Hare's theory and others like it should give us pause. We can acknowledge that ethics might be difficult to do because it takes professionals to carefully examine and recognize many or most of the moral factors and principles involved, and then apply the principles correctly using both. However, there is no reason to assume that professionals do not use the same principles as everyone else to make their decisions. Instead of asserting multiple tiers of rules that are inaccessible to few other than the elite, the simpler hypothesis is that there is only one moral theory and code applying to all persons. Everyone's moral tools are identical in this view. The difference between professionals and nonprofessionals is how well each applies theory and code, not that they have different rules. One group has less skill at their exercise than the other, in much the same way a journeyman can utilize carpentry tools, but not as well as the master craftsperson.

The deficit in academic ethicists' outreach to a community demanding more ethics training and information has not gone unfilled. Applied ethics in medicine, business, research, technology and a large variety of other fields have seen rapid expansion, especially after the unethical, illegal activities of Enron, Worldcom, Adelphia, and other businesses.

The trend's unfortunate part is a number of those teaching, writing, or talking about these issues seem to know little ethical theory. Ethics is an extremely difficult subject area because it affects so many people in so many different, important ways. It requires master craftspeople to teach others how to become their own masters. Since academic philosophy departments have not been producing as many adequately trained applied ethicists as they should, many who style themselves as practical ethicists have been exposed to only shallow introductions to classical philosophers and their work. In fact, some applied ethicists have advanced degrees in related associated areas, such as medicine, that, do not provide the skills needed to examine an ethical issue fully. Without teachers fully cognizant of the field, students fail to obtain the required skills to make good decisions, but rather receive material characterized by its weak grasp of the underlying ideas. Instead of the richness of over 2,500 years of development by some of the best minds in humanity, thoughtful people are offered information or advice based upon one or two page characterizations of a person's lifetime work. The result is that too many applied ethicists are trying to teach on the cheap, which makes it impossible for thoughtful individuals to have the best tools they need to make good decisions for themselves and their communities.

Of course the problem's answer is to guarantee that applied and theoretical ethicists are fully trained both in theory and application. They will then have a balance allowing them to provide adequate tools for people to use in their lives. Fortunately, there are those such as Loretta Kopelman and Tristram Engelhardt, who fulfill this goal, but there need to be many more as ethics becomes more recognized as vital to a thriving society and integrated into professional and private lives.

Until ethics can be worked out completely and practically, the best course of action is to create a moral code that everyone can use in all her decision making, including those touching upon technology and transgenic organisms. In developing the practical moral code used in this book, I will incorporate not only the ideas of Stevenson and Gert beginning this introduction, but those from others such as Gary Comstock, Bernard Rollin, Lilly Marlene Russow, and others as well. Paul B. Thompson, for example, reminds us that regardless of their origins, the framework of applicability of ideas must be global (Thompson 1995, p. 13). Why? Because people all over the world have to use it to make decisions that affect other people from all over the world. J. Baird Callicott argues that philosophy – and I take it that he includes ethics under that umbrella term-should avoid being a socially irrelevant, academic ivory-tower endeavor by pursuing its original purpose as one of the most potent forces of social change (Callicott 1999, p. 27). New technology, for example, will have enormous impact on what our society will be, and we have to be able to understand and deal with the inherent changes in such a way that we leave society better off than it otherwise would have been.

As Gert pointed out, people utilize complex moral codes in their everyday lives, however, just because a moral code is used does not entail it is a simple matter to discover and explain it (Gert 1998, p. 6). This fact should not surprise us. Our brains are in use all the time that we are alive, but neuroscience is still in a relatively early stage of informing us about how brains actually function. Ethics is no different, although it does not lend itself to physical science the way the brain does. Neuroscientists can stimulate or suppress brain activity in certain sections, and then record the responses in a variety of ways, including but not limited to MRI's. Morality, on the other hand, is not a physical entity or reading. Rather it is a set of beliefs, rules, and emotions un-examinable in the same way "hard" science employs. Hence, although the following practical moral code might be true in that it captures one part of what ethics really are, it cannot be established in the same manner as showing scientifically the light is on in a room.

What I intend to do in this work is to capture as much of what people actually think about morality in the moral code as can be done without making the code inconsistent, such as classifying the same action as morally right and morally wrong. I will attempt to create and defend a practical moral code that can be used by any thoughtful person anywhere in the world to solve moral problems caused by technological advances in all areas of their lives. The code might have a vagueness problem because it will not classify all morally right actions, but it is guaranteed to provide a useful tool to discover and defend at least one morally right action in every situation.

Before beginning the difficult task of creating a practical moral code, I first want to introduce the topic of transgenic organisms, which will serve both to help develop the code and as a way to test its usefulness.

Genetically Modified Organisms (GMOs) and Transgenic Organisms (TOs)

Bioethics, technology ethics, and agricultural ethics are extremely diverse subject matters dealing with some fundamental needs of human beings, animals and the environment. The three disciplines cover everything pertinent to technological innovations in biology, technological progress, and more narrowly, the production of crops and animals generally used for food consumption by humans and other animals, as well as clothing, shelter, fuel, and a variety of other needs. In addition, there are overlaps in the ethical issues each area faces. Pesticides, transgenic organisms, bovine growth hormones, animals for food production and their treatment, and human/non-human chimeras are but a few of the moral controversies encountered by the three.

The main ethical controversy addressed in this work is technology and transgenic organisms. More specifically, the morality of the creation, production and marketing of transgenic organisms are examined in light of a practical moral code, which itself forms the larger share of this work.

But what are transgenic organisms? Roughly, TOs are created by splicing parts of one organism's DNA into that of another organism to produce desired traits that the recipient organism did not previously possess. Generally, the two organisms are from different species, and sometimes from different kingdoms, such as bacterium and corn. The traits are selected on the grounds of how they will improve the recipient's characteristics. For example, golden rice was developed using bacteria and daffodil genetic material spliced into the recipient rice's DNA. The resultant rice can provide consumers with a percentage of the daily Vitamin A needed to prevent blindness and death. What makes this such an intriguing discovery is that no variation in the rice species could produce any Vitamin A prior to Golden Rice's creation.

The term "transgenic organism" is used in place of the more popular, but misleading "genetically modified organisms" or "GMOs." This nomenclature justification is that the first term more accurately captures the process' essence than does the more general GMO. After all, humans have been modifying organisms ever since the former first had an impact on the environment. Animals have been domesticated, and bred to be more productive and easier to care for than their ancestors. Crops have been drastically altered to produce more of what humans need and want than were found in their non-artificial ancestors. For example, natural teosinte has considerably fewer kernels than does its modern corn descendants, which are the result of human organized breeding. To facilitate clarity, I will use "transgenic organisms" to help develop the practical moral code and to talk about bioethics, biofood, and making ethical decisions about new technology.

A Groundwork for Ethical Decision Making and Civil Discussion

Before analyzing the ethics of transgenic organisms in general, it is necessary to set the groundwork for fair, civil ethical discussions. Ethical debates, as all other types, require that certain parameters be scrupulously maintained by those discussing the issue at hand. All debates are supposed to be useful in that they are intended to advance the amount of information available, allow people to participate in the marketplace of ideas, and eventually to help find ethical solutions to problems. The solutions, of course, will be unlikely to be approved unanimously, but should be acceptable to the majority of reasonable people using reasonable decision procedures to come to reasonable beliefs. To achieve the ethical debates' ends, several guiding principles must be adopted by all involved, including but not limited to, the realization that reasonable people can reasonably disagree on issues and both still be correct in their beliefs. As long as the available evidence supports them, contradictory beliefs held by different people about what is ethical are legitimate because there is often more than one answer to a moral dilemma, such as to what is the best life or what should a person do in a particular situation.[4]

The Principle of Charity is designed to advance discussions as efficiently and respectfully as possible. The principle first requires all debaters to assume that every participant is a reasonable person who is trying to say something important to everyone else. Of course, sufficient evidence might arise to show the person need not be listened to, but merely not liking what he has to say is insufficient to justify disregarding his input. The principle's second step is to strengthen, if necessary, what the person said by adding more evidence, reformulating it, or otherwise improving it. Many people are unable to present their ideas and arguments as efficiently as the most skilled debaters. However, if we are serious about finding the best solutions to problems and respecting individuals as persons, then we are obligated to help them make their best case, even if we vehemently disagree with it. If we do not, we waste valuable resources having to debate weaker arguments, when better ones are available. Analogously, we would disarm a person with a handgun before we expend effort addressing a person with a pea-shooter. The charity principle is therefore practical. The principle of charity's third step is to evaluate the improved idea, position, or argument to find its strengths and weaknesses. In current conditions of political and social polarization, many people have forgotten that criticism is supposed to be useful. In order to fulfill its true purpose of advancing knowledge, both reparable and fatal flaws have to be identified in conjunction with the positive components. How else would someone know if a defect is so severe that it renders the argument unusable? The principle's fourth and final step is to put aside unreasonable bias and draw a conclusion based upon the evidence available. If the argument goes against

[4]All philosophers will be familiar with these principles.

one of our most cherished beliefs, then it is time to rethink our justification for holding that belief and all other beliefs, principles, rules, etc. the belief has supported or caused us to adopt. For example, in the fourth chapter, it will be argued that the eastern world's assumption that intrinsic value exists in all things is a better starting point for developing a theory of value than the western world's requirement to prove something is valuable in and of itself before it can be ascribed that worth.

The principle of charity is consistent with moral duties to create better societies. By implicitly demanding efficient discourse and respect for all individuals' input, the principle fosters the best atmosphere for problem solving. For example, instead of focusing on primarily emotion driven arguments, as happened when transgenics were attacked using the "Frankenstein myth" fears, the best real arguments can be developed for decision making (Rollin 2006, pp. 131, 135). In addition, even if a problem's solution is not one championed by an individual, he is more likely to understand and acquiesce to its implementation if he knows his view has been understood and adequately incorporated into the debate. Furthermore, by using the principle on his opponents' views, he is more likely to see that they are as reasonable as his. This perception will allow him and others to work together in solving problems facing their community better than if he had merely rejected his opposition's arguments as being the result of a stupid or unthinking mind.

Another requirement of adequate ethical debate is the old rule to answer questions of meaning before tackling questions of truth. That is, it is impossible to know if someone is saying something true or false before understanding what the person actually means by her statements. For example, going into a restaurant, one might hear the assertion, "That man is hot." However, it is not clear what the sentence means. It could mean what old timers, such as myself, would immediately think, viz. the speaker believes that the man feels the room's high temperature exceeds his comfort level. The statement could also mean that the man is attractive to the individual making the assertion. Without knowing which interpretation is accurate, it is impossible to state the truth value of the proposition.

Although this example is not important to ethical debates as such, it does illustrate the need to clearly define terms so that everyone understands what is being said. In many scientific and public debates people simply talk past each other (Zimdahl 2006, p. 13). That is, they use the same terms, but never realize that different definitions are intended. Since it is a waste of time for people not to understand what everyone else is saying when we are trying to find solutions to vital problems affecting society, it is never legitimate to assume that everyone knows what is meant by words such as "good," "bad," "right," "wrong," "permissible," "forbidden," and so on. These ideas can denote vastly different things at different times for different people. In order to have our ideas understood, the burden is on us to ensure we make the effort to explain how we define our terms. We cannot complain that someone has misunderstood us if we have not provided adequate information for her to use.

It would be helpful to see how the principle of charity and requirement for clear definitions of terms works in the transgenic organisms' context. There have already been heated discussions about what to label the result of mixing TOs with non-TOs which can serve as a case study.

Ethical Reasoning and Use of Terminology

In any relatively young[5] ethical debate over an emerging technology, the first problem to surmount is agreeing to the terminology used in the discussion. One of the transgenic nomenclature difficulties is agreeing upon what to call the mingling of transgenic and non-transgenic organisms, especially for seed crops. To date, no mutually satisfactory term has been adopted because of the seemingly intractable nature of the various concerned parties' positions. As a result, progress on the moral issues of the debate, such as whether or not transgenic organisms are morally bad in and of themselves, has been hindered. Before supporting my suggestions of "mixed" and "unmixed," I want first to mention and examine the current terms used by either side of the debate and show why they are inadequate.

Defective Terms

The test of an adequate term is how practical the term is for the circumstances in which it is used. More precisely, an adequate term captures, as much as it can be captured, the essence of what is being referred to, is readily understandable by those intimately involved in the debate -and hopefully the general public as well- and does not have an illicit positive or negative emotive impact on the listener.[6] For example, in order to advance the abortion debate, emotive and unrepresentative terms such as "babies" and "fetuses" have to be given up for a more neutral term such as "unborn." One benefit of the latter word is it requires debate participants to present their best arguments and evidence, instead of primarily making appeals to emotion. I will use the pragmatic test to evaluate the terms currently utilized by both sides of the controversy over transgenic organisms.

"Contamination" and "Pollution"

First, opponents of TOs have been employing the terms "contamination" and "pollution" to describe the state which results from the mixing together of transgenics and non-transgenics (Davies 2004, p. 71). [7,8] One of Greenpeace's background papers, for example, states that, "Spring planting in the Northern hemisphere has started, and with it numerous seed contamination scandals have broken loose in the United

---

[5]In comparison with issues such as racial equality and abortion, the transgenic organism debate is in its infancy.

[6]Defining terms and concepts using Conceptualistic Pragmatism will be discussed in more length in Chapter 3.

[7]For example, see the G.M.O. ALERT at http://www.organicvalley.com/member/forum-gmo1.html

[8]A Netscape search on the terms "transgenic" and "contamination" yielded 13,900 hits, and "transgenic" and "pollution" yielded 14,801 hits. Most of the sites were against transgenic organisms.

States...,Canada and in Europe" (Greenpeace 2007). Moreover, when responding to France's refusal to tear out 4,500 hectares of "polluted" corn, Dominique Voynet, French Minister for Land and Environment, said France should "show that she doesn't spread GMO contaminated seeds, whether authorized or not" (Walgate 2000, p. 1). Finally, on the Friends of the Earth Europe's website, the section headings include "Contamination in the field", "Contamination of our food", and "Contamination around the world" (FoEE 2006). Generally, when the opponents of transgenics employ words such as "contamination" and "pollution," they want to imply the negative, i.e., transgenics are contaminating or polluting the non-transgenics, rather than the reverse.

Both "contamination" and "pollution" should be rejected because they have immediate, negative emotive connotations, which may illicitly influence people to be biased against transgenics.[9] There seems to be something bad about an organism contaminating another, much like a paper mill willfully dumping untreated waste water into a pristine stream. What is emotively entailed by opponents' terms is that there is something morally bad in and of itself about TOs, even though no rational evidence has been provided to support such a conclusion.

Emotive terms tend to lower the evidentiary standard required for fair moral debate. Consider the impact the word "contamination" has on the following argument of David Vetter, an organic farmer from Nebraska.

> The bill for tests that revealed the contamination of [Vetter's] corn crop ran to $450, including a scan for StarLink that turned up negative, he says. It cost him $1,500 to evaluate a load of corn worth $4,000...At the very least, he says, Monsanto, Aventis and others in the biotechnology industry should pay these costs. (Schubert 2001)

If the transgenics are contaminating non-transgenics, then it follows in this argument that TOs are not pure or good. They wrongly destroy the "genetic integrity" of the organic seeds (Schubert 2001). In other words, there is something morally bad in and of itself about transgenic organisms. Of course, it is still an open question as to whether transgenics really are morally bad, but the emotive terms make the

[9]Many opponents to transgenic technology fall into the trap of using value laden language. Richard Hindmarsh and Geoffrey Lawrence, for example, state "And what about the public: does it have a say about the genetic manipulation of life, or is this just a scientific enterprise that we should leave to the elites?" (Hindmarsh and Lawrence 2004, p. 26) The last line introduces the idea of class warfare that is unsupported by any evidence. John Gray uses Fidel Castro's advocacy of biotechnology and the poor regulation of nuclear weapons as indicative of biotechnology regulation to generate an argument from fear (Gray 2005, pp. 27, 30). Sonja Schmitz likens biotechnology to colonization and invasion (Schmitz 2005, p. 59). Although it might be convincing to people who already believe in this type of conspiracy, it will do nothing but stop those who the authors most need to convince from listening to them. Proponents of transgenics have also been known to use the same tactic. Ronald Bailey states that the actions and statements of Vandana Shiva, Mae-Wan Ho, and Benedikt Haerlin of Greenpeace's European anti-biotech campaign are part of their disdain for the poor (Bailey 2002, pp. 34–8). The problem is that the instances cited do not establish these three people have any such feeling. Making matters worse is that by focusing on extreme positions, Bailey does not address many thoughtful people's real concerns.

less-than-careful reader much more likely to accept Vetter's conclusions about the existence of an injury.

Furthermore, Vetter's assertion that the seed companies should pay for testing is emotively, not rationally, supported by the terminology he employs. On the average, people do not like polluters or those who contaminate what was once pure. If a company enables its bad seeds to pollute others' good seeds, then most people would agree those who suffer from the corruption must be compensated by the polluter. Hence, many people would be inclined to agree with Mr. Vetter's demand for reimbursement based solely on the grounds of his labeling the seed companies as polluters.

However, the use of "contamination" and "pollution" begs three central questions. First, are transgenics morally bad in and of themselves or less valuable than non-transgenics? Second, is the mixture of transgenics and non-transgenics bad, and do transgenics actually pollute the non-transgenics or is it the reverse? Third, if the mixture is bad, then who should bear the costs of testing, maintaining separation of TOs and non-TOs, or compensation for mixing the two types of seeds, and why? Rather than merely taking the expedient route of appealing to the listener's non-rationality by utilizing emotionally charged terms, what is required to further the debate, and possibly reach some sort of consensus, is to answer the begged questions, and then provide adequate justification for the responses, as will be done in Chapters 2, 4 and 5. If it is shown that transgenics actually are detrimental in all the relevant ways, which I do not believe they can be, then they may be labeled as contaminants -but not before. Hence, until use of the terms is adequately justified, the emotionally charged terms "pollution" and "contamination" should not be employed.

"Adventitious Commingling"

From looking at available sources, it is clear that the terms, such as "adventitious commingling" and "adventitious presence," employed by those who favor transgenics are not as popular as "contamination" or "pollution." For example, a simple Netscape search on "GMO" and "adventitious commingling" only produced 64 hits, while "GMO" and "pollution" yielded 11,902. However, since the agencies employing the former are powerful policy makers, including The European Parliament and the United States Department of Agriculture which heavily influence the debate, their terms should be evaluated for their usefulness.[10]

While "adventitious commingling" does not appeal immediately to emotions as do those from the other side, the term is defective on two other grounds. First, "adventitious commingling" is a mastery of confusion for many people. In order

10For examples, see the; the meeting summary of the Fifth Plenary Meeting of the Advisory Committee of Agricultural Biotechnology at http://www.usda.gov/agencies/biotech/acab/meetings/mtg_8-01/su..., and FY03-NDSU Extension Service Program #203-Cropping Systems in the 21st Century at http://www.ext.nodak.edu/progplan/203%20FY%2003.htm

to understand what adventitious commingling is, even those most knowledgeable about the debate have been forced to turn to a dictionary. Although requiring people to use a dictionary is not wrong in and of itself, it does show a certain rhetorical clumsiness on the part of those who coined the term. Due to the expression's relative obscurity, many members of the public probably would not bother to learn more about the issue's complexities because they feel the debate is beyond their grasp. If our goal is to usefully educate citizens and advance the debate on TOs, then it is our duty to provide clear and understandable terms and arguments for them.

A more important defect is the fact that "adventitious commingling" fails to capture the referent's essence. "Adventitious" means unintentional or accidental, which renders the whole term unable to perform the function for which it was intended. By explicitly limiting it to incorporate only unintentional commingling, all intentional or negligent commingling has been excluded. Perhaps, for example, a farmer or grain elevator operator *intends* to commingle the two organism types because he is tired of what he believes is organic farmers' unjustified complaining. The result cannot be an adventitious commingling, even though the resulting state is identical to the one which would have arisen if the commingling had been unintentional. In order to refer to the second type of commingling, we will have to make matters needlessly more complex by coining the term "intentional commingling."

Moreover, negligent commingling, which also results in the same states of affairs as the unintentional and intentional, would not be an adventitious commingling. Suppose there is a lazy grain elevator operator, who inadequately cleans his elevators and machines, even though it is foreseeable there will be mixing of the different seed types. The result, according to the definition, is not an adventitious commingling, but negligent commingling. Now we have three terms referring to the same resulting state of affairs, *viz.* the mixture of TOs with non-TOs, when one would better serve the interest to include more of the public in the ethical discussion.

Furthermore, the real dissimilarity between intentional, unintentional, and negligent commingling is an unhelpful moral difference rather than a descriptive one. When we talk about intentional, negligent, or unintentional, then we are actually focusing on moral responsibility, the morality of an action, situation, or something similar rather than the mere fact that two types of things have mixed together. Unintentional may mean no culpability on the part of the agent, for example, while intentional mixing might entail the agent is responsible for the results of his action. However, what matters in the ethical debate over TOs is the fact that transgenics and non-transgenics have been mixed, not what the mental states of the person or persons who did the mixing were. The latter is a different moral issue which should be addressed on it own.

For the sake of usefulness, since all we want to do is to talk about the states of affairs in which transgenic and non-transgenic products are mixed, we should use only one, non-emotive, publically accessible term rather than three.[11]

[11] On the same grounds, "unintentional presence" must be rejected as inadequate. A second sufficient ground for rejection is the fact that the term seems to be able to equally refer to a mugger on a dark street or any person we do not want to meet.

Unmixed and Mixed

In place of the less useful terms, my suggestion is to adopt the perfectly workable adjectives of "mixed" and "unmixed" to describe the state of mixed transgenic and non-transgenic organisms and the states of unmixed transgenic or non-transgenic organisms, respectively.

My suggested words satisfy the three conditions for an adequate term from above, *viz.* they are emotionally neutral, understandable and accurate. First, the terms have no illicit emotive impact. If a group of seeds is unmixed, it merely means that the seeds are all of the same type, regardless of whether or not they are non-transgenic or transgenic. Seeds being mixed entails solely that there are at least two types of seeds in the group. In addition, neither "mixed" nor "unmixed" implies a mixture or uniformity of seed is somehow better or worse on mere emotive grounds.

One beneficial result of using emotively neutral terms is that people who oppose transgenics or non-transgenics must now focus on better arguments to justify their positions, rather than relying too heavily on the listener's emotion. For instance, a person must present a proof showing that the existence of transgenics is bad, instead of letting the labeling of it as a contaminant perform too much work. As the arguments are developed, it will become clearer that many of the assumptions made so far in the debate are unjustified, which will call for a re-evaluation of beliefs. For example, by eliminating "adventitious commingling" from the discussion, then the assumption which some make that mixing transgenic and non-transgenic organism is innocuous would not be accepted as a given. Proof is required that any thoughtful person can understand. Perhaps, in the long term, what now seems to be a fight between individuals with intractable positions will progress into a consensus-at least on some points such as on terminology.[12]

Since the suggested words are readily understandable, the second condition of an adequate term is satisfied. Mixed is a combination of different types of things, while unmixed means all the parts are homogenous, for example. Anyone conversant with the transgenic debate, as well as the general public, will be able to immediately understand the terms.

Finally, by capturing the essence of the referent, the suggested terms satisfy the third condition of an adequate expression. In order to be as clear as possible so that everyone understands what claims and arguments are being made, we are trying to find terms accurately describing only the states of affairs in which transgenic organisms are mixed with non-transgenics. "Mixed seeds" means there is a mixture of the two seed types, while "unmixed seeds" merely means there is only one type of seed in the group. If organic farmers want to maintain purity, for example, then they want to keep their seeds from mingling with transgenic or non-organic seeds. If proponents of TOs are not concerned with mixed seeds, then they are not concerned with

---

[12]From my experiences at workshops and focus groups, the first likely consensus will be that transgenics are not morally bad in and of themselves.

a commingling of seed types, although they might be worried about the percentage of seed types in the mixture.

Furthermore, there are two benefits in making the mental states of the agents who mingled the two types of seed irrelevant to the terms. First, the referent is more accurately captured by the term. Since all that was under discussion was the actual state of affairs of the mixed seeds, it does not matter what the agent intended, should have foreseen, or did not intend. Second, the terms are more efficient. We are only concerned at this stage of the debate with the morality of the mixture of the two types of seeds, not the different issue of the morality of the agent who did the mixing. By making the mental states of the agents irrelevant, we are freed from the duty of actually discovering whether the mixing was unintentional or not, which allows us to focus solely on the relevant matter, *viz.*, the mixed seed. Hence, for both of these two reasons "mixed seed" is a far better term to use than "adventitious commingling."

From this point onwards, the same care needs to be taken with any word or term used to discuss transgenic organisms or any type of technology. It is only when this is done that the public will be able to solve its problems efficaciously and avoid pointless strife.

A Very Brief Book Overview

The purpose of this work is twofold. First, some of the most important legal and moral issues in the transgenic organism debate, such as labeling, market concerns, and trade agreements, will be addressed and some solutions formulated and defended. Second, and more importantly, a practical moral code will be developed for use in transgenic debates and for any controversial issue facing society, especially for technology. The code is not intended to nor can it identify all morally right and wrong actions facing individuals, but it can find at least one morally right alternative in any situation. Moreover, it will be based upon how people actually do their moral reasoning.

The first chapter begins the work of identifying moral principles and values people from all walks of life have used to make their ethical decisions. The groundwork is based upon sociological principles and USDA sponsored surveys as well as five of the most influential moral codes some professional groups have adopted, including the National Commission that wrote the Belmont Report, and four professional organizations. The first chapter also starts turning the raw data and ideas into a consistent decision making procedure with consistent moral principles. The second chapter develops and refines the code's – the Practical Moral Code (PMC) – two normative principles – Reasonable Person Utilitarianism (RPU) and a Quasi-Categorical Imperative (QCI). Chapters 3 and 4 develop PMC's complex, hierarchical axiology to be used in transgenic and other technology debates. Much of the disagreement in various moral issues does not stem from the use of different normative principles of right or wrong action but from the values people attribute to various things, e.g. people, animals, plants, and the environment. If we are clearer on both principles and values, then the controversies are not guaranteed to vanish, or at least, be more manageable. However, those engaged in the debate should be able to understand and respect other viewpoints better, and perhaps, know more about their own. To develop an axiology, I will first reject the simplistic definitions of the unnatural and the misuse of evolution, and then in Chapter 4, formulate a complex, hierarchical value system. The fifth and final chapter provides an overview of some of the arguments for and against transgenic organisms and the first applied uses of PMC. Included in the final chapter are the applied moral issues of whether to create transgenics, labeling, traceability, and market access levels. These four issues I contend are among the most contentious in the transgenic debate, but reasonable solutions that every thoughtful person can appreciate can be found. That is, the conclusions I derive will be acceptable to some and anathema to others, but every rational person on both sides of the debate should, at bare minimum, understand why another reasonable person can justifiably arrive at those conclusions.

Chapter 1
Applied Groundwork for a Practical Moral Code

1.1 Introduction

We should take seriously Gert and Stevenson's claim that philosophers in blind pursuit of their theories and principles are at risk of becoming ineffectual. For example, many ethics articles are too theoretically academic rather than practical. Generally, the author chooses a controversial moral topic, and then applies one or more principles to which she is particularly drawn. As a result, the issue is evaluated in light of the moral codes that academics, for the most part, have adopted, rather than those the people more directly involved in the situation would use. Although the articles tend to be well written, they carry little weight where people are making real world decisions about what to do.

A practical moral code based upon the work of ethicists, philosophers, sociologists, and other social scientists, as well as what thoughtful people believe in and apply in their lives is a necessity for three important reasons. First, it will be practical in legitimate decision making that is understandable and justifiable to all reasonable people. Second, it will be something people are more likely to use. Third, because of its universal acceptability, it can help lead to social decisions and a better society. I will address each below.

Moral codes based on how ethics is actually done by individuals and communities must be practical because they are intended for applied individual decision making. That is, people use them to figure out what actions they are morally required to perform or refrain from doing, what type of people they should be, what thoughts they should have and so on. Impractical theories and principles such as many forms of consequentialism are impossible for people to utilize. If they are unsure of how to apply the theory or principle correctly, then it does not help them to choose, much less justify their choices to others.

Impractical codes also pose a danger to individuals and society. If agents cannot know with some level of acceptable certainty what is right or wrong in real world

---

An earlier version of the practical moral code based on USDA survey results appeared in Gary Goreham, George Youngs and my "Practical Moral Codes in the Transgenic Organism Debate." It has been extensively updated since that time.

D.R. Cooley, *Technology, Transgenics and a Practical Moral Code*, The International Library of Ethics, Law and Technology 4, DOI 10.1007/978-90-481-3021-4_1,

situations, then they might begin asking if there is anything to know in the first place? In other words, without being able to recognize our actual duties, which actions are right or wrong, what it is to be a good person, then why should we assume there are any moral truths of any kind? It is a fundamental fact that rational belief in a proposition requires some sort of adequate evidence; otherwise belief in something's existence seems to be merely an article of faith, or a non-rational or irrational belief. If someone proposes an impractical moral code that does not allow people to know what to do, then he will be unable to prove rationally that the moral code classifies one way or another in any situation. If we want to establish that there is a morally right path in any particular situation, then we must accept that codes producing unknowable classifications must not be impractical.

Second, moral codes should be practical because moral debates need resolution for their own sake and those of the community. Practical codes allow us to use the moral ideals and reasoning processes all of us share universally to discuss the issue and try to find consensus at least an understanding that reasonable people can reasonably disagree will help people live their lives well and create a better society. Issues such as abortion, euthanasia, technology and transgenic organisms cause a great deal of strife in people's lives, groups, and communities. After rigorous discussion, there must be solutions that can heal and move us to new issues needing the community members' attention and energies, otherwise nothing will get done and the community and its citizens cannot pursue their highest good of flourishing.

Mark Sagoff's work on local civic engagement in environmental problem solving as illustrated in his Quincy Library case study serves as one justification for why a "bottom up" rather than "top down" approach is more appropriate in developing both an axiology and practical moral code. In Quincy, California, there was a dispute between environmentalists, local officials, and the timber industry over the dispensation of three national forests. A desperate deadlocked situation resulted in which each local faction was fruitlessly expending its resources while tearing apart the community. Resolution through consensus was achieved only when local citizens realized how bad things had become, and then worked together without outside provocateurs to find an ethical solution to their dilemma. The democratic result was a plan in which each side sacrificed part of its goals at the same time its primary desire was satisfied. Using arguments and this example, Sagoff proves that a democratic approach to problem resolution, especially if those engaged are the same individuals who will be directly affected by the decision, is more practical than having outside agents/experts intervene (Sagoff 2004, Chapter 9). In fact, when outside factions, such as the Forest Service, environmental groups, and logging businesses interfered in Quincy's process to pursue their own interests, many of which were benefitted by keeping the problem alive, the community plan was derailed (Sagoff 2004, pp. 225–7). Although external agents and entities are not always harmful, the damage and distraction they can cause should make us look for more practical, local venues through which people solve their own problems using moral codes based upon their universal and local ethical beliefs, values, and principles.

Many bioethicists promulgate a democratic decision procedure for technology issues (Korthals 2002 and 2004; Light 1996 and 2002; Rollin 1996; Thompson 2007). Rollin argues that governments have an obligation to poll the public for any risks it perceives, no matter how farfetched the concerns are, which are then addressed by scientists in terms any layperson can comprehend (Rollin 1996, p. 93). Michiel Korthals' applicationism or deliberative ethics, which seems very similar to Andrew Light's methodological pragmatism, requires that "all sorts of consultation between consumers and producers (and other parties, such as governments)" (Korthals 2004, p. 52). The justification for deliberative ethics is the fact that consumers and producers are manifestly unable to acquire adequate information or power necessary to protect autonomously their own interests, while there is a dearth of entities that will do a proper job of it for them. From being involved in the various decision processes, consumers can guide technology to a state more conducive to their well-being. By adopting Korthals' approach, two chasms can be bridged. The first is the epistemological distance that has grown between consumers and producers as food production became more specialized and divorced from the everyday experiences of most consumers (Ibid., p. 153). In developed societies, very small numbers of the populace grow or raise the foodstuffs used by the rest; hence, both have a poor understanding of what the others do or think. Deliberative ethics overcomes the estrangement by integrating each group into decision making bodies. The second chasm is a schizophrenic division between consumers and citizens (Ibid., p. 155). Consumers are defined as individuals active in markets who act according to their personal preferences, while citizens are individuals participating in the political arena who act with others according to their common preferences (Ibid.). Obviously, these disparate foci are incompatible unless sensitively overcome. Deliberative ethics empowers consumers by allowing them to have greater control in the market besides merely whether or not to buy a product that is produced by what can sometimes be viewed as Others.

Deliberative ethics is one aspect of a current revival of pragmatism in technology and ethics circles. Instead of being concerned about the end results of ethical decision making, pragmatic ethics focuses upon the decision making itself. In other words, it is "more process than product oriented" (Keulartz et al. 2002, p. 15). Deliberative ethics ensures that all affected stakeholders have input in the final result, whatever it may be. Pragmatic ethics also involves "substantive interventions" that break stagnation and entrenchment in ethical debates by creating new moral vocabularies, bringing novel perspectives or ideas, or otherwise moving the process forward through the introduction of some new element into the debate (Ibid., pp. 15–16). In a partial justification for his methodological pragmatism, Andrew Light claims that one must be a pragmatist in order to function well in bioethics – and by extension – technoethics. The reason why is based upon the assertions that:

1. bioethics is a social activity,
2. the value of ideas is ultimately weighed in terms of their value in practice,
3. the reliance in all approaches on past experience, and
4. bioethics is always aimed ultimately at influencing policy (Light 2002, p. 85).

Using principles or codes that fail to recognize these fundamental facts result in possibly impressive abstract arguments and conclusions, but does not guarantee that any problems will be solved in the real world (Gremmen 2002, p. 101). Laying the groundwork for one way to discover defensible, reasonable solutions to actual problems – the next chapter's PMC – is what this work is striving to achieve.

Although I am not at the moment, nor might ever be, ready to adopt a form of academic pragmatism, it is important to acknowledge its usefulness in various ways. First, it focuses on solutions to problems using what is real in the world rather than taking flights of philosophical fancy to complex, non-existent realms. Second, it frames ethical discussions in more inclusive terms than merely having battles between two or more conflicting moral principles. "The moral arguments that are used [in academia] belong to a specific practice and may lead to problems in understanding members of other practices" (Gremmen 2002, p. 101). The major disagreements between those opposed or in favor of transgenic organisms, for example, organic and transgenic producers, can be understood in terms of how each group practices animal husbandry or farming. Once the various involved practices are appreciated, then it will be easier to obtain some form of consensus on fitting solutions. These two benefits of pragmatism are considerable and cannot be readily dismissed or minimized.

There are several severe drawbacks to pragmatism that will give anyone pause. First, without some sort of fundamental principle to evaluate a practice's moral legitimacy, then how can one practice be set aside in favor of another? The issue is not generally clear in the pragmatists' writings for the examples they use to support pragmatism tend to be all morally legitimate. Gremmen talks about clashing views over how horses are treated as a conflict between practices people have for pets and those they have for natural resource management (Gremmen 2002, p. 101). Either practice is acceptable to reasonable people because we as reasonable people allow either in our societies. Since both views are morally permissible to hold, it does not ethically matter which one a person adopts.

However, in more difficult dilemma cases, the inherent defect with pragmatism becomes apparent. Suppose the incompatible practices have different moral statuses. In the case of societies that condone women's abuse, reasonable people want to adopt an ethical practice, and reject and condemn an unethical practice, but there is neither incentive nor justification for doing so. For example, in May 2008, 15 women in western Kenya were burned to death by a rampaging mob on the grounds that the women were witches. Although the conflict between the mob's superstition based practice and the evidence and equality practices of any reasonable person should be resolved in favor of the latter, with pragmatism, there is no mechanism that can be used to support such a decision as there is in consequentialism, Kantianism, or another Realist theory. They are merely conflicts of practices. The result is rampant relativism unless pragmatism can go against its central foundation and incorporate some essential principle that allows for conflict resolutions for these situations.

Of more practical concern is how are we to do what those who want a more democratic or pragmatic approach to decision making want us to do without becoming too counter-intuitive? Decisions about technology development and deployment affect many different people in many different communities; thereby, making each person a stakeholder. If everyone must be consulted for their concerns and have each of them addressed by the proper authorities, no matter how unlikely these consequences are, then a great deal of technological innovation will be delayed if not prevented by the task's sheer magnitude.[1] No company would ever develop new technology because it would be unlikely to have a net profit. Second, there is the danger posed by the tyranny of the majority. If all stakeholders have to be consulted, and the stakeholders are uniformly technologically and scientifically ignorant, then they will stifle innovation in thought and deed (de Tocqueville 1966, p. 235). Many times, the stakeholders as a whole are against innovations because it frightens them, they do not believe it is in their interests, or for some other irrational reason, assuming adequate evidence is being ignored by them. In these situations, those in the minority have no avenue of appeal since democracy is the final arbiter. But this cannot represent what ethics is all about. It offends us to think that minorities' interests have to be sacrificed to the majority's sometimes irrational will. In order to reign in tyranny, there must be some objective principle, such as justice and flourishing, that takes priority over democracy.

Given the unwieldy requirement to consult all affected stakeholders, it might be reasonable to limit information gathering to representatives of the populace as a whole. At the very least, the costs involved would decrease dramatically if some smaller set would be consulted.

This approach's benefits are enticing, but its difficulties arise when considering who should be selected to represent the stakeholders. If the stakeholders as a whole are ignorant of science, then should the representatives represent the majority or those know enough about science to make more informed decisions? If the latter, then they might not represent the majority of the society; hence, there is none of the empowerment desired by pragmatists and others. It is merely doing what a small group believes to be right or listening to what they think is important rather than allowing the community as a whole significantly to affect negotiations and decisions. Although it might be a bit better to make decisions this way than merely allowing scientists, producers or similarly conflicted parties do it, there still might be a bias toward science that the community as a whole rejects. If the more democratic representation is adopted, then decisions can be made based on ignorance rather than evidence. Given that governments are supposed to be pursuing the best interests of the society and its citizens by promoting and maintaining the flourishing of each, decisions of this type will tend to retard government efforts to achieve its goals. In addition, there is the problem of the tyranny of the majority here as well. The only change between asking all stakeholders and representatives who mirror the views of all the stakeholders is merely a difference in the group's size; not its practices, ideas, and beliefs. Hence the result will be the same.

[1] How would future generations, who will be impacted by the technology, be consulted?

A practical way out of this conundrum is for reasonable people to utilize a practical moral code such as the one developed in this work, which incorporates RPU and QCI or similar principles. RPU examines what is best overall, which will tend to be identical to society's interests, while QCI requires proper respect for all people involved. To fulfill the latter duty, those affected should be consulted as long as so doing satisfies PMC. Therefore, the benefits of pragmatism and democratic consultation can be achieved with PMC without simultaneously being required to take on pragmatism's drawbacks.

To begin developing the practical moral code for all forms of technology and ethics, the central focus is on the sets of theories and rules governing ethical conduct gleaned from surveys conducted under a United States Department of Agriculture grant for the study of the social, economic, and ethical impact of biotechnology, especially transgenic organisms. In depth, structured interviews were conducted with a range of individuals, including farmers, legislators, clergy, government officials, scientists, agriculture school administrators, and agribusiness officials. Their responses to the interviewer's questions not only indicate that versions of the standard moral principles of consequentialism, Kantianism, and justice are used in decision making procedures, but in certain cases, environmental ethics as well. Moreover, the principles employed in biotechnology discussions are general enough to be utilized by the human subjects for other moral issues as well.[2]

This chapter's structure is broken into six main sections. First, the definition of terms employed in the arguments and discussion are stipulated. Second, several professional ethical codes, including the Belmont Report's code for physiological and behavioral research on human subjects, are examined. They will later prove a valuable comparison to those adopted by the individuals involved in the transgenic organism debate. Third, the results from the USDA study are stated and formulated into the most plausible moral principles. Fourth, the most reasonable versions of transgenic moral codes are compiled, and then examined for internal and external consistency. Fifth, in an attempt to be as inclusive as practical, the Common Moral Code is devised from combining portions of the professional and survey participants' moral codes. Finally, it will be shown that although it provides insight into necessary moral factors in any adequate moral decision procedure, the Common Moral Code is evaluated and found wanting. The code's lack of clarity poses serious problems for interpreting it, much less employing it to help resolve moral dilemmas. In the end, it is obvious that if applied ethicists are concerned with settling controversies, such as the transgenic organism issue, then one of their tasks is to formulate a clear, practical moral code that appropriately incorporates the ethical principles people use in their moral decision procedures. However, before proceeding further, it is important to first stipulate definitions for the terminology that will be employed in the later discussions.

---

[2]Gary Goreham's "Ethical Perspectives on the Transgenic Organism Debate" unpublished PowerPoint presentation of research data at North Dakota State University, 2002: 1–10.

1.2 Definitions

Since the USDA survey's respondents used principles in their codes which rely upon more basic moral theories, it is vital to start with the latter and progress to the former. Moral theories are the most fundamental level of ethics. For the purposes of this work, a moral theory "provides an overall framework for specifying ethical norms and interpreting ethical concepts" (Shamoo and Resnik 2002, p. 12). For example, consequentialist theories are fundamentally about the consequences of actions, which can be spelled out in terms of doing the best one can or producing good and avoiding evil (Holmes 2003, p. 125). Therefore, all principles derived from the theory will evaluate only the consequences of actions, rather than the action's antecedents or the action itself. Furthermore, meta-ethical definitions of moral terms, such as good and evil, will be based upon the framework of consequentialism, e.g., evil is defined as pain for hedonic consequentialism.

On the second level of ethics are normative principles, which evaluate the morality of actions, people, or things (Shamoo and Resnik 2002, p. 12). Among the other possibilities, a moral principle could be a rule of right behavior. An action rule can incorporate the necessary or sufficient conditions for a right action. For instance, a utilitarian principle asserting that an action is morally right if and only if the action maximizes utility[3] states both the necessary and sufficient conditions of permissible actions. However, a moral principle can also be more limited in scope. The rule asserting that an action is morally right only if the action maximizes utility is just as much of a moral principle as the former. The only difference between the two is that the latter states a necessary requirement of morality rather than both necessary and sufficient conditions. Further moral principles evaluating goodness or badness apply to classifying mental states, individual characteristics, and other states of affairs.

Finally, the set of moral principles that a person uses in her decision making procedures is her moral code. Although it is possible that all of her principles fall under one moral theory, such as consequentialism, there is no reason that they must. It is perhaps likely that people use multiple moral theories to support their moral principles (Shamoo and Resnik 2002, p. 20). For example, a consequentialist moral theory can help justify the moral principle to maximize utility. Kantian moral theory can be used to support a principle to never treat anyone as a mere means. Justice moral theories validate justice principles, and so on. It is helpful to see how different groups of people developed their moral codes from several different moral theories.[4]

---

[3] Utility is defined as the result of subtracting the value of all of the evil consequences produced by an action from the value of all of the good consequences produced by the action.

[4] Audrey Richards' work on primitive human societies supports the idea that there are basic universal characteristics to ethical systems (Richards 1969, pp. 23–32).

1.3 Professional Moral Codes

The five selected professional moral codes are interesting to compare to the study participants' codes and the Combination Moral Code for two reasons. First, even though 23 of the 25 USDA survey participants do not have an upper level education equivalent to those of the professional society members, most of whom have the highest degree attainable in their field, when considered as a whole, they have adopted a very similar set of principles and ideas to those of the professionals. Second, it is intriguing to compare the two types of codes to see how the groups' different perspectives and value foci influenced their moral rule sets. Overall, the professional societies tend to be more precise than the USDA participants in their statement of ideas and principles, but the former do not include the more emotional components of ethics that the latter identify and incorporate into their codes. These two results entail that the six moral codes might have broad, universal features which groups and communities can use to evaluate the morality of technological developments and to help set public policy. I will begin with the professional society codes of behavior, and then proceed to the Belmont Report's set.

Each moral code of the four professional societies has elements of at least one of the three main theories or moral ideas used in most professional writing on ethics, *viz.*, utilitarianism, Kantianism, and justice, while two also include versions of virtue ethics. The elements derived from these four and the Belmont Report codes will be turned into the Professional Moral Code by which to compare the code created from the responses of the USDA participants.

Let us begin with two societies closely related to transgenic organisms and bioethics: the American Society of Agronomy (ASA) and the American Society for Biochemistry and Molecular Biology (ASBMB). ASA's Statement of Ethics is relatively short with six behavior prescriptions, mostly dealing with the professional conduct of work. However, the fourth rule states that members shall "Demonstrate social responsibility in scientific and professional practice, considering whom their scientific and professional activities benefit, and whom they neglect" (ASA 1992). Although the requirement is very broad, it does incorporate forms of Kantianism and social utilitarianism in ASA professional duties. A version of social justice can also be gleaned from the material, but the rule is a bit too vague to provide a strong case for such a claim.

The American Society for Biochemistry and Molecular Biology's code has more complexity and provides more depth than that of ASA. ASBMB asserts that professionals should have the "ultimate goal of advancing human welfare." Included in this duty are the sub-obligations to promote and follow practices that enhance the public interest or well-being and comply with government and institutional regulations, "such as those ensuring the welfare of human subjects, the comfort and humane treatment of animal subjects, and the protection of the environment" (ASBMB 1998). Although it appears as if the code is heavily dependent on utilitarian ideas and principles, the professional duties to trainees require the former to respect the vulnerable population by being good mentors and creating and maintaining a working environment that encourages cultural diversity.

Two other groups involved heavily in technology, the Association for Computing Machinery (ACM) and a consortium of the US Department of Commerce, Russian Chamber of Commerce and Industry, and US–Russia Business Development Committee, have more complex codes than either ASA or ASBMB but still share common features with them. The consortium's *Basic Guidelines for Codes of Business Conduct* requires business members to respect the partners and participants in a shared business venture, as well as minority shareholders' rights. In addition, members are expected to live up to the trust placed in them and "endeavor to earn a reputation for integrity, competency, and excellence." The consortium not only advocates the utilitarianism of a capitalistic economic system, but in pursuing profit, every buyer and seller should simultaneously respect those affected by their actions, especially if a trust relationship has been developed. What might be most interesting is the virtue ethics component that deals with characteristics of the person rather than her actions. Not only should she do the right thing, she must also acquire certain beneficial character traits.

The ACM, a computer science professional organization, has general moral imperatives to curry and maintain certain virtues, and well as acting in a useful, socially just way. The relevant portion of the group's ethical code is below.

1.1 Contribute to society and human well being including a local and global safe natural environment, protect human rights, respect cultural diversity, and meet social needs. Minimize negative consequences from technology.
1.2 Avoid harm to those affected by the technology. Recognize that well-intended actions might lead to unexpected harm, which professionals have an obligation to minimize or mitigate.
1.3 Be honest and trustworthy.
1.4 Be fair and take action not to discriminate.
1.5 Honor property rights. . . .
1.7 Respect the privacy of others.

In addition, computer professionals are obligated to treat their employees in certain ways including instituting and maintaining non-discriminatory practices, committing to worker health and safety, creating effective systems for consultation with employees on employment conditions and other issues affecting employees, and providing forthright information and policies affecting workers. The global community is also considered in the ACM document, unlike the other three codes. Professionals must be sensitive to the local populations' concerns, have good communications with them, abide by all applicable environmental laws and regulations, and show tolerance for people of other cultures, races, beliefs, and countries. Although utilitarianism is more than hinted at in this code, the main moral idea seems to be the Kantian respect for human beings and some social justice principle of those with power and resources being responsible for caring for those lacking one or more.

From these codes, several interesting conclusions can be drawn. Each has an element of utilitarianism, generally directed at advancing human welfare. Second, respect for persons is clear in the ACM, ASBMB, and consortium's rules, while

ASA's mandate to determine who is benefited and neglected by research and how to treat others in the field establishes it has a version of Kantianism. There is also a virtue ethics in both the consortium and ASBMB's documents. Not only are people supposed to act in certain ways, they are obligated to acquire specific virtues or beneficial dispositions to act in certain ways. Finally, both ASBMB and ACM have conduct rules regarding protecting or sustaining the environment and the creatures living in it. Since these elements have been found to be practical to the professional societies and others, they will be part of the Professional Moral Code developed using the Belmont Report's central material.

The Belmont Report was chosen as the primary moral code for two reasons. First, the Belmont Report's code has moved beyond the narrower realm of medical and behavioral research to become one of the fundamental codes in bioethics. Tom Beauchamp and Leroy Walters use their variations of the Report's three moral principles as "the beginnings of a framework through which we can reason about problems in bioethics" (Beauchamp and Walters 2003, p. 21). Adil Shamoo and David Resnik claim that principles identical to the Belmont Report's "are supported by more than one ethical theory and agree with commonsense intuitions" (Shamoo and Resnik 2002, p. 15). Finally, many government agencies in the United States make the Report part of their evaluation of research proposals and activities.[5] Hence, the practicality of any moral code developed without reference to them is in severe doubt.

Second, the Belmont Report's moral code incorporates three of the required elements of an adequate theory: treating all people as they deserve, utilitarianism, and justice (Rachels 2003, pp. 191–202). Gary Comstock thinks that the "high-level principles of justice, beneficence, and autonomy or utilitarian calculations of costs and benefits" are important enough to combine with narratives to form his theory in Life Science Ethics (Comstock 2002, p. 8; 2000b, pp. 7–8). Since the moral principles found in the Belmont Report's moral code have been recognized by leading theoretical and applied ethicists as necessary components for applications outside of medical and behavioral ethics, they provide ample groundwork for a moral code that combines the theoretical and applied ethics worlds' best.

The Belmont Report has what the National Commission, its creator, labels as three moral "principles" which govern research ethics: Beneficence, Respect for Persons, and Justice. Although the Report does not make Section 1's distinction between moral theory and moral principle, it is clear that such a division does exist in the document. Beneficence is actually the moral theory of consequentialism; Respect for Persons is Kantianism; and finally, Justice is Justice. The Commission first provides the framework for specifying ethical norms and interpreting concepts before stating the moral principles, which is a rational approach to take. If they had not laid the ethical theory groundwork, the moral principles would have been vaguer and more ambiguous than I will later claim them to be.

---

[5]United States Department of Agriculture, http://warp.nal.usda.gov/awic/legislat/nasa.htm and United States Food and Drug Administration, http://www.fda.gov/oc/ohrt/irbs/belmont.html

The Belmont Report's Beneficence theory supports a combination of two principles: non-maleficence and beneficence.

> Two general rules have been formulated as complementary expressions of beneficent actions in this sense: **[non-maleficence]** do not harm and **[beneficence]** maximize possible benefits and minimize possible harms. (NCPHSBBR 1979)

Although the Report claims the two consequentialist principles are harmonizing, it is obvious that there will be situations in which both principles cannot be satisfied. For example, sometimes, in order to maximize utility, it might be necessary to harm an evil person so that good people can benefit, as in the case of justified self-defense (Miller 2003). However, even though it is clear that the principles are not always complementary, using the Principle of Charity as a guiding interpretation principle, it will be assumed that if an action will satisfy both principles, then there is a very strong *prima facie* case that it is a moral action.

The Respect for Persons theory is Kantian in nature, with all that Kantianism entails.[6] As the Beneficence theory before it, the Respect for Persons supports two principles:

> The [moral theory] of respect for persons...divides into two separate moral requirements: the requirement to acknowledge autonomy and the requirement to protect those with diminished autonomy...To show a lack of respect for an autonomous agent is to repudiate that person's considered judgments, to deny an individual the freedom to act on those judgments, or to withhold information necessary to make a considered judgment, when there are no compelling reasons to do so. (NCPHSBBR 1979)

Given the fact the Belmont Report's authors use "moral requirement" in their explication, it follows that treating individuals as autonomous agents and protecting persons of diminished autonomy, if that situation arises, are necessary rather than sufficient features of moral actions. Hence, the two principles must be incorporated into the moral code that has other conditions as well (Miller 2003). The combination of all the requisites, including the two apiece from Beneficence and Respect for Persons, will be sufficient for morality.

Finally, the Justice theory in the Belmont Report is left underdeveloped, as happens in many cases in which people attempt to incorporate justice principles into their codes of ethics.[7] The National Commission states that:

> There are several widely accepted formulations of just ways to distribute burdens and benefits. Each formulation mentions some relevant property on the basis of which burdens and benefits should be distributed. These formulations are **(1)** to each person an equal share, **(2)** to each person according to individual need, **(3)** to each person according to individual effort, **(4)** to each person according to societal contribution, and **(5)** to each person according to merit. (NCPHSBBR 1979)

---

[6] See Kant's *Metaphysics of Morals* and *Groundwork of the Metaphysic of Morals*. My interpretation of what it means to respect persons as ends in themselves will be sketched out in the second chapter.

[7] See Shamoo and Resnik's *Responsible Conduct of Research*, p. 16.

Although justice as a moral theory seems a natural candidate for adoption, its complexity and principles makes it very difficult to integrate into a moral code. First, there are four different types of justice from which to choose: Retributive, Compensatory, Distributive, and Reward. Even if the type of justice is stipulated, there are further divisions possible, as in the Belmont Report's five justice rules, which incorporate elements of Egalitarianism, Socialism, Capitalism, and Libertarianism. The result is often one or more of the five principles will contradict one of the remaining principles (Miller 2003). For example, giving a person what he needs, according to Socialism, might not be what he is entitled to based on his individual effort, social contribution, or merit, according to Capitalism. Babies who require transplants have not contributed in any of the proper capitalistic ways; therefore, they should receive nothing. However, principle two states that the baby should receive according to his need which in these circumstances is great. In these moral dilemmas in which the Distributive Justice principles contradict each other, instead of providing the reader with a practical conflict resolution procedure, the Commission merely states that the principles are guides to right behavior.

In order to try to make the justice principles more useful in guiding moral agents, it is necessary again to use the Principle of Charity in interpreting this area of the Belmont Report's moral code. It will be stipulated that if we can satisfy all five principles with an action, then we *prima facie* ought to perform that action, provided that all other alternatives satisfy less than the five principles. Most of the time, however, in those situations in which we only can satisfy fewer than five principles, we are required to satisfy at least one of the five.

Gathering together all the strands spun from the data sources, a plausible Professional Moral Code for right action can be formulated. Included are the three elements – utility, respect for persons, and justice – that appear throughout the five codes.

1.3.1 Professional Moral Code (ProfMC)

An action, A, is morally right only if

1. A either produces no harm or maximizes utility, in those cases in which both cannot be done,
2. in doing A, the agent treats all individuals affected by the action as autonomous individuals, which entails that no agent's considered judgments are repudiated, no individual's freedom to act on those judgments is denied, and information necessary to make a considered judgment is not withheld, when there are no compelling reasons to do so,
3. in doing A, the agent protects those with diminished capacity if they are affected by her action,
4. A satisfies at least one of the justice principles provided that all five cannot be satisfied.

5. A protects or does not cause undue harm either to animals or the environment, and
6. A fosters or maintains virtuous characteristics in the agent of A, including but not limited to honesty, trustworthiness, integrity, competency, fairness, and respect for cultural diversity and everyone affected by the agent's actions.

Although satisfying all six conditions is necessary for right action, it does not follow that an action which does so is morally right. It might be the case in the particular situation that there are further requirements which must be met. However, that being said, it is safe to assume that in the majority of cases, satisfying the six in actuality is sufficient. For the purposes of developing a practical moral code, the weaker claim that meeting all six conditions makes an action *prima facie* morally right is stipulated.

1.4 USDA Study: Methodology and Results

Now that ProfMC has been identified, laying the groundwork for the moral codes of the USDA sponsored study can begin in earnest. The study was conducted to determine the ethical principles, guidelines, or ideas used by proponents and opponents of transgenic plant crops in North Dakota. Since the state's economy is heavily agricultural and particularly dependent on wheat production, there is considerable debate over the advisability of introducing transgenic wheat into the state. Thus, the bioethics of transgenic crops is a very salient, state-wide issue. Personal, in-depth interviews were used as the study method in order to determine the nominal categories of responses one would typically expect to find. The researchers did not attempt at this stage of the study to determine the relative percentages of the population who adhere to these nominal categories of responses.

1.4.1 Sample

Purposive and snowball sampling techniques were used to select respondents for the study. Four criteria guided the purposive sampling. First, the respondents were to be North Dakota residents. Second, respondents were sought who are knowledgeable about transgenic plants. The intent was not to seek data on the distribution of opinions across the general public, but to understand, in-depth, how opinion leaders constructed their opinions. Third, leaders were sought from a diverse array of professions to maximize exposure to different viewpoints and to examine the relationship, if any, between professional commitments and positions on transgenic plant crops. Respondents included conventional, biotech, and organic farmers, research scientists, clergy, business leaders, legislators, environmental organization leaders, and representatives of state regulatory agencies. Fourth, both supporters and opponents of transgenics within each of these occupational categories were sought. Interviews were conducted initially with notable individuals who met these four criteria. At each interview's conclusion, respondents were asked who they would recommend should also be interviewed, thus providing a "snowball" sample. This two-fold sampling method – purposive and snowball – resulted in 26 in-depth, structured interviews that lasted between one and two hours each.

1.4.2 Interview Instrument

The interview questions were organized into three sections: ethic guidelines regarding transgenic plant crops, agro-biotechnology as a social problem, and the potential for consensus among proponents and opponents of transgenic plant crops. Following are the questions used to determine the ethical guidelines used by the respondents: "What is your position on GMOs? What is your rationale for your position? What information do you use to support your position? How did you come to your position? What ethical principles have shaped your position? In what way do you see your position as the logical result of these ethical principles?"

1.4.3 Findings

The interview data can be succinctly broken down into the four moral theories and the dependent principles found below. Some of the rules show necessary or sufficient conditions for morally right actions. Others are rules of evaluation, which must be taken into account in classifications of actions or things, but are neither necessary nor sufficient conditions on their own. The evaluation principles merely state the weight of moral evidence in a situation (e.g., the more natural an object is the better the object is). In parentheses, next to each principle is listed whether or not the respondent thought the principle as a necessary or sufficient condition or as a way to evaluate the value of evidence, and if the respondent was for, against, or neutral toward transgenic organisms and their introduction into the market. In those instances in which more than one side of the debate used a principle, it is labeled as neutral to signify that any position in the debate can employ it.

The moral theories and their related principles can be gleaned from the following five ethical values that were held by those who favored or opposed transgenic organisms in commodity crops. These ethical values include environmental ethics/nature, respect for personal autonomy, social and economic distributive justice, beneficence, and nonmaleficence. Each will be described and illustrative quotations from the interviewees provided.

One problem with creating a plausible moral code out of two or more sufficient principles is that contradictions will arise between the rules, as has been seen in the Belmont Report's justice principle prior to the stipulation that at least one of the five principles had to be fulfilled, rather than requiring all of them to be satisfied at the same time. Inconsistencies occur when an action which is classified by one sufficient principle as morally right is classified by another rule as morally wrong.

In order to eliminate inconsistencies, the development of a practical moral code that avoids as many internal contradictions as possible is required. It is best to say that, unless clearly stated as a sufficient condition, each principle the respondents utilize is merely a necessary condition of moral behavior or weighs the situation's moral factors/evidence. In what follows, all of the necessary and evaluation principles will be incorporated into a moral code.

1A) Environmental Ethics/Nature

One ethical value that emerged from the responses was that living, working, and farming in harmony with nature, including the physical, biological processes of nature. The respondents described this theme as a deontological principle carrying a moral obligation. Other related concepts included creation, life, land, and the environment. These concepts were not always differentiated by the respondents, however some nuances were noted. For example, "land" carried economic or production implications, whereas concepts like "the environment," "environmental health," "natural resources," or "natural resource management" had implications of the human/nature connection.

Three variations of the nature ethic were noted. The first variation was the autonomy of nature above that of human domination and control. Some believed that attempts to dominate or control nature result in problems both for nature and for humans. This notion is exemplified by an organic farmer who said:

> When I look at our organic farm it's a constant balancing act. When there is a pest problem, we look at the system that nature uses and try to figure out why it's a problem for me. Nature just has this wonderful way of constantly finding ways to correct things. There is an inherent wisdom in nature; for lack of a better term I use 'wisdom.' There is an integrity to that system. For instance, when we go in and try to correct a problem with an external correction, often times we end up with bigger problems than what we started with.

Life is autonomous even above human domination and control. One farmer pointed out:

> We didn't author life. It's not ours to own. That's like saying I can own you. Can I own you? Certainly we can own seed. We can own the field that we are growing the seeds on, but we don't own the entire crop. We don't own the wider essence of that particular crop. But with transgenetics, when you genetically modify something and patent it, that genetic material can't be contained. It co-mingles with non-transgenetic crops. Then where does ownership end?

Furthermore, humans should not and cannot dominate nature. A legislator stated:

> You kind of get a healthy respect for what the way things are and the soils that are required to support crops and of nature's way. I think that's kind of led me to believe that this is something that ought not to be tampered with. Things are made this way for a reason, and granted, trying to impose their will and trying to improve their lives with a tampered-with nature is a little bit beyond the line.

A second variation was the inherent goodness and value of nature relative to that which is human-made. An organic farmer stated,

> Those [genetically modified] genes were not there before. They are manmade genetic combinations, and since they weren't in the natural environment before, I view that as contamination.[8]

The interviewees noted that diversity is good. A clergy person pointed out that:

> Nature abhors uniformity. Industrialization is based on uniformity. We don't know what kinds of long-term consequences we're going to have dealing with this biology. Europeans have this thing about Frankin-food. I'm not sure that I would. I don't have that same kind of fear, but what I do have is a concern about is what happens to our genetic pool. Will a GMO so dominate our genetic pool of a particular crop that we will essentially lose the genetic history of a particular crop that we will essentially lose the genetic history of that crop?

A third variation on the nature theme was land stewardship. A legislator said:

> I think land ethics are concerned here. Thinking about land as a long-term investment – not even thinking about it as an investment, but as a long-term part of your heritage.... The way we've been doing things all along has maintained the land.... I think for those GMOs that allow you to use less chemicals, till the land fewer times, and conserve moisture are positives.

The environmental ethics as expressed by some USDA survey participants bases morality in part on the intrinsic value of nature or "that morality is part of the natural order of things" (Holmes 2003, p. 92). Nature or the natural has value in itself, which is equal, if not superior, to that of a person. Hence, the theory is much like that of Kant's Respect for Persons, with the significant exception that nature plays the same role as that of persons in Kant's theory. It follows that, in order to do what is moral under environmental ethics, nature should be respected in as close as appropriate a way that any person is to be respected (Reiss and Straughan 2001, pp. 65–7). Furthermore, acting ethically requires acting in accord with nature or the natural order of things. Those actions which are incongruent to the natural order are wrong, while those in accord have prima facie moral justification for them. The principles that can be derived from the interviews are:

1B) Environmental Ethics/Nature Principles[9]

i. Autonomy of Nature: It is morally wrong to own or alter the fundamental parts of nature – e.g., DNA (Sufficient condition, Anti-transgenic).

---

[8]The participants who pursued an organic lifestyle would likely agree with Schmitz's contention that biotechnology is too concerned with generating products "whose sole purpose is to benefit and sustain industrial agriculture" (Schmitz 2005, p. 60).

[9]I have divided the participants into three groups: Anti-Transgenic, Neutral, and Pro-Transgenic. Those who are in the Anti-Transgenic group are generally opposed to the technology. Those in the Pro-Transgenic group are generally in favor of the technology. Finally, those in the Neutral group have no strong feeling either way.

ii. Inherent value of nature: The natural is always morally better than the artificial. The more artificial something is, the worse it is. The more natural something is, the better it is (Evaluation principle, Anti-transgenic).
iii. Stewardship of nature: An action is morally right only if the action assists or does not harm the flourishing of nature and the environment (Necessary condition, Neutral).

2A) Respect for Personal Autonomy

A second ethical value was personal autonomy. One variation on the personal autonomy ethic was freedom of choice. It was expressed by those who were concerned with farmers' freedom to choose (or the loss of the freedom). Farmers feared the loss of freedom of choice in that they may lose markets, their ability to save seeds from harvest for future replanting ("brown bagging") because pollen mingling renders them unusable, or their right to save seeds because corporations have patented the genes that mixed with their crops. A conventional (non-organic and non-biotech) noted:

> You can do what you want to do with it. You can plant it again next year or you sell it. You can go somewhere else and buy something else. It's a freedom that's taken away from a farmer when all of a sudden he can't make the choice on saving or seed or buying new seed. Often a farmer, there are plenty of farmers who buy new seed every year, our friends over just west of us do every year, they buy they think it keeps a stronger seed. But we've always had that choice. You save it or you buy it. So that choice is taken away...I guess it [developing one's own crop varieties and saving seed] is a sort of sense of freedom that it's yours. It's a natural freedom.

Another variation on the personal autonomy ethic involves individuals taking personal responsibility for their lives and their choices. It was described by the respondents both in terms of consumers' freedom of choice and in terms of farmers' planting decisions. A legislator stated:

> I feel very strongly that people have some responsibility for their own situation, for their own actions and that sort of thing. I know it may sound kind of goofy but you know, if you don't like the American food supply, go to a different store. I don't mean leave the country, but I think people make choices all the time.

Further,

> If farmers are really concerned about these national markets, we shouldn't have to pass a law that makes it illegal for them to use gene-spliced materials. It's their business. If they want to give away those markets that they have been putting money into building and they are not worried about it. . . . I sort of stand back and say, 'Hey if they all want to jump over a cliff, I can't hide all the ropes.'

The principles that can be derived from the interviewed supporting the moral theory of Kantianism or Respect for Personal Autonomy derived from these types of interviews are:

2B) Kantianism/Personal Autonomy Principles

i. Freedom of Choice: An action is morally right only if in doing the action, no individual's freedom of choice is unjustly limited (Necessary condition, Neutral).
ii. Responsibility: It is morally obligatory for an agent to take responsibility for the consequences of her actions, unless doing so presents an overwhelming burden for the agent. (Focus of this principle is on the obligation of the agent rather than on what makes actions right or wrong, Neutral)

3A) Social and Economic Distributive Justice

The third ethic was that of social and economic justice. Four variations emerged. The first variation was personal, social, and economic rights. It involved both rights, equity, and fairness for individuals as well as the loss of freedom they believe farmers and consumers may suffer as a result of the monopolistic actions of corporations. Illustrative of this variation was a comment made by a legislator:

> I think that no one should be able to own the rights to a natural process even if it is an altered natural process that's life. But I think farmers have had success in the U.S. and around the world in developing crops and growing crops and I think they should be able to retain the rights to maintain their seed and grow them again the next year without having to rely on chemicals or a certain monopoly to provide those seed to keep the farm growing, keep the food supply going.

Similarly, an organic farmer said:

> It's a question of how to you regulate an industry and balance everybody's rights in this.

A second variation on the social and economic justice theme was an equitable structure of agriculture. Respondents commented on the integration and consolidation of the agro-food system. Central to these systemic changes is the role that agribusiness corporations unjustly play in the ownership both of agricultural inputs and genetics. They believe these changes place family farms and consumer products at risk. A clergy person said:

> If you look at the Levitical law particularly in Old Testament theology, Levitical law basically says that food needs to be produced in justice, distributed in justice, prepared in justice, and eaten in justice.... We're coming to this question in an Old Testament way. Are we doing justice in our food system? Are we providing a solid nutrition for the people? If we're not, then we're not doing justice to the consumer.

An equitable distribution of genetics was a third variation on the social and economic justice theme. Concerns were raised about whether or not genes should be owned and patented by a private corporation and if the concentration in ownership precludes others from access to genetic material. Some respondents saw free access to genetic material as a basic right that should be held by the public. For example, an organic farmer said:

> My biggest objection to bio-technology in agriculture is that taking our crops that our foods that are based on out of the realm of the public's control and putting it into the control of

> aiding for-profit corporations. I very much see control of food and seed and our genetic heritage as the issue. And I find it unacceptable for anybody to control something as basic as food.

Another organic farmer echoed the same point:

> They are talking about 'viro-piracy' or 'bio-piracy' where corporations come in and patent crops that the native people have been using for thousands and thousands of years. All of a sudden they no longer have the right to save seed from those crops and to replant those seeds. This is going on beyond the issue of trans-genetics. The germ plasm has traditionally been in the public realm. Nobody owned it. It was there for the public good, and now we have a situation where our genetic heritage can be taken out of the public realm and put in the hands of private, for-profit corporations. We are going to see people being disenfranchised.

The fourth variation on the social and economic justice theme was the equitable distribution of food. Concentration of input ownership in the hands of a few agribusiness corporations was a concern voiced by a broad array of respondents, both those who favor and those who oppose transgenic crops. Seed genetics is the very basis of food production; hence, the control of genetics results in the control of food. A legislator expressed his concerns as follows:

> I think there is definite concern, at least on my part, that one or two companies having control of the entire food supply is a dangerous thing.... I'm concerned that having one or two multinational companies having control of the food supply from start to finish is dangerous.

From another vantage point, a clergy person observed that maldistribution, not lack of production, is the cause of hunger. Thus, he argues that the agribusiness claim that transgenic crops will reduce hunger is false:

> Who are the beneficiaries or what are the benefits right now of GMO? I've looked at the arguments. One of the arguments is that it will provide new base and increase food supply, and thereby be able to deal with hunger. Well those of us that have been dealing with issues of hunger know that hunger has never been an issue of enough production. Hunger has always been an issue of distribution, and the economic systems of distribution, not the production systems, have created the hunger issues.

The moral principles that can be derived from the Distributive Justice interviews are:

3B) Justice: Social and Economic Distributive Justice Principles

i. Rights-Libertarianism: People have a moral right/entitlement to do as they want, as long as doing so does not unduly harm others. It is generally morally wrong to interfere with that right (Sufficient condition, Neutral).
ii. Capitalism: Benefits are to be distributed according to the contribution an entity makes toward achieving its group's goals. The value of the contribution is determined by a free market (Sufficient condition, Neutral).
iii. Socialism: Benefits should be distributed according to need, while burdens are distributed according to abilities (Sufficient condition, Neutral).

4A) Beneficence

The fourth theme that emerged from the study pertained to the beneficial results of biotechnology. The comments stressed the utilitarian or consequentialist impacts of transgenic crops. The respondents noted that there were both risks and benefits of transgenics. Whether or not the benefits outweighed the risks and whether or not the risks were acceptable typically differentiated between the proponents and opponents of transgenics. One legislator weighed the risks and benefits as follows:

> [GMOs are] one of the biggest areas that we can increase the food supply without increasing the **cost**. GMOs, in fact, increase disease resistance, increase disease tolerance, eliminate diseases, and that sort of thing. I think that if the whole GMO thing had looked more like a **benefit** to the starving zillions on this planet by increasing disease resistance within our diet and that sort of thing I don't think it would reach the same resistance.

Similarly, a biotechnology corporation official said:

> As you start looking into this as far as what is the relative **risk**, as somebody once told me, 'There is no such thing as zero.' There is always some **risk**, but I think the **risk** is to a level that, in my perception, is so low that it's probably perceivable [only after] everything has been checked off.

Many of the respondents, particularly those who favored transgenics maintained a strong belief that human progress will come through scientific and technological achievements. Applied scientific research can be used to improve the everyday, practical quality of life, as noted by a crop scientist:

> The only principle I have is one of **practicality**.... It's a question of weighing the advantages and disadvantages of new technology, [whether it] is an overall improvement to or society or not.... An improvement for society could be that we have better economy. It could be more convenience for farmers. Higher profitability for farmers. It would include keeping the food supply safe. Keeping our water supply safe. I don't know if that's an ethical principle, but just that its advantages outweigh disadvantages from a practical stand point. I wouldn't oppose to a new technology just because it's new.

The optimistic view that scientific research in the field of transgenics was reinforced by a legislator who favors transgenics:

> I think plant breeding for food production is ultimately pretty important and I have quite a bit of faith in the demonstration of science to increase productivity. That's my support for it.

The same sentiment was echoed by another legislator who opposes transgenics:

> I'm a believer in science to improve the quality of life for us folks.

5A) Non-maleficence

The final theme, non-maleficence, could be characterized as the "Above all else, do no harm" principle. In fact, several respondents referred to the Precautionary Principle. It was described by a clergy person as follows:

> GMOs fall into this category in terms of environmental concern that we have to use the precautionary principle. That is, before now, we assumed it's safe until proven otherwise.

> The Precautionary Principle says we need to investigate all sorts of possibilities to assure ourselves that it will be safe in the future, not that it appears to be safe today. We need to provide a regulatory framework that sides on precaution and hesitancy rather than open the doors and go full speed ahead on these kinds of things.

The results of Beneficence and Non-maleficence can be combined together to produce the principles that follow:

4B and 5B) Consequentialism: Beneficence and Non-maleficence Principles

i. Beneficence: An action is morally right only if the action produces benefits (Necessary condition, Neutral).
ii. Non-maleficence: An action is morally right only if the risk of harm is minimized (Necessary condition, Neutral).

There are five interesting features of the respondents' principles to note before proceeding to the formalization and evaluation of the different moral codes. First, each of the derived principles is either a neutral principle or it is used solely by those who are opposed to transgenic organisms. Furthermore, the anti-transgenic principles are exclusively found under the environmental ethics, which indicates that opponents are more likely to use environmental ethics principles against the morality of transgenics than supporters are to apply the same moral theory and principles to establish the morality of the technology.

Second, from the responses to questions, the survey participants clearly have adopted four moral theories to support their moral principles, instead of ProfMC's five. The four, which are also included in ProfMC, are versions of consequentialism, Kantianism's respect for persons, various justice theories, and environmental ethics. The main differences between the two are ProfMC has a virtue ethics the USDA respondents do not address and ProfMC's environmental ethic is oriented toward protecting the environment not for its own sake but because human beings need it.[10]

Third, there are two main differences between the ProfMC's Respect for Persons rule and that of the survey participants. Unlike the former's principle, the respondents clearly believe that respecting an individual requires a willingness to accept responsibility for one's actions. Hence, for an agent to respect himself and others, he must be willing to take responsibility for his action and its effects. Furthermore, the participants clearly do not mention two of ProfMC's Respect for Persons principle's clauses: to repudiate that person's considered judgments and to withhold information necessary to make a considered judgment, when there are no compelling

[10]For some of the classical formulations of Natural Law Theory, which seems to have some connection to environmental ethics, see Aquinas' "Creatures Have Their Own Activity" in *On the Power of God*, q. 3, a. 7, and Aristotle's *Nicomachean Ethics* and *Physics*.

reasons to do so. The result is that ProfMC is much stricter than the respondents on what counts in respecting a person as she deserves.

Fourth, as in the Belmont Report, the justice principles derived from survey data cannot be made consistent in certain cases because each principle is sufficient. Each distributive justice principle disperses social benefits and burdens in a virtually contradictory way to the remaining two principles. For instance, if benefits are distributed according to the Socialist principle, then Capitalism will say that dispersing goods in this manner is wrong, and that the resulting situation is morally worse than would have occurred under Capitalism. As was done for the Belmont Report and ProfMC's justice principle, to rectify the situation in the best manner possible, it is stipulated that the agent performing the action has to follow at least one of the distributive justice principles to perform a right action.

Furthermore, the notion of an equal distribution of benefits does not find a place in any of the participant's interviews while being the first justice rule on the list for the Belmont Report. One conclusion that is supported by Egalitarianism's absence is people – at least in the transgenic organisms debate – do not use such a principle in their moral reasoning or to guide their actions. Hence, it will not be included in any of the three moral codes.

Fifth, the surveys' consequentialist moral theory and its two principles are always consistent with each other, unlike the two consequentialist principles from the Belmont Report. In the latter, when both principles could not be satisfied in the same situation, the agent could choose between doing no harm and maximizing utility. Sometimes, however, maximizing utility required doing some harm. This was rectified in ProfMC. From the respondents' principles, a person is to minimize harm, while producing at least some benefit to someone. Therefore, they do not include the problematic principle to do no harm. This is a superior feature of the USDA survey respondents' code over the Belmont Report's. However, in Section 1.6, the respondents' two consequentialist principles will be proven to be inadequate in those cases in which the agent is faced with only a set of alternatives in which no one is benefited and the best the agent can do is to minimize the evil that is done.

1.5 Formalization and Evaluation of the USDA Study's Moral Codes

From the principles the respondents adopted to govern their decision making processes about transgenic organisms, it is possible to formulate the moral codes that each side of the debate uses. While these sets of rules might be too inclusive for a particular participant's view, it is useful to explicate the code for the particular group, and then analyze it for its merit in the debate. If the code is defective, then it should be modified or abandoned. Furthermore, a common moral code might be able to help with a consensus on the issue of transgenics, provided that the principles and theories are not too divergent or the people holding them are willing to alter their rules to make their sets of principles consistent.

A) The Anti-Transgenic Code (ATC):
An action, A, is morally right if

1. A assists or does not harm the environment or nature,
2. in doing A, no one's freedom is unjustly limited,
3. the agent(s) of A are willing to take responsibility for the consequences of A,
4. A satisfies one or more of the justice principles,
5. A produces benefits for at least one person or entity,
6. the risk of harm from doing A has been minimized,
7. A does not entail the ownership of alterations or ownership of the fundamental parts of nature, and
8. A is as natural/non-artificial of an action as A can be.

B) The Pro-Transgenic Code (PTC):
An action, A, is morally right if

1. A assists or does not harm the environment or nature,
2. in doing A, no one's freedom is unjustly limited,
3. the agent(s) of A are willing to take responsibility for the consequences of A,
4. A satisfies one or more of the justice principles,
5. A produces benefits for at least one person or entity, and
6. the risk of harm from doing A has been minimized.

C) The Neutral-Transgenic Code (NTC):
An action, A, is morally right if

1. A assists or does not harm the environment or nature,
2. in doing A, no one's freedom is unjustly limited,
3. the agent(s) of A are willing to take responsibility for the consequences of A,
4. A satisfies one or more of the justice principles,
5. A produces benefits for at least one person or entity, and
6. the risk of harm from doing A has been minimized.

One interesting feature to note is while the three codes have the first six principles in common, the Anti-Transgenic Code is different from the latter two. The ATC is the most restrictive with eight principles, while the Neutral-Transgenic and the Pro-Transgenic are identical.

Even though it would produce a uniform code of morality, it is impossible to include the last two environmental ethics principles in the Pro-Transgenic Code because the rules are designed, almost by definition, to classify the creation or introduction of transgenics into the market as morally wrong. Given this fact, it might be impossible to build a consensus between the anti-transgenic people and those who are pro-transgenic. What is stranger is that consensus might not be possible between the neutral–transgenic people and the anti-transgenic people, while consensus seems to be *prima facie* likely for the neutral and pro-transgenic people because they share the same moral code – at least according to their responses to this survey about biotechnology.

If those opposed to transgenic organisms eliminated the environmental ethics principles ATC7 and ATC8, then there would be an increased chance for consensus on a resolution to the conflict – or at least a better understanding and empathy for others positions. There is good reason to reject the two principles on rational grounds – both lead to unsupported or implausible conclusions.[11] First, ATC8 states "A is as natural/non-artificial as it can be," which means that a right action is required to be as lacking in human participation as possible. However, the principle evaluates human actions, which means that a human action must eliminate the human factor as much as possible in order to be ethical. That is an unreasonably high standard of acceptability. Furthermore, since all human actions are artificial or human-made, then it follows under the principle that all human actions are morally wrong. Hence, the implausible result it that no one, including those who oppose transgenic organisms, can ever do anything that is ethical.

On the other hand, if what are actually being evaluated are the consequences of human actions, such as human-made object creation, then all products of human action will be morally bad precisely because they result from human action. Organic farming is no less bad or wrong than farming transgenic crops on these grounds for both require a great deal of human interaction. Possibly, if people did nothing but eat what they gathered from the environment, then they would act in the only manner in which they can act morally according to ATC8. This implausible result would entail anything a human does that is not at the base subsistence level would be wrong, including developing medical technologies that fight cancer, polio, tuberculoses, or malaria. Moreover, given that absolute naturalness or the lack of human intervention is the best situation possible, it follows from ATC7 that many humans in our overpopulated world might have to cease existing in the best manner possible for the environment. This issue will be addressed in more detail in Chapter 3.

Rule ATC7 is also problematic, mostly due to the obscurity of what it entails. First, "A does not entail the ownership of alterations or ownership of the fundamental parts of nature" means that owning a part of something with intrinsic value, such as a person, is morally wrong because it necessarily disrespects the value of the entity. However, it is not obvious that nature is valuable in itself in the same way or to the same degree as a human person. If it is a choice between owning a person or part of nature, such as DNA, for example, many would say that the former is always

[11]There are other ways of interpreting what "natural" means when the respondents use the word. One standard, but vague, definition is to equate natural actions and products to activities and products that are harmonious with or beneficial or neutral to the flourishing of nature. If an action is out of harmony or prevents flourishing, then it is morally wrong. Morally bad products are unharmonious or detrimental to nature. Christine Pierce, in *Immovable Laws, Irresistible Rights: Natural Law, Moral Rights, and Feminist Ethics*, analyzes various ways of defining the principle. Chapter 3 will address this issue in depth.

prohibited, while the latter is not.[12] Unlike the environment, people have autonomy, which makes them an end in themselves deserving respect. There are those who would go against such a strong intuition and deny it, but the burden of proof is on them to show either the obscure – that nature is an autonomous entity – or that some other characteristic it has gives it the intrinsic value.[13]

Perhaps a more telling argument against including ATC7 in a moral code is that it forbids clearly right actions. If the principle proscribes owning genetic information, then it misses the point of why owning people is morally wrong. Owning an intrinsically valuable entity is far different from possessing and adequately controlling information about the entity. For example, a person might own information about Elizabeth Taylor, but it does not follow that she owns Elizabeth Taylor. Even if some entity was capable of owning all the information about the human genome, it would not follow that it also owns the human genome.[14] The former is morally permissible, while the latter clearly is not based on the value of the moral agent involved.

If principle ATC7 means that owning physical objects such as the actual DNA double helix is wrong, then it follows that owning any seed, plant or animal with DNA is morally wrong, since the cells of the object have DNA in them. Furthermore, owning anything dead that has DNA, including but not limited to the food that a person eats, will be unethical. Now it might be a very good thing that no one owns anything that has DNA in it, but adopting that position would entail the undesirable destruction of all arguments against pollen drift damaging a person's organic crops, for instance. If no one owns the crops, then they cannot claim or sue for damages to property no one owns. For practical and consistency reasons, it would be better at this stage to eliminate ATC8 and ATC7 from any practical moral code in favor of one that is agreeable and useful for public policy and decision making.

1.6 Formalization of the Combination Moral Code

In order to formulate a moral code that will be more practical than the Professional Moral Code and clearer and more general than that of the Neutral-Transgenic Code, it is first necessary to evaluate each code in turn. Not only should the irreparably defective principles from each be identified and discarded, but the value of the remaining must be discussed in terms of how they best fit into the Combination Moral Code. Once again, the codes are:

---

[12]See the 13th Amendment of the Constitution of the United States of America.

[13]See Rita C. Manning's contention that the earth is a living body (Manning 1992, pp. 127–9, and 132).

[14]The patenting or ownership of information, processes, cell lines, or living organisms other than human beings is one ethical issue, while moral access to the patented or owned material is another. Although it might be morally permissible to patent or own these sorts of things, it might be impermissible to deny access to some people if certain conditions obtain.

The Neutral-Transgenic Code (NTC):
An action, A, is morally right only if

1. A assists or does not harm the environment or nature,
2. in doing A, no one's freedom is unjustly limited,
3. the agent(s) of A are willing to take responsibility for the consequences of A,
4. A satisfies one or more of the justice principles,
5. A produces benefits for at least one person or entity, and
6. the risk of harm from doing A has been minimized.

The Professional Moral Code (PROFMC):
An action, A, is morally right only if

1. A either produces no harm or maximizes utility, in those cases in which both cannot be done,
2. in doing A, the agent treats all individuals affected by the action as autonomous individuals, which entails that no agent's considered judgments are repudiated, no individual's freedom to act on those judgments is denied, and information necessary to make a considered judgment is not withheld, when there are no compelling reasons to do so,
3. in doing A, the agent protects those with diminished capacity if they are affected by her action,
4. A satisfies at least one of the justice principles provided that all five cannot be satisfied,
5. A protects or does not cause undue harm either to animals or the environment, and
6. A fosters or maintains virtuous characteristics in the agent of A, including but not limited to honesty, trustworthiness, integrity, competency, fairness, and respect for cultural diversity and everyone affected by the agent's actions.

In evaluating the merit of the two codes, it is apparent that each has its individual strengths and weaknesses. Both are remarkably similar, although they do have significant differences. First, ProfMC2 and ProfMC3 more practically deal with the Kantian imperative to respect all persons as valuable in themselves, while NTC's rule to not illicitly limit an individual's freedom is much easier to fulfill. The benefit of the former is that it better captures more of the necessary conditions of how to treat the individuals affected by the agent's action than does the latter. When dealing with respecting people it is always better to error on the side of caution and use higher standards than to set the standards too low and do something that is easy but also wrong.[15]

[15]This argument will be fleshed out more in Chapter 4.

Second, ProfMC3 features an element which NTC does not have, *viz*., a requirement to protect those who cannot protect themselves. Since transgenic organisms will affect a great number of people in the world who have little to no political voice or power, especially in the developing world, it is more than reasonable to have a principle to protect the vulnerable than completely exclude them from consideration in decision making.

Third, the justice principles in both codes are virtually identical in their vagueness and problems, which have already been noted. We will spend no further time on them here.[16]

Finally, the ProfMC's first principle has a choice between producing harm and maximizing utility, in those cases in which both cannot be done. Since harm will be done by all of the alternatives open to agents in most cases in which people have to make difficult choices, the "do no harm" clause does no real work. Consider what happens in the case of immunizing a child. In one set of alternatives the child is harmed by the pain from the inoculation needle, in the other, the child is harmed by the disease. Even though it is clear that the right act is to inoculate the child, ProfMC1 says it is not because it violates the condition to do no harm. This result is not unique to childrearing. Implementing business decisions in regards to the creation and introduction of a transgenic organism or other biotechnology is going to harm someone no matter what policy is decided in the end. Those who oppose transgenics will be harmed if the decision goes against them, while those who support them will be harmed if they are banned. In order to keep the spirit of the "do no harm" clause without the devastating drawback of not being able to do anything ethical in situations in which all alternatives produce some harm, then a practical maximization of utility clause must be adopted. It would be best to revise ProfMC1 by eliminating the first condition and leaving the second.

However, NTC4 and NTC5, the combination of which is merely a version of consequentialism, are insufficient substitutes for ProfMC's revised first principle. Principle four of NTC places its standard very low by stating that at least one person must benefit from the action in order for the action to be moral. But there will be circumstances in which an alternative will produce great benefit for many with low risk, while another alternative has the same low risk but only benefits one person. NTC would be unable to distinguish between the two alternatives, while ProfMC's revised first principle would say that the former is superior. Clearly if it is a choice between the two actions, *ceteris paribus*, the better with its much greater benefits is *prima facie* right. As a result, the second part of ProfMC1 is the consequentialist principle that should be adopted. It merely says to do the best one can in those situations governed by the moral code.

If the remaining principles from each of the two moral codes are merged, the result is the Combination Moral Code found below:

---

[16] I have argued briefly in "The CIOMS's Distributive Justice Principle: A Reply to Dr. Benatar" that an adequate distributive justice principle would merely be the combination of a form of utilitarianism and a version of Kant's second Categorical Imperative (Cooley 2002a, pp. 11–13). The Practical Moral Code will develop just such a set of principles.

Combination Moral Code (CMC)
An action, A, is morally right only if

1. A assists, protects, or does not unduly harm animals, the environment, or nature,
2. in doing A, the agent treats all individuals affected by the action as autonomous individuals, which entails that no agent's considered judgments are repudiated, no individual's freedom to act on those judgments is denied, and information necessary to make a considered judgment is not withheld, when there are no compelling reasons to do so, and the agent(s) of A is willing to take responsibility for the consequences of A, provided that it is not too great a burden for the agent,
3. in doing A, the agent protects those with diminished capacity if they are affected by her action,
4. A satisfies one or more of the justice principles,
5. A maximizes utility, and
6. A fosters or maintains virtuous characteristics in the agent of A, including but not limited to honesty, trustworthiness, integrity, competency, fairness, and respect for cultural diversity and everyone affected by the agent's actions.

The resulting code is not only a valuable resource in the debate over the morality of transgenic organisms, but can be used in all aspect of bioethics and technology development. After all, the Combination Moral Code incorporates practical moral principles based on commonly held moral theories at the same time it requires individuals to follow the general moral rules of conduct.

1.7 Problems for the Combination Moral Code

Although the Combination Moral Code is better than those from which it was formed, there are still substantial problems to overcome. First, even though no one would disagree with the very general moral ideas behind the six principles,[17] the latter are too vague and abstract at this time, which renders any actual application of them suspect. For example, the first principle of CMC requires that animals, the environment, or nature be either assisted, protected or not unduly harmed, but provides no clue as to what it means to assist, protect, or unduly harm animals, the environment, or nature. It might be the case that my exhaling carbon dioxide harms nature because there is already a surplus of the compound which cannot be converted back into its key elements by plants. This is one of the reasons that CO_2 sequestration is becoming more important, especially in those nations relying heavily on coal burning power plants. Does my creating excess CO_2 mean that my

[17] No one disagrees that it is necessary to respect other agents, pain is to be avoided, social justice should be pursued, and so on. The disagreement begins when individuals discuss how to perform those tasks. For example, a Libertarian would have a very different approach from a Socialist, although both would agree to the basic idea that justice is something people ought to pursue (Miller 2003).

breathing is morally wrong? What of my other activities that in infinitesimal ways contribute to global warming that will harm some animals, some parts of the environment, and some aspects of nature? Do all human actions in some way cause some harm to non-human entities and things? Are these all classified as morally bad or wrong? The other principles of the Combined Moral Code are equally vague and need more precise definitions than can be gleaned from the USDA respondents or the professional code statements.

The definition of terms and principles cannot be left to each individual trying to apply CMC in a decision procedure. The lack of clarity is a problem because opponents might very well satisfy all six rules of CMC, but have vastly different understandings of each. Someone who is unconcerned with the environment might believe that an action which does not fully destroy nature fulfills the first principle, while a deep ecologist, such as George Sessions, would take a much more restricted view.[18] Given the rules' vagueness, it would be impossible to prove which one is right and which is wrong, if either.

The second problem is that each principle can be internally or externally inconsistent. CMC's fourth principle allows individuals to choose which distributive theory of justice she wishes to fulfill. However, as mentioned before, the distributive justice principles are inconsistent with each other. Giving a person what she needs might not be the same as giving her what she deserves. If an individual cannot contribute in the proper way under Capitalism, then she receives no benefits, even though her needs might be great. Socialism, on the other hand, requires that the agent get what she needs, even if she cannot contribute. Hence, the two theories are contradictory.

Furthermore, in many circumstances, it might be impossible to fulfill all the Common Moral Code's six principles at the same time. For example, in a complete set of alternatives for a particular situation, utility maximization could entail that nature be harmed in some way. Perhaps, plants must be sacrificed in order to save the lives of sentient animals, and to preserve the plants, the only alternative is to sacrifice the animals. In these situations, according to CMC, it is impossible to do anything that is ethical. Other examples could be described to illustrate different conflicts, but it is relatively clear that moral codes with more than one principle – if they are not properly finessed in their development – are likely to have incompatibilities (Miller 2003). In order to manage or eliminate conflict situations, a justified process for determining which principle or value, if any, has greater weight in the context is needed. However, that is a question for Chapter 4.

Finally, the most significant problem the Combination Moral Code encounters is that its principles might not correctly capture morality. Kant, for instance, rejected any form of consequentialism in his Categorical Imperative (Kant 1956, p. 62). For Kant the outcomes of actions never matter. Rather an action is ethical if and only if the agent of the action fulfills the conditions of the Imperative, i.e., not treating anyone as a mere means, acting on a rule autonomously, and/or being able to

---

[18] See *Deep Ecology* (Peregrine Smith: Layton, UT, 1985).

consistently will the generalizable form of the maxim governing the situation as a law of nature.[19] If Kant is correct, then the other principles in the moral code would misclassify actions, unless by luck, they made the same classification as the Categorical Imperative. The same type of argument works against each of the five principles of the Common Moral Code. Just because there is consensus among many different people from various walks of life on the principles, it does not follow that the principles are correct. For example, if we polled slave owners from Rome, 19th Century America, and parts of Africa today, we very well might find they all agree to the principle that it is morally permissible to own slaves. However, the fact of the matter is that it is morally forbidden to own other people for a number of obvious reasons.

The most the agreement in principles can tell us is that there probably are universal moral considerations that must be taken into account in ethical decision procedures, but not exactly how that is to be done. Furthermore, the Common Moral Code and its principles can give us some insight into what is needed for a moral code that most, if not all, of the people in moral debates can adopt to help structure the dialogue of the discussion. After all, having everyone agree to a common set of moral principles makes it much easier evaluate and understand each others' position, although they do not have to concur with it.

1.8 Conclusion

Even though, the Combination Moral Code faces severe problems, it is an excellent way to try to begin bringing debate factions together to engage in constructive dialogue. If further progress is to be made in the reaching consensus on the issue, one of the first steps must be to understand the moral codes that each side is using. Besides explication, part of that understanding requires knowing how the codes work. It seems from the professional codes and respondents' interviews that there is a large sharing of moral theories and principles, but until the moral code is sufficiently clarified, the potential and social effects of transgenics understood, and the factual disagreements about how much scientific evidence is required to show that a risk is acceptable are settled, there will be little hope of reaching consensus on the morality of transgenic organisms or their introduction into markets.

In the next chapter, I will begin the process of creating a practical moral code from the groundwork of CMC and the other codes. The groundwork will also prove valuable in Chapters 2 and 4 when the axiology – a theory of value – required by the practical moral code is developed.

[19] See either *The Metaphysics of Morals* or the *Groundwork of the Metaphysic of Morals* for the various formulations of the Categorical Imperative.

Chapter 2
A Practical Moral Code

2.1 Introduction

Thomas Jefferson once said that ploughmen are more likely to find the right answer to moral questions than professors. Why? Because the former have not been led astray by overly abstract, theoretical rules and ideas, which are impressive in their elegance, complexity, and nuance, but unable to capture true morality. By eliminating formal education as a criterion for moral knowledge, Jefferson rejected the elitism of other philosophers, such as Plato, Aristotle, Mill, and Hare. As long as moral agents are not distracted by irrelevancies, ethics is something each of them can do correctly.

Although Jefferson's claim of self-evident truths here and in the Declaration of Independence is suspect, he does correctly identify a universal trait of personhood. Namely, everyone has a moral instinct and love of humanity much in accord with Hume's idea of sympathy/empathy, which enables them to do the right thing at the right time for the right reasons. If a practical ethical code is to be discovered and elucidated, then this universal sense must be incorporated in a meaningful way into the end product. The new role of professional ethicists eventually might be to take the common ideas, and then develop a moral code based upon them – rather than developing their own code, and then requiring people to adopt it.

A combination of the main tenets underlying utilitarianism and Kant's Categorical Imperative can provide an adequate moral code for technology, if it is defined adequately through a reasonable person. The resultant principle not only prescribes all people to respect the autonomy of each individual affected by their actions, they must also try to bring about what at least one reasonable person would reasonably believe is most likely to produce the best consequences of the alternatives open to the agent. Furthermore, the code incorporates a value system of weighted interests in favor of the agent performing the action to eliminate the possibility of a duty to sacrifice her best interests for relatively minor gains by others.[1] Overall, while striving for consistency and rigor, the code captures much of what people actually use when making ethical decisions in tough cases.

[1]This value system or axiology will be developed in Chapters 3 and 4.

D.R. Cooley, *Technology, Transgenics and a Practical Moral Code*, The International Library of Ethics, Law and Technology 4, DOI 10.1007/978-90-481-3021-4_2,

2.2 Utilitarianism

Standard act-utilitarianism focuses solely on the consequences of an action to determine if the action is morally right or wrong. The value of the action's antecedents or the action itself is irrelevant to the action's morality. A person might have evil thoughts, intend to do what is wrong, yet might actually perform a right action because the act produces at least as good results as any alternative open to the agent at that time. The counter-commonsense result is that if a terrorist intends to create a transgenic organism deadly to all human life, but instead finds one beneficial to treating serious illnesses, then according to act-utilitarianism, if no other action would have produced better results, the terrorist action is not only morally right, but morally required of him as well.[2] Morality for this theory and principle is too dependent on how things work out in the end, which includes forces beyond the control or foresight of any moral agent affecting the action's moral value.

Since I want a practical ethical code, I contend that the action's actual consequences are irrelevant to the situation. Seeing as no one can know all the consequences of many trivial actions, much less those of significant ones, only the agent's reasonable under-the-circumstances-perception of one action's *probability* being better than the alternatives is necessary to make the act right. The moral agent is obligated to evaluate the alternatives open to her at a particular time and select from them an action that, given the experiences of reasonable people and the situation's constraints, is likely to maximize utility. Under this theory, an agent who attempts to bring about the likely best action can begin to fulfill her obligations even if the act proves not to maximize utility. Hence, unlike standard act-utilitarianism, unforeseeable consequences cannot alter the utility calculation so that an action's moral status changes regardless of whether or not the agent could have foreseen or controlled them in any way.

There are several reasons why we should adopt this more relativistic version of consequentialism instead of standard act-utilitarianism. First is the fact that although we are manifestly unable to determine absolutely which consequences will occur, we can still act morally with a strong degree of probability. Consider some real life circumstances people encounter – at times, we face situations in which the only alternatives open to us are not ones with which we would normally desire to be confronted. At these times, we are obligated to choose what appears to be the least bad thing out of a set of terrible choices. Suppose, for instance, Mary's supervisor asks her to create a transgenic wheat species designed to cross pollinate more easily than the normal 6% of conventional and organic plants. Given her extensive knowledge, Mary reasonably believes that this wheat type will be devastating for organic and conventional farmers' global markets, while providing little benefit to the transgenic producers. In this situation, the utility produced by Mary's rejecting

---

[2]The details of how to make this case work out so that utility is maximized might seem unlikely, but so many of the counter-examples to utilitarianism are unlikely to happen. The reason they are so effective against the theory and principle is that they actually do occur.

the request appears to be better than her agreeing to develop the product. Her obligation is to do what she thinks is the best she can. To try to prevent the organic and conventional farmers' livelihood destruction when there is so little gain to anyone else – though it may not actually be the best thing she can do – appears better by far than honoring her supervisor's order.

Second, even if we were able to know what the future holds, it is impossible to perform a complete cost/benefit analysis for many actions. For instance, significant acts have too many consequences for any agent to accurately evaluate. Consider the 1914 assassination of Archduke Ferdinand in Sarajevo. Although other factors had more influence on starting World War I, killing the heir to the Austro-Hungarian throne created such tensions that the major European nations declared war as a consequence. As everyone knows, wars generate a great deal of results both great and small in positive and negative value. The Russian revolution was caused in large part by WW I, which in turn brought about realignments in political alliances, which in turn helped instigate WW II and the Cold War. Every negative and positive consequence of that fateful encounter must be included in the assassination's utility calculation – every pain state, pleasure state, death, birth, happy life, unhappy life, good thing, and bad thing. Even if this information was available to any person, which it cannot be given the causation' complexity and the impossibility of recording every bit of datum, no person could accurately perform the cost/benefit analysis. Other actions such as choosing a career, marrying, having a child, will generally have fewer results produced, but still have consequences which occur long in the future and affect people in ways that we can never know or about which we cannot speculate reasonably. Therefore, in major decisions affecting people lives in important ways, it is impossible to know with any type of certainty what one should do.

Some actions are relatively trivial, which means they have a few consequences without much value to them, but being trivial does not necessarily mean their analysis is easily performed. For instance, choosing which shoe to put on first is generally unimportant, but it is still classifiable as morally right or wrong as any of our actions are. That is, because we always have an obligation to act ethically and we have alternative actions from which to choose, according to standard act-utilitarianism, we must maximize utility even in the case of insignificant actions.[3] Most people would say that in normal circumstances putting on either shoe first is morally permissible, but act-utilitarianism does not. If there is even the slightest difference in utility, the agent must select the best action. Suppose putting on the left shoe before the right

[3]There are some who argue that some actions, such as trivial ones, are morally neutral or that trivial actions for the agent are unclassifiable, while trivial actions affecting other people are classifiable. The problem with making a third classification beyond that of morally right or wrong is that it is difficult to draw the line between what counts as a morally neutral action and one that can be evaluated. Should the line be drawn by each individual agent? If so, then the line could vary widely from one person to the next depending on what each thinks is insignificant. The same problem holds for any relativistic demarcation, including but not limited to what societies or human persons as a whole think unimportant. In order not to create additional controversy and to incorporate the common moral sense idea that we are obligated always to act ethically, all actions are classifiable as morally right or wrong, and obligatory, permissible, or forbidden.

has a small amount of negative value that reversing the order does not, while all other values are the same. Let us plausibly assume that the reason for the negative value is the person is accustomed to putting on the right shoe first, and changing her habit causes the most minor annoyance to her. However, regardless how small the difference is, for act-utilitarianism, the agent is obligated to put the right shoe on first because it maximizes utility, while the other alternative does not. The problem being drawn out here, of course, is an epistemological one. The action is right, although the agent cannot reasonably believe it to be the only right action for she does not have the ability to measure such small differences. Hence, although the actual calculation is simple if the agent knows the relative values generated by each alternative open to her, she cannot practically perform the computation because she cannot acquire the necessary data.

What makes matters worse for standard act-utilitarianism's practicality is that the moral agent would have to analyze each alternative action for every situation in which she has to make a choice before she knew which one is best or tied for the best. First, if the agent cannot readily perform the calculation for one significant action, doing it for alternative actions increases the complexity and makes it harder for her to know what to do in the circumstances. Second, there might be an infinite number of actions open to the agent at a particular time. Suppose for instance, Mary wishes to leave her lab at the end of the day. She can exit in a variety of ways: through the door, the window, the air duct, and so on. She can also leave under her own power, through the assistance of another, e.g., leaning on another, or completely relying on the actions of others, e.g. someone could carry her out. If she departs through the door under her own power, she could do it by walking, running, skipping, slithering, and so on. The point in making this long list of alternatives is to illustrate that an agent can perform actions in a variety of ways. What makes each alternative different can be a very minor alteration, such as how high to lift her foot when crossing the threshold. Since there can be an infinite number of changes, there are an infinite number of alternatives open to the agent at any particular moment. Hence, although act-utilitarianism classifies all of the actions open to each agent each time she can act, we, as people without supernatural insight, are unable to know what our actual duties are in most cases. The result is that even though we can not know an action's moral status, we can do what is right and wrong, and thus, be blameworthy or praiseworthy for actions that accidentally work out in utility's disfavor or favor, respectively.

The question that arises at this juncture is can any adequate moral principle or theory create an unbridgeable gap between what is actually morally right or wrong and reasonably believing or knowing what is morally right or wrong? Some forms of consequentialism mistakenly ask us to see into the future to be able to select the best act, which we cannot do in the absence of a one hundred percent reliable scrying device. However, in the real world, we do not need to select the genuine better alternative, in order to do what is required of us. Even Mill, who seems to think certain elite individuals can eventually use a form of standard act or rule utilitarianism to evaluate actions or other human activities, acknowledges that average citizens often make correct ethical decisions.

> [M]ankind must by this time have acquired positive beliefs as to the effects of some actions on their happiness; and the belief which have thus come down are the rules of morality for the multitude, and for the philosopher until she has succeeded in finding better. (Mill 1988b, p. 25)

As Mill implies, we have general rules-of-thumb assisting us in making our ethical decisions. Bernard Gert argues that there are ten imperatives that all people as rational persons use in moral decision making and actions (Gert 1998, p. 158). Under no interpretation can Gert be considered to be a utilitarian, but his rules can be examined as rules-of-thumb. With only the slight alteration from Gert's "rational" to my "reasonable" people, Gert's code ranked from most to least important is:

1. Do not kill
2. Do not cause pain
3. Do not disable
4. Do not deprive of freedom
5. Do not deprive of pleasure
6. Do not deceive
7. Keep your promises
8. Do not cheat
9. Obey the law
10. Do your duty. (Gert 1998, pp. 159–217)

In the absence of being able to see into the future, these rules, based on past experience and probabilities, usually allow us to do the right thing, and are part of how we actually decide to act as we should. As I argued in the Introduction, since these rules are so useful while the "real" standard act-utilitarian principle is not, then it would be best to try to eliminate the impractical.

In order to keep the appealing notion that doing one's best must be part of at least one morally right action in all situations *and* make it a practical principle, standard act-utilitarianism must be changed to something more useful, and therefore more reasonable. Instead of morality depending on all the actual consequences of actions, it makes practical sense to incorporate only the consequences at least one reasonable person would foresee in the particular situation into any utility calculation.[4] Put more formally, the reasonable person principle is:

Reasonable Person Utilitarianism (Do the best you can.):

> An act is morally right only if under morally identical conditions, a reasonable person would reasonably believe the utility of the consequences of the act will probably be as great as any alternative to the action at that time, where utility is the result of subtracting all of the evil produced from an action from all of the good produced by the action.

---

[4]The reasonable person standard has been adopted by tort law to determine if an agent's action negligently caused damages to others (Robinson 1972, p. 178).

Proper consideration means the agent has taken into account the probable consequences as permitted to her by external factors, e.g., the amount of time and information she has, and internal factors, e.g., the agent's reasoning ability of the situation. If the agent is not hurried, for example, then she will have more time to determine probabilities, thereby generally coming up with more insight for the decision procedure. The judgment to carry on or stop researching a particular transgenic organism or other biotech item could be classified differently according to the amount of time available. If Mary is forced to choose in one hour, then she might select an option she would not otherwise have selected if she had been given more time for reflection. However, the decision made in the short time period is as legitimate as one made in the longer time period. Regardless of the external pressures the moral agent is under, as long as she selects what at least one reasonable person would reasonably choose under the same conditions, she has satisfied one morally right action condition.

Internal factors are also important moral decision elements. Each agent's mental faculty is composed not only of the actual ability of the mind to reason, but the relevant information the agent possesses as well. In general, an agent with more pertinent information will have more insight for her choice than one with less. Furthermore, an agent with greater faculties will also generally be able to collect data and evaluate situations better than an agent with less ability; thereby making her decision more insightful in two different ways.

Of course, these two internal factors, as well as all the others, entail that many alternatives at a particular time and situation can be permissible. The acts of those who are less able to perform the proper reasoning may be moral for them, when the same act in virtually the same situation would be immoral for an agent under slightly different constraints. In addition, those with one set of morally permissible, justified beliefs may have dissimilar moral actions from those with another set of permissible, justified beliefs. Unless it is an obvious classification, such as torturing innocent people for the agent's personal amusement having less utility than leaving them alone, or saving a child at no cost to anyone being better than letting her die, reasonable people often disagree about which action is more valuable. Organic producers, for example, believe that organic products are ethically superior to transgenics, while transgenic producers disagree. Given this form of utilitarianism, they could both be right provided that there are sufficient differences in the circumstances in which they make their decisions. Reasonable people can also change their minds without losing their reasonableness. Hence, there is an agent-centered relativity at work in RPU that not only recognizes the differences in human persons and their abilities to reason, it also acknowledges that in many situations there are a large number of morally permissible actions open to an agent.

2.3 What Is a Reasonable Person?

Since both the Practical Moral Code's RPU and Quasi-Categorical Imperative (QCI) heavily depend upon the notion of a reasonable person, it is necessary to take time here to stipulate some necessary conditions for being a reasonable person.[5] Reasonableness is a very complex concept and requires a great deal more attention than I will give it, but what is said here should give an adequate idea of what is required of a person in order for her to be reasonable.[6]

In general, the definition of reasonable person must be linked both to having an ethical "good" life for the person, herself, and trying to make the world a better place for other individuals, groups, living things, or society (Sagoff 2004, p. 2). A reasonable person is someone who makes the best judgments she can given her overall knowledge and understanding given the situation's constraints. The best judgments are those that are, in each state of affairs, the most likely to achieve the sustainable flourishing of herself and help at least some others with intrinsic value to achieve or maintain their own flourishing.[7] The goal is to increase overall worth by instantiating as many values as is practical given these constraints.

2.3.1 The Reasonable Person Standard in Tort and Criminal Law

The reasonable person standard is used in tort and criminal law as a way of humanizing reasonableness for juries (Johnson and Gunn 1994, p. 233), but the term's definition remains vague at best. All agree that the standard must be as objective as appropriate in the given circumstances, otherwise it would be too variable to be practical, would encourage defendants' fraud and deception to mislead the jury, and would not serve as an incentive to make all people more likely to exercise as much care and skill as they can in all their decision making and actions (Abraham 2002, pp. 54–5). Although we can list the desired functions, what it actually means is murky or outright inconsistent.

Sometimes the reasonable person is described as an individual with merely ordinary prudence; for other definitions, she is one with reasonable prudence (Keeton et al. 1984, p. 174). There also appears to be a number of reasonableness levels and several audiences depending on the term's intended function and who is applying it. A police officer with expert knowledge, for instance, might evaluate an action differently from a juror or other member of the public. The citizen might think that the force used to subdue an individual excessive, while the officer believes

[5]It might be simpler to make the reasonable person into an economist who is an expert in cost/benefit analysis. However, I take seriously Sagoff's rejection of economic science as sufficient for adequate decision making. Economics "cannot measure the benefit, value in use, or the utility an object provides" (Sagoff 2004, p. 7). The reasonable person takes all relevant values into account.

[6]My standard for a reasonable person may appear to be too high, but, in order to reduce the probability of action misclassifications based upon some defect of the individual, I would rather set a higher standard than one that is too low.

[7]Gordon Graham adopts a contemporary Aristotelian line when he states, "the biology of a thing should be so ordered as to promote and maximize its flourishing" (Graham 2002, p. 181).

it to be reasonable. Hence, the standard, although we would like for it to be universally objective, actually is relative to who is doing the evaluation, her set of pre-existing beliefs, personal experiences, and other factors that will influence her decision.

The vagueness and ambiguity problems do not end at identifying reasonable people, but find their way into how people should behave. In the law, how a reasonable person would act in a particular situation has at least four ways of being established: Factfinder Determination, Judge-Made Standards, Legislatively Determined Standards, and Judicially Declared Standards Based on Legislation (Johnson and Gunn 1994, p. 234). Generally, the Factfinder Determination is used which allows the court or the jury to make the decision about reasonableness using their own standards (Ibid.). The other three are attempts to codify a definition that can be universally applied, but because complete precision is impossible for these word types, there is an inherent degree of interpretation in each case.

In Judge Spotswood W. Robinson's opinion in *Canterbury v. Spence*, the reasonable person standard appears to be based on how the average individual involved in that situation would act (Robinson 1972).[8] How much information an average patient needs to know about risk for a particular procedure, for example, should be the guide for how much information the patient's physician should disclose to her. If most people are uninterested in a relatively infrequent, minor side effect, then the physician is under no obligation to tell the patient about it, unless some other intervening factor is at play, e.g., the physician knows the patient has an idiosyncratic concern about the side effect. And this is the juncture at which we should begin asking what duties the physician or any fiduciary has to find out these unusual personal details, which will again involve the reasonable person standard that was vague in the first place.

Legal textbooks help clarify the reasonable person standard to some degree, but never define it with the clarity desired by many. In fact, Keeton et al. state, "The utmost that can be done is to devise something in the nature of a formula, the application of which in each particular case must be left to the jury, or to the court" (Keeton et al. 1984, p. 173). Besides being as objective as practical, at the very least, the reasonable person must be prudent and careful for the situation and its circumstances (Ibid., p. 175). In addition, in order to have a true understanding of the situation, the reasonable person has the same relevant traits as the actual actor who makes the decision. If the person has a physical disability, then so does the reasonable person who must choose based on what the actual person can do at that time and place. The actual person's mental capacities will also be part of the reasonable person evaluation, although to what extent is controversial because of the inability to measure the infirmity, unlike that prevailing for physical conditions (Ibid., pp. 176–8; Abraham 2002, p. 58). A physical scar is simple to perceive, while mental ones are not. Finally, the person's ability to foresee risks deriving from

---

[8]From US Court of Appeals, District of Columbia Circuit, 10 May 1972, 464 Federal Reporter, 2nd series, pp. 772–96, West Publishing Company.

his actions is also a factor in what a reasonable person is (Ripstein 2001, p. 105). If the actual person in those circumstances could know that a hazardous consequence would follow from a particular action, then a reasonable person would know it, and take adequate precautions to prevent or minimize attendant harms (Abraham 2002, p. 59).

What the reasonable person need not be is *the* average or normal person, *the* average or normal reasonable person, or a reasonable man or woman. The first definition is problematic because it depends upon the society in which the person exists. The better educated the society, then the higher the standard. The worse educated the society, then the lower the standard. Speaking of the latter, an average or normal person in a particular society might be an exceedingly ignorant or reckless individual who cares little for what happens to others, or who bears some other defect in her moral reasoning processes. But if the average person in her society has the same characteristics, then being this type of person or acting based upon these rather bad qualities is what a reasonable person would do. However, reasonableness cannot mean the same thing as base stupidity. Therefore, to protect those who deserve it, respect everyone affected by our actions, and bring about a better world, a higher standard than that is called for.

The average or normal reasonable person standard is superior to the first for it makes the average depend solely on the class of reasonable people, but it is still defective. The actions of children, the mentally infirm, and others without the same average adult's advanced cognitive abilities and reifiable potentialities would be evaluated on the same level as those adults. In consequence, their actions are adult yet they cannot make adult decisions. This creates a fiscally unpleasant situation for their caregivers. Their guardians would be partially responsible for whatever bad action the less competent person intentionally or negligently performed, thereby creating an incentive for the guardians to prevent or monitor the less competent individual in her risky behavior. Although the definition and standard is appealing on egalitarian grounds and might increase security by maximizing moral hazards to the competent, so that they in turn minimize risks to others, not all people should be treated the same. Since it is beyond their capacities to be reasonable people, individuals who are not fully realized moral agents should not be held to such a standard.

Furthermore, the average reasonable person might come to a very different conclusion from another reasonable person, without the minority view being wrong. The mere fact that an opinion is held by enough reasonable people does not in itself make it superior to those held by other reasonable people. If all reasonable opinions lead to flourishing but use different means to achieve it, then *ceteris paribus*, there is no moral reason to favor one over another. In fact, allowing for different reasonable people to make up their minds to find divergent reasonable solutions to problems fosters a better society, with a more engaged populace and dynamic marketplace of ideas than always relying only upon the average.

Finally, on its mere face, the reasonable man or woman standard is less than helpful due to its unnecessary limitation caused by morally irrelevant individual characteristics (Moran 2007, p. 1). No one would disagree that women can be reasonable and hold reasonable beliefs about what should be done different from those of the reasonable man (Moran 2007, pp. 199–201). The same is true for men and the reasonable woman (Moran 2007, pp. 277–8). In addition, there are a plurality of masculinities and femininities that defy any attempts to universalize them. The reasonable man, for example, is often equivocated with the "hegemonic masculinist's" ideal of a white, heterosexual, Protestant, employed, urban man (Kheel 2008, p. 37). By narrowing the reasonable man class too far, all other types of reasonable men, be they of different races, sexual orientation, religions, etc., are excluded from the standard for no other reason than simplicity at best, and illicit discrimination at worst. So who decides who the reasonable man is? An identical problem arises for the reasonable woman standard.

Moreover, claiming that there is such a thing as "the" reasonable person is unjustified given that there are many different types of reasonable people who can legitimately hold different beliefs, act in different ways, make different judgments, etc., and do it all in a reasonable way. Although incorporating this complexity into a definition makes the reasonable person standard more difficult to use, it is more practical to capture reality than to force an easier, fictitious solution that will not help us achieve the beneficial ends we all desire. For example, if we are using a hegemonic masculinist's reasonable man standard, we can find an answer by merely asking someone fitting the description what he thinks. However, what he believes is only part of what may be permissible to believe. Other options can be legitimate.

My solution to the need for an objective standard that allows for the subjective is to list the necessary characteristics of reasonable people, which if instantiated simultaneously in a person, makes the individual a reasonable person. The standard's objectivity can be found in possessing these general features. The standard's subjectivity arises from making the characteristics capable of being instantiated in different ways, as will be seen below. The result is that there might be as many different reasonable people are there are different people, since all people are capable of achieving the necessary features. At the same time, the relatively high standard will eliminate those who intentionally or negligently do not reify all nine features of a reasonable person.

2.3.2 The Reasonable Person's Necessary Characteristics for this Moral Code

A reasonable person has nine traits he uses to accomplish the two general, overall goals of having a good life and making the world a better place. Roughly, they are that the person recognizes what morality is all about, accurately applies her maxims to each situation she encounters, tries to make herself and others better as long as doing so sacrifices nothing of comparable worth, adopts reasonable goals, knows that reasonable people will have different reasonable views at times, strikes the proper balance of emotion and reason in decision making, correctly analyzes and uses the data available to her, is more reluctant to impose risk on others than she is on herself and takes an appropriate precautionary approach in uncertain times. I will address each of these below.

First, the most important component of a reasonable person is her recognition of what morality is all about. A reasonable person accepts and evaluates situations using the idea that morality is:

> a creative, cooperative enterprise whose end is to better the world by trying to realize in ourselves and others nurturing goods such as caring, considerateness, compassion, sympathy, and love. (Holmes 2003, p. 217)

As in Hume's moral theory, the Practical Moral Code requires the existence of emotions allowing us to care about and for others affected by our actions. Unlike Hume's, all interpersonal interactions' primary goal is to cause the world to be a better place by helping each person improve his human empathy, making himself more virtuous by increasing his frugality, appreciation, temperance, self-development, dedication, benevolence, generosity, justice, or any of those other virtues that foster intrinsically valuable entities' flourishing[9,10] and appropriately respecting all intrinsically valuable things, as will be discussed in the next section.

Second, a reasonable person is able to correctly apply his rules of conduct to whatever situations confront him. He will understand how his maxims work and when they are appropriate for the situation. Suppose, for example, a farmer's cattle have been infected with hoof and mouth disease, and he is faced with a dilemma between two alternatives, each of whose utility is negative. He can either destroy all his animals or send them to market to attempt to recoup some of his investment. The first will destroy his financial health, but the second will harm many more people and animals. In this particular situation, the farmer would recognize that he is obligated to perform the lesser of the two evils, if he is using a utilitarian theory to classify the actions. He must annihilate his herd in order to prevent other creatures in the food chain from being infected.

Third, the reasonable person wants to make himself and others better people provided that doing so does not surrender something of greater moral value (Singer 2006, p. 255). The reason he does this is because a reasonable person is not satisfied with the status quo for himself or another, when he or the other can be a better person (Kant 1956, pp. 90–1). When being good is available at a reasonable price, the reasonable person understands being satisfied with less-than-good is a defect. However, if he sacrifices something of comparable moral worth in order to try to become better, he cannot make himself better. The loss of something greater than the increase in value from the betterment of self or another necessarily entails that the situation is worse than it was. It is neither reasonable nor rational to prefer a pure loss of value for inadequate recompense.

Fourth, a reasonable person adopts realistic goals and the means to achieve them, based upon the evidence available, with the understanding these goals may require

---

[9]These virtues are from Peter Wenz's work (Wenz 2005, p. 197).

[10]I believe that Aristotle's definition of the human good as activity of the soul in accordance with virtue can be made consistent with this view of ethics (Aristotle 1941b, 1098a).

some work to achieve them. He does not set too high an end because it is unreasonable to work for an unobtainable goal when there are achievable ends to be had. To pursue world peace instead of helping a person in dire need living next to you, when you can only do one thing, for example, is to act unreasonably. There is little chance of success for the former goal, while the latter is easier and more likely to achieve good. Furthermore, the reasonable person will not set too low a goal because it would be too easy to bring about. The ease in which the end is met would not work toward making him better; his potential will be underdeveloped and his talents will rust (Kant 1956, p. 90). Setting one's ends at a very low threshold does not require the person to improve, merely to maintain the status quo, which in turn leads to stagnation and the prevention of new ideas that can be beneficial to the individual, those for whom he cares, or society as a whole.

The reasonable person's fifth characteristic is that she realizes that other reasonable people can legitimately interpret the same information in different ways from her and come to different conclusions than she has. Although they might not change her decision, she is willing to listen and respect other people's views because she knows she can learn from them and that they have legitimate conclusions. As Mill states in *On Liberty*, "But on every subject on which difference of opinion is possible, the truth depends on a balance to be struck between two sets of conflicting reasons" (Mill 1988a, p. 104). In other words, rarely does one person know the whole truth on her own, and even if she does, her view must be "fully, frequently, and fearlessly discussed, [otherwise] it will be held as a dead dogma, not a living truth" (Mill 1988a, p. 103). In the vast majority of decision cases, there is a correct results' set containing many members from legitimate evaluation processes. If a person performs one of the actions from the permissible action range, then he has acted reasonably, even though it might not be something a different reasonable person would have done for legitimate reasons important to her.

In conjunction with the third, fourth, and fifth characteristics, although there can be reasonable disagreement, a reasonable person will select solutions that can accommodate divergent viewpoints when doing so is practical (Thompson 2007, p. 105). Suppose, for example, that two groups have disparate views on labeling biotechnology. One wants it to be marked, while the other does not. If the dilemma can be resolved by satisfying each group's primary goals and desires, which does not entail that each gets everything the group wants, then the reasonable person would adopt this course of action over those in which a balance is not struck.[11] Since flourishing is better achieved in groups, communities, and societies in which consensus building is a standard way to resolve problems, it is sensible to adopt this decision procedure as well.

Sixth, a reasonable person is more than merely a rational person (Ripstein 2001, p. 7). As with reasonable people, rational people are those who correctly evaluate

---

[11] As will be seen in Chapter 6, the labeling issue can be settled by the simple expedient of labeling being paid for by those who desire it, while those who do not want labeling need not do so for their own products.

evidence in a framework of realistic parameters, and are willing to change parameters or their evaluation provided there is new information available justifying this alteration. Furthermore, they are ready to seek new information rather than merely waiting for it to come to them. Finally, they set sensible goals and the means to achieve the goals. However, rational people are not necessarily reasonable people. For example, a hedonic egoist can be rational and rationally pursue his long term self-interest even though others are severely harmed by his behavior. The fact that he disregards what happens to them in pursuit of his own interests makes him better able to achieve his goals within his egoism framework, but it disqualifies him from being reasonable.

Missing from the rational persons standard are necessary emotions and feelings. David Hume effectively argued that reason is not the sole component in ethics. Desire and emotions play vital roles as well. In fact, if desires and emotions did not exist, our ethics system would not be what it is. We might not even be able to understand what an ethics stripped of moral sentiments or emotion is like because we would have nothing by which to reference it. It would be along the lines of not being able to understand what colors are if we have been blind from birth because we have never perceived color, therefore we have nothing we can use to understand the concept of color.

To prove that desire and emotions are required for morality, Hume asked if fully rational persons would prefer the destruction of the world to the pricking of their thumbs. Suppose we are creatures who have no feelings at all, including a desire to survive or care for anything, including our family members or ourselves. We operate using pure reason alone, much as Mr. Spock is said to do in the *Star Trek* series.[12] Hume states if we are uncaring because we are using pure reason alone, then we will choose neither. That is, if I do not care about myself or others, have no feelings about pain being bad, or have any of the desires human persons actually have, then how can I prefer one alternative over the other? It would not matter to a fully rational person which event came about because he would never be motivated to choose, since motivations are basically desires a person wants fulfilled. Hence, given that ethics is meaningful to us, we have ethical systems, and we can choose between the two alternatives, then it must be the case that desires and emotions of some type are necessary to our ethical systems.

Hume states that sympathy is the universal moral sentiment or emotion allowing morality to exist.[13] Sympathy for Hume is approximately what we currently call "empathy."[14] That is, there is a fellow feeling of care we have for all other humans just because they are human. No one wants another person to be harmed needlessly, and feels bad about it when she discovers occurrences of it. Everyone also has

---

[12]Kant does not believe that moral sentiments have any role to play in normative ethics (Callicott 1999, p. 102).

[13]Callicott claims that these feelings and moral sentiments can vary but there is a range of normalcy for all human beings (Callicott 1999, p. 108).

[14]Kheel calls empathy "the culmination of many small acts of attention" (Kheel 2008, p. 227).

some good feeling when she learns about something positive happening to another, such as a hard working individual receiving some form of adequate reward for her actions.

According to Hume, the sympathy feeling can be stronger or weaker relative to situational conditions. First, the fellow feeling is more powerful the closer in care relation the person is to the observer. For instance, we care more for our own family than we do the neighbors. We care more about neighbors than fellow citizens who do not live near us, and we care more for fellow citizens than we do for foreign nationals not residing in our country. Moreover, the greater similarity a person has to us, the more we feel for him because there is a greater connection between us. The less like us, the less we feel. If the person is vastly different from us, then we tend to feel even less because we do not have enough in common to strike an emotional note in our minds.

There are a number of other factors that can affect our sympathy for others. We tend to feel stronger about infants than adults. Young children tend to stimulate more feeling than older folks, which can be seen by society's negative reactions to crimes committed against them in comparison to the same crimes committed against adults. For example, it is unfortunate for an adult to be beaten by another adult, but a tragedy if the same is done to a child. The "yuck" factor most people have as an instant reaction to such news is stronger in the latter case.

We also feel better or worse for individuals depending on how much of their story we grasp. If we can place ourselves in their position or make the individual more real to us by knowing more about her, then she becomes someone we can sympathize with more. This fact relates to one of the demonization problems often encountered in current debates. If an opponent is demonized, then he is not a person any more. By eliminating those features that stimulate the natural sympathy everyone has to other persons, it is easier to distance emotionally oneself from the particular person, which in turn allows the person to be treated in a manner inconceivable for those for which we care.

Finally, physical distance and time between our existence and others influence the feeling of sympathy we have for them, but we will always have some for any entity we consider a person. A tragedy occurring next door is more real to us than one happening halfway around the world. Furthermore, an event from today is much more influential on our emotional responses than something that occurred many years ago. Harms to people during World War II, for example, create greater feeling on the part of those who lived during that time than it generally does for those who were born long after it. The reason, of course, is that the former experienced the war when it was happening, while the others can only reference it indirectly. However, the more a person can put himself into the position of the person who has experienced similar goods or evils, the greater the feeling of sympathy he has for the latter.

Reasonable people have empathy or sympathy for those affected by their actions. Hence, the main difference between a rational person and a reasonable one is that a reasonable person incorporates all morally relevant considerations, including emotions and feelings, into his evaluation processes rather than some smaller set.[15]

Returning to the rational part of being a reasonable person, a reasonable person correctly analyzes the value of the data available to him for the particular situation in the particular time frame. That is, he gives the evidence he has the proper weight of being reliable, certain, and so on. The information not meeting the particular standard he has legitimately adopted is discarded, while that meeting it is used in the appropriate way for the purpose he has selected. For example, when making decisions of vital importance, he will use only the best data. For more trivial matters, additional information can be included in the set. If he should receive new relevant information, then he will re-analyze the situation to see if his position should be altered accordingly (Rescher 1983, pp. 120–1).

Proper data analysis requires an understanding and use of what rational, irrational, and non-rational beliefs are. After all, data evaluation depends in part on the individual's abilities, standards and beliefs. Difficulties stem from the fact that "assessing the rationality of beliefs about what is probable is more difficult than assessing the rationality of beliefs about what is actual" (Lehman 1995, p. 7). A partial reason for the increased complexity is that some people have better cognitive capacities in certain situations than others. A doctor in a medical emergency situation, for example, is generally able to come to a better supported conclusion than someone without the same background.

In addition, there is the problem of cognitive conservatism, which causes people to hold on to beliefs even when confronted with overwhelming contrary evidence and in some cases, allows them to turn that evidence in favor of the belief (Smith 1994, p. 153). In fact:

> Human history indicates that people will maintain their beliefs not only in the face of apparently contrary evidence but even when those beliefs have severely disagreeable and disadvantageous consequences for them-not to mention for many other people. (Smith 1994, p. 152)

For one case in point, although the evidence for evolution is overwhelming, a Creationist interprets fossil records and other geological data much differently from a scientist. For example, to fit its truncated existence time line, the $27 million Creation Museum, which opened in May 2007 in Petersburg, Kentucky, claims that dinosaurs and humans lived simultaneously. In the TO controversy, as evidenced by a perusal of their website offerings, it is clear that some of the anti-transgenic groups have not modified their assertions against the technology regardless of the positive scientific evidence that has been generated since the issue arose and much less was known about transgenics' risks.[16] It is almost as if the dangers have become part of

---

[15] Another difference is that rational goals may not be reasonable ones because of the differences in ethical considerations included in the decision process.

[16] An activity is risky or an object poses a risk if and only if there is a probability that the activity or object will lead to some harm or other negatively valued outcome (Lehman 1995, p. 22). The

a religious dogma that is keenly felt by the adherents but not critically examined. Sites in support of transgenics sometimes share the same sort of overwhelming faith rather than reason-based beliefs. However, if we are in search of solutions to moral problems and interested in our own and others flourishing, we have to use the best evidence available to us.

Regardless of their background, all reasonable persons hold that irrational beliefs may not permissibly be the basis for a moral decision, that rational beliefs are better than non-rational beliefs for knowledge, and that non-rational beliefs play a vital role in practical morality. As stipulated here, rational and irrational beliefs are based upon evidence, while non-rational belief is primarily generated by non-rational mental processes. A person has a rational belief when he believes a proposition primarily due to the objective evidence the person possess for the proposition's truth outweighing the objective evidence the person has that the proposition is false. A person's belief is irrational if he believes a proposition is true despite the fact he possess sufficient evidence to show that it is false. Finally, a non-rational belief is one that a person believes primarily because of some emotion or desire. In cases in which there is equipoise in regard to the evidence to a proposition's truth or falsity, a person can non-rationally and permissibly believe that the proposition is true or false depending on how the person feels about the proposition. For example, believing in a god who is infinitely good and powerful is irrational once the Problem of Evil argument is known by the individual, but belief in a god with less than those two characteristics is non-rational and permissible until it is proven otherwise.

One reason why taking evidence seriously is so important in debates about policies affecting communities can be seen in a current controversy over wind turbine technology in western New York State in the United States. Among the potential turbine risks listed by opponents are turbine and gas well fires caused by lightning strikes, catastrophic blade failure sending pieces of blade scything off up to 400 m from the tower, ice flying off the blades and hitting people, birds dying from colliding with the tower, wires and blades, and the blade noise creating the same sort of disease said to afflict some people living too close to major airport runways. On the other hand, proponents cite wind turbines' proven record of profitability for owners, turbine farm landlords, and local communities, and the reduction of fossil fuel use, pollution, and reliance on foreign resources. Given the evidence of extremely low risk probability and the much greater dangers posed by well established technology, e.g., cars kill more birds and people than do wind turbines, the question is why would anyone believe that these extremely remote risks are sufficient to neutralize or overwhelm the potential benefits? If this is the only evidence available to reasonable people, then it is irrational to think that wind turbines are more dangerous than common place technology.

In the developed world, especially the United States and European Union both of which have greatly benefitted from it, it sometimes seems odd to meet people

definition's broadness entails a great deal of things are risky or have risk to them, but it captures the essence of what risk is.

who fight against technological developments in energy, biotechnology, agriculture, and elsewhere. Technology and science has been proven as a potent force of positive change in the world, even causing some religions to rethink their fundamental assumptions (Edwards 2002, p. 12). The US became the world power it is, in part, because of its development and adoption of technological solutions to perennial problems, such as feeding its populace and providing sufficient healthcare. And it is clear by even a causal perusal that those who embrace technology generally acquire greater goods and power for themselves. Why is it then that people so vehemently resist certain biotechnology and other technology types and products?

The answer can be teased out of Calvin Coolidge's statement that "When people are bewildered they tend to become credulous." Obviously, one reason for being befuddled is ignorance. This condition can be the result of intellectual sloth, requisite information not existing to be obtained, or an inability to find or understand information once it is exists. People are also poor judges of risk, in part, due to generally held beliefs that technology is either completely safe or dangerous with nothing in between, that nature is benevolent while the artificial is dangerous, and that it is possible and appropriate to abolish risk entirely in technology (Sunstein 2005, pp. 36–7). In the case of food biotechnology, even after public discussions, news coverage, and a wide variety of academic and other information becoming readily available, the public familiarity with transgenic food – 40% – was less in 2006 than it was in 2001 – 44% (Mellman Group, p. 2). Without adequate understanding of issues, it is much easier to sway people by inciting negative emotions.

In addition, there are probably certain universal psychological traits which make people comfortable with the technology with which they grew up, but allow them to feel intimidated or threatened by new science they do not understand or know how to use.[17] From personal experiences and anecdotal evidence, it seems as if people do have this characteristic.[18] They assume their lives were much safer or simpler when they were younger and wish to freeze development at the *status quo* of that time either from a sense of laziness not to learn how the new technology works, or desire for what they are comfortable with and what they consider to be known by them. In addition, we should never forget that the more dramatic the technological change in the status quo, the more dramatic the response to it (Rollin 2006, p. 131). Again, it is a matter of one's accepted, comfortable lifestyle seemingly being threatened by technology that one does not readily grasp.

Adding to people's bewilderment is science's failure to predict accurately certain outcomes, as well as scientific research being demagogued by the media and others beyond what the findings actually support. Rollin goes so far as to claim that the "public confidence in scientific reassurances has precipitously diminished as a result of an apparently endless list of scientific prognostications gone afoul" (Rollin

[17] Jessica Hutchings has done some excellent work on the Maori's opposition to transgenic technology. Many of the objections raised to TOs also surface in the Maori study, including TOs violating Maori culture and causing spiritual and moral offense (Hutchings 2004, p. 181).

[18] These are never great sources of evidence to establish a claim, but are useful to indicate a direction in which to begin to find an argument.

2006, p. 5). Hindmarsh and Hulsman accuse scientists of too zealously applying the "reductive genetics laboratory approach" to transgenic organisms and their consequences once they are released into the general population (Hindmarsh and Hulsman 2004, p. 54). Robert L. Zimdahl states that science and scientists have lost credibility due to the perceptions that they can be unduly influenced by social, economic, and political pressures and the negative consequences that occurred even after some scientists had assured the public that they would not happen (Zimdahl 2006, pp. 138–9, 150). In other words, many members of the public believe that scientists are either incompetent or driven by unethical considerations to skew their work in ways that benefit only those controlling it (Rollin 2006, pp. 5–6). Their findings, research, and technological developments automatically become suspect in the minds of many and decisions are made based not on the evidence but on the scientists' perceived character traits (Thompson 1995, p. 37).[19]

The result can be hysteria from the general public and people who should know better. In 1999, Vandana Shiva accused the US of treating storm victims as guinea pigs to test a high nutrition mixture of corn and soy meal for its effect on human populations. The trouble with her allegation was that it was unsupported by any evidence of duplicity. In fact, there was no need to test the mixture because US citizens had been eating it for years without any harmful effect (Bailey 2002, p. 34). Anyone knowing the situation would have immediately grasped that it made no practical sense to test it on storm victims. In consequence, all that was accomplished was scaring people away from a product Shiva thought dangerous that distracted from the efforts to help them and others. Acting in this manner is clearly inconsistent with anyone's flourishing, although it does get attention.

Beliefs based on the fear of new technology are fallacious for two reasons. First, although people believe their lives were simpler in the past, life has always been difficult relevant to the exigent circumstances. Consider how those who preceded them thought cars and other motorized devices were too complicated and frightening and longed for a return to the day of horse drawn conveyances. In Fargo, ND, for example, before a motorized vehicle could enter the city, someone had to run ahead of it to warn people that it was coming. Second, technology-phobes assume that if something is familiar to them, then they know all relevant information about it, but these beliefs are based upon unsupported non-rational or irrational desires rather than on a full understanding of the item and the consequences of its use. For example, people utilize cars, telephones, microwave ovens, and so on, without knowing how they actually work. However, just because someone makes do in general without understanding all established technology's risks does not entail that the technology is safe

[19]The general consensus to rectifying this situation seems to be for greater engagement of the public in science and scientists in non-scientific areas affecting their work. Scientists should consider the consequences of their work on society and elsewhere and take advantage of the "cooperative rationality and social wisdom in the social enterprise" before they begin their research or create new technology (Zimdahl 2006, pp. 23–4).

or all risks are known, e.g., lead in paint was widely used in residential applications until it was found to be a significant contributor to lead poisoning in children.[20]

Scientific evidence is helpful in eliminating ignorance and judgments but, in a world that respects individuality and individual autonomy, should never be taken as the only information relevant to many decisions in the public or private realms (Fincham and Ravetz 1991, pp. 131–4). There is always the danger of an excessive reliance on science as the sole or primary basis "on which we construct our complicated society" (Teitel 2002, p. 25). For some, science becomes a religion that answers all questions in life, such as those found in ethics. For others, science is a totally value neutral endeavor; so, scientific and technological advances should only be evaluated objectively within the field and not on external subjective considerations (Fox 1999, pp. 1, 5, 9; Rollin 1996, pp. 16–17; Teitel 2002, p. 25). There is also the danger of misusing science or scientific credentials. Some scientists have manipulated trials to produce the results they seek, such as what seems to have happened in the Pusztai transgenic potato and Ermakova transgenic soybean studies (Bailey 2002, pp. 38–9; Preston 2005). More egregious are multiple paperbound and internet citations of a linkage made by Traavik in 2004 between Philippine villagers suffering from respiratory, intestinal, and skin problems and exposure to Bt corn pollen.[21] If there actually had been a cause and effect relationship, then the news would have been devastating to Bt corn, and possibly, transgenic technology as a whole. However, after three years no valid evidence has been printed supporting Traavik's conclusion. The conclusion that must be drawn is that the mere fact a scientist asserts a claim does not make the claim true nor always adequately supports it enough to make it a rational belief.

Proper care needs to be taken to sort out which scientific information is valuable and which may be reliably placed aside. The problem is made more difficult by the fact so many sciences are involved in decisions about technology, business, and public policy.[22] However, practical measures can be adopted and used. First, the source of the information should be considered (Lehman 1995, p. 10). If a scientist has a particular stake in the development of a technology, then her claim about its potential should be evaluated in light of the conflict of interest, although the conflict does not entail that the information is illegitimate. The same situation applies to those researchers who oppose the technology and their findings. The informations best source, of course, is a neutral or independent party when such is available, but if one is not, then transparency is vital for all conflicted parties to maintain. The published or publicly available information should make clear any and all actual or apparent conflicts. In addition, scientific claims should be judged according to how well they follow the scientific method for the particular area, if they are consistent

[20] As will be seen in Chapter 6, conventional and organic produce can be much more dangerous than transgenic goods, even though many people have used the former for many years.

[21] See Fox (1999, p. 124).

[22] Bernhard Glaeser claims that in human ecology the sciences of biology, geography, psychology, sociology, anthropology, medicine, geology, mineralogy, botany, zoology, and ecology are involved (Glaeser 1995, pp. 7–9).

with generally accepted "facts" in the field, if they, the method used for testing them and the data supporting them have undergone rigorous peer review, and if the data supporting them have been confirmed by replication. A claim that has satisfied all five criteria will be better than any claim lacking at least one.[23] However, given science's nature, the evidence will never allow anyone to make an absolutely certain claim.[24,25]

Reasonable people take all evidence types seriously, and use them to evaluate beliefs. They know that it is irrational to reject all possible forms of technology and that some objections to technology are rational because there is always the possibility of some bad consequence resulting from its creation or use (Lehman 1995, pp. 200, 205). When evaluating available information, the reasonable person will know that there are conflicts between certain types of beliefs and will weigh the beliefs according to their respective evidentiary weight and moral permissibility. Generally, if a rational and an irrational belief are contradictory, then the rational belief defeats the irrational one. For example, if a person is opposed to any possible form of biotechnology when the person is aware that not all biotechnology is problematic, then the person is operating under irrational belief states (Lehman 1995, p. 200). She has an obligation to alter her thinking and behavior appropriately otherwise she is not acting as a reasonable person. If there is a conflict between rational and non-rational beliefs, then the issue becomes more difficult to resolve. However, since emotions are subjective by nature, and rationality is universal or at least understandable if people know the paradigm by which the person made her decision, then for the sake of practically solving moral dilemmas, rationality trumps non-rationality. Decisions, when possible, should be based on evidence rather than emotion. When evidence for the truth and falsity of a proposition is lacking, a non-rational belief is morally permissible with the condition that if relevant evidence becomes available, the person will take it seriously, and then alter her belief states accordingly.[26]

[23]Ralph Nader thinks that openness, vigorous peer review, and intolerance of commercial repression are sufficient for good science (Nader 2002, p. 48).

[24]It would be useful for findings in controversial areas to be made more available to the public. One of the greatest problems to finding solutions to such issues is a lack of reliable information.

[25]PB Thompson argues that science and scientists have a moral duty to "develop a better conceptualization of risk and to engage the broader public in conversation and deliberation on the types of science that they undertake" (Thompson 2007, p. 304). Fulfilling the obligations would eliminate some of the unwarranted fears about technology so that people could make decisions based on evidence rather than mere feelings.

[26]Many people believe in psychic powers such as mind reading and premonitions (BBC News 2007). There also seems to have been a rejection of science and an acceptance of mysticism such as that of crystals and Wicca (Rollin 2006, p. 5). Such beliefs are permissible for the individual if and only if there is no evidence readily available that debunks the beliefs. In the cases of psychic powers, the fact that no legitimate scientific test has ever found evidence for their existence is more than sufficient to show that belief in them is irrational.

There are specific types of evidence a reasonable person uses to evaluate situations for what is morally permitted, required, or forbidden. Reasonable people seek out and apply information about the:

1. External world society's rules, practices, and customs.
2. Rules and responsibilities associated with specific roles the agent is playing at the time.
3. Claims that others have on the agent and the agent has on others.
4. Maxims growing out of previous judgments that the agent has made, in order to maintain ethical consistency.
5. In conflict situations, what is right on balance.
6. Which consequences are important and their value, as well as the value of other relevant things.
7. Which if any mediated consequences count and which do not. (Holmes 2003, pp. 215–17).

Once again, a form of relativism is clear in RPU and the reasonable person standard. Since people have made different judgments in their past, for example, one agent might come to a conclusion different from another who has the same information but a different background. In other circumstances, two agents, in virtually identical situations can have different moral duties based upon the society in which they find themselves. One society might require certain ethical behavior other societies forbid. For example, a man might be required to give up his bus seat to a woman in one society, but doing so would be morally offensive in another. However, a reasonable person will find and evaluate the seven types of evidence available in the circumstances prior to making an adequate determination as to which further action the agent may or should do.

The eighth characteristic of a reasonable person is that he is more reluctant to impose risk on others than he is on himself because he thinks people should decide for themselves how much risk they are willing to take.[27] In addition, as the risks increase for an action with reduced chance of offsetting benefits for those affected by his actions, he becomes more reluctant to choose or to perform the action. Again, the reasonable person's concern is to let people make their own decisions about actions that affect them in significant ways rather than acting paternalistically toward them. Severe harm and death tend to be the boundary on his decisions in these situations. No matter what the possible benefits from an action, if someone is likely to die as a result of the action, then the reasonable person is unlikely to take the risk (Rescher 1983, p. 122).

[27] As Reiss and Straughan recognize, when risk and safety raise questions about responsibility, accountability, and justifiability, they become matters of moral concern (Reiss and Straughan 2001, p. 53).

Nicholas Rescher also argues that reasonable people do careful cost/benefit risk analysis for alternatives open to themselves and others. According to Rescher:

> For any given level of benefit, people are prepared to tolerate a greater level of risk for activities that rate more highly in point of being: voluntary; avoidable; controllable; familiar (i.e., not particularly striking, memorable, shocking); well understood; not dreaded; not potentially disastrous; remote (not immediate or near-term). (Rescher 1983, p. 123)

In other words, when an action's advantages are low and none of the seven factors above are instantiated or likely to be, a reasonable person is likely to believe the action unreasonable and is justified in drawing this conclusion. As the benefits or potential levels of the seven factors increase, the action's reasonableness increases. Highly risky activities can be sensible as long as the benefits are sufficient and the action is completely voluntary, avoidable, controllable, familiar, well understood, not dreaded, not potentially disastrous, and remote.

Ninth, related to the desire to minimize change from that which one is comfortable, reasonable people take a precautionary approach in unknown situations. They believe that it is better to leave things as they are than to alter them, if there is no reason to believe things will be better as a result of the change. In favoring the status quo, reasonable people are very conservative. They will act according to what they know, which has provided them with the evidence allowing them to generate their general rules-of-thumb about beneficial behavior, rather than venture into the unknown, which relies upon speculation and insufficient data.

However, although all reasonable people consider precaution in making their decisions, they are unlikely to use most versions of the Precautionary Principle. From even the shallowest review of the literature on transgenics, it is clear that there are multiple interpretations of the principle rather than one to which all people refer. What is held in common to all of them is the intuitively appealing idea of maintaining the *status quo* until certain risks from the product have been identified and dealt with adequately. As stated above, all reasonable people would adopt such a belief.

Although it poses problems, it must be acknowledged that arguments and suggestions for precaution and a Precautionary Principle are, when viewed as general ideas, very appealing. In the face of uncertainty, we should be careful to avoid upsetting the environment and harming those with interests when doing so is not necessary. Even Cass Sunstein, no fan of the Precautionary Principle, states that the higher the magnitude of harm that might occur from technology, then the less evidence is needed for its probability (Sunstein 2005, p. 115). This means that decisions about technology creation and adoption rest upon relativistic circumstances. Furthermore, rules to allow us to achieve this goal seem reasonable on first face. For example, Kerry Whiteside advises us to:

1. Set up research programs whose purpose is to gather further information about the risk and test successive hypotheses about it.
2. Institute long-term environmental and health monitoring. If doubts remain about some technology, then labeling and traceability should be required.

3. Multi-disciplinary expertise to evaluate the technology should be orchestrated.
4. Reinforcing the independence of regulatory bodies.
5. Systematically favoring "clean" technology.
6. Building larger safety margins, devising backup safety systems, and putting emergency plans into place.
7. In the most potentially serious and uncertain risk cases, precaution can mean banning a technology or strictly minimizing its use. (Whiteside 2006, pp. 53–5).

When ignorance of overall outcomes is all that anyone has to evaluate in situations in which both the impacts and probabilities are unknown, then the Precautionary Principle should take action to "anticipate, identify, and reduce the impact of 'surprises' of products ... promotion of robust, diverse and adaptable technologies and social arrangements to meet needs" (Harremoes et al. 2002, p. 217). All of these suggestions would be agreed to by all reasonable people – much as they would concur that we need to be kind to others – but the difficulty, especially when it comes to regulation, is determining what each of these means and what it entails. For example, how much is enough for Whiteside's sixth suggestion? What are "robust, diverse and adaptable technologies"? More importantly, who gets to decide the answers to these questions? The same concern applies to all seven of the suggestions as well as the Precautionary Principle in general: who gets to choose how to fulfill these?

There is a great deal of variation on what entities are covered by the principle and how its basic terms are to be defined. Both the Rio Declaration's principle 15:

> where there are threats of serious or irreversible damage, the lack of full scientific certainty shall not be used as a reason for postponing cost-effective measures to prevent environmental degradation.

and European Communities Treaty's versions, for example, mention only risks to the environment, while the Commission of the European Communities has expanded its principle's scope to include human, animal and plant health (CEC 2000, pp. 9–11). Poul Harremoes et al.'s Precautionary Principle closely mirrors the Rio Declaration but adds that stakeholders use a cooperative approach to take precaution when there are potentially serious or irreversible threats to health or the environment (Harremoes et al. 2002, p. 4).[28] Of course, the issue of what is covered is vital to making informed decisions and evaluating claims about actions and obligations. If only the environment counts, then actions minimizing risk to it, while increasing danger to humans, animals, or plants, are permissible. On the other hand, a principle which includes the environment and any of those three groups would classify the same act as wrong. Since, businesses, government officials, and others need to have clear principles on which to base their decisions and actions, and there

---

[28] Sunstein takes a similar approach when discussing regulation. According to him, cost/benefit analysis is important, but democracies can choose to do what is important to them even when it is not cost efficient (Sunstein 2005, pp. 129–30).

is no consensus on which rule and interpretation to employ, there is good reason to reject the Precautionary Principle on the grounds of insurmountable vagueness and ambiguity unlikely to find consensus solutions.[29]

The more serious problem is that the Precautionary Principle can be impractical and gives inaccurate ethical classifications in many cases. According to Greenpeace, the principle means that based on the available evidence, caution should be taken for all activities that might harm human health or the environment even if the full extent of the harm has not been scientifically established (Greenpeace International 2007). Other interpretations of the principle are just as draconian. The Precautionary Principle "would more actively encourage that minimal or no adverse interference occurs with the nonhuman world from development practices or technologies" (Hindmarsh and Hulsman 2004, p. 56).[30]

Besides providing informative examples of the detrimental impact of technology that should have been identified and dealt with earlier on than was indeed the case, Harremoes et al.'s justification of a strong Precautionary Principle sheds light on two subjects. First, the book amply shows an example of well-intentioned authors who do not fully grasp their argument's actual defects. The editors state that there are no uncontroversial "false positives" on which all can agree (Harremoes et al. 2002, pp. 3–4). A false positive is an example in which people raised a hue and cry about a technology only to discover that it was not as hazardous as feared. At the same time, false negatives – in which a product or activity was wrongly thought to be harmless – there are in plenty (Ibid., p. 3). It never seems to occur to the authors to question why one class would have obvious members, while the other is empty. And this is what brings us to the second subject on which light is shed: The reason that there can never be an uncontroversial false positive is simply because all technology has inherent risks that will affect health or the environment that anyone can use to trigger use of the Precautionary Principle.

Although the immediately preceding claim might seem overly strong if not extravagant, it is well borne out by the evidence. Medical procedures, drugs, and other technology, for example, are tested repeatedly for morbidity and mortality before being allowed into the market, yet some of it has been proven to have negative, unforeseen side effects. Even those products about which there is little concern have had long term negative impact on both human health and the environment. For example, it is more than clear that humans have altered homo sapiens' genome by using medicine and other technological innovations that allow those with physical

[29] Philip Davies requires that TOs not be released until there is sufficient evidence to establish that they will not have a significant effect on the environment, but never states how much and what kind of information would be enough (Davies 2004, p. 75). Given the rest of his claims about the possible dangers, it appears that he requires certainty that TOs will not pose a threat which is a scientific impossibility.

[30] Judy Carman argues that all transgenics should undergo extensive animal trials in the same way that drugs are tested (Carman 2004, p. 90). Given her acknowledgement that there is a lack of funding and interest to investigate transgenics and the virtual impossibility of connecting transgenics to disease, it is clear that the evidentiary burden will be too high to satisfy.

detriments, such as weaker immune systems, to survive and reproduce rather than die off quickly (Jonas 1981, p. 247; Taylor 1998, p. 24). Therefore, the beneficial effects of competition in non-artificial environments have been reduced, if not outright eliminated, in order to save those to whom we feel some moral obligation. In fact, modern technology has altered the very nature of human action (Jonas 1981, p. 23). Moreover, with the discovery and use of medicine to eliminate smallpox, malaria, and other often terminal diseases, we have increased the world's population and the demand for food, energy, and other resources. The electric light development would be thought to be a boon to humanity and others, but it is proving detrimental to both humans, animals, and the environment. In one study, bright lights are shown to correlate to breast cancer rates twice as high as areas without them. Furthermore, herbivores that eat less under moonlight to avoid predators also limit their eating near urban development because of the night glare from electric lights. The result is chronic underfeeding for them (Harder 2002, pp. 16–18). Cars are detrimental to lives – the accidents caused, road rage, etc. – to communities – by causing people to be isolated in their cars from each other and of course, to the environment with the effects of paving, greenhouse gases, and all other negative consequences of such technology. In fact, these are not isolated cases. All technology, because it alters the artificial and natural environments and uses resources, will have negative impact on some individual or class of things. There will always be something that is used up or no longer receives what it did before the technology was developed, thereby harming it. The Precautionary Principle has no obvious mechanism in place to evaluate objectively which technology should undergo serious study and for how long. That is, since there is no way to know which technology will have negative impact and to what extent because they are unknown by definition, then it would have to apply to all technology.

Oddly enough, maintaining the status quo will fare no better with extreme Precautionary Principles than forging ahead with technological innovations. In its stronger forms, the principle permits nothing because it is uncertain which course of action will achieve the safety level desired by proponents (Sunstein 2005, p. 14).[31] That is, all behavior entails risks; hence, if risk is to be prohibited, then no actions can be permitted. Since it violates the practical view that, granted all human actions and the products they use or create have some risk involved, we can act ethically anyway, it follows that excessively strong versions of the Precautionary Principle are to be rejected.

The strongest versions of the Precautionary Principle are clearly defective, but what about more moderate versions? The EU's more moderate political version of the principle is to achieve a high level of protection "when there are reasonable grounds for concern that potential hazards may affect the environment or human, animal or plant health, and when at the same time the available data preclude a

[31] Gordon Graham argues along similar lines when he states that a Precautionary Principle based on the possible catastrophic results caused by adopting biotechnology leads to a situation in which no action is permissible. Simultaneously, accepting and rejecting the technology is forbidden because each has a possibility of creating a catastrophe (Graham 2002, p. 130).

detailed risk evaluation" (CEC 2000, p. 9). The EU has argued that because of their inherent character, transgenics require "particular scrutiny" and a stringent precautionary approach (WTO 2006b, 4.502).

The European Union established a risk assessment procedure for each proposed transgenic organism in Directive 2001/18/EC. In order for a transgenic to be approved for release into the EU's market, transgenic producers and companies must:

1. [Prove the TO passes] theenvironmental risk assessment, which includes studying the short and long term effects of the product on the environment and human health from cumulative and long term use, [which requires]
   a. Identification of any characteristics of the [TO] which may cause adverse effects,
   b. Evaluation of the potential consequences of each adverse effect,
   c. Evaluation of the likelihood of the occurrence of each identified potential adverse effect,
   d. Estimation of the risk posed by each identified characteristic of the [TO], and
   e. Determination of the overall risk of the [TO].
2. Implement mandatory post-market monitoring, including collecting data on long-term effects associated with the interaction with other transgenic organisms and the environment,
3. Provide mandatory information to the EU's public,
4. Implement labeling and traceability at all stages of the placing on the market-farm to fork rule,
5. Go through mandatory consultation with the Scientific Committees,
6. Consult the European Parliament on decisions to authorize the release of the candidate TO, and
7. Pass through the Council of Ministers for the authorization of the TO by a simple majority. (EC 2003a, p. 2)

Since the EU uses the Precautionary Principle to interpret the conditions, producers and companies must show that each product is safe, rather than there not being any unreasonable risks associated with it. The result for producers and others is providing information about the cumulative and long term direct and indirect effects on human health becomes very difficult and expensive in terms of time and resources. Given the complexity of the process, its evidentiary requirements and the lack of uniform administration of EU customs law, it is unsurprising that it takes over six years for some TOs to pass through the EU's regulatory system (USTR 2003, p. 112; USTR 2006, p. 236).

The EU's principle is weaker than that of Greenpeace because of the former's reasonable person standard, but it still gives inadequate guidance for making decisions. After all, reasonable people can reasonably disagree about risk. What is classified as unwarranted risk in one person's evaluation might be acceptable to another. Given this fact, when applying the EU's principle, it is unclear which reasonable person is supposed to win the argument on approval or marketing of transgenics. If Monsanto, one of the largest players in transgenic technology, states that reasonable grounds do not exist while an EU politician who knows nothing about the science says there are reasonable grounds, then does the EU's Precautionary Principle allow the product to be delayed, labeled, or banned from the market? In practice, the answer is a very strong affirmative as evidenced by the EU's *de facto* moratorium on TO releases imposed before 2003 and the restrictive regulatory framework in place today. Any imagined concern about risk is often counted as equally legitimate to contradictory beliefs well grounded in scientific evidence (WTO 2006b, 8.5–6). As a result, technological progress becomes virtually impossible in any contentious area where reasonable people reasonably disagree.

Besides ambiguity and vagueness, another shortcoming that Precautionary Principle supporters seem to share universally is an ignorance of personal identity and future generations. Whiteside, for example, argues in part for his version of the Precautionary Principle on the grounds that the damages and their effects from new risks can take years to become evident and last for generations, and that there is no way to judge technological policy adequately because we do not feel its affects immediately (Whiteside 2006, pp. 33–4). In fact, many believe that present generations might get the benefits of technology released into markets in the near future while future generations such as our grandchildren's children have to pay the costs (Whiteside 2006, p. 34).[32] However, even if technology should alter our environment so that we cannot survive as we currently do in it, it does not follow that we have done anything wrong to those who come after us, as can be seen in Derek Parfit's work.

In *Reasons and Persons*, Parfit considers claims about non-identity, future generations, and whether bringing someone into existence is a harm or benefit to him. I will first explicate several of Parfit's ontological and meta-ethical claims in regard to humans, which can be applied to any entity that has an interest.

Parfit is an essentialist because, according to him, human identity in general and for each person in particular has certain necessary features. One required characteristic is the time at which conception occurred and the genetic material used in the event.[33] This claim put more formally is:

> (TDP) If any particular person had not been conceived within a month of the time when he was in fact conceived, he would in fact never have existed. (Parfit 1992, p. 352)

---

[32]Here is where we begin to see the non-rational desire to maintain the status quo for our species, society, environment, and biosphere.

[33]Jones states that, "If that egg and that sperm do not unite, he cannot come to exist" (Jonas 1981, p. 249). Unlike Parfit, Jones places too much emphasis on the impact a person's genetic endowment has on the person's identity (Jonas 1981, p. 248). Kripke makes a stronger claim that both an object's origin and the substance from which it is made are essential to the object (Kripke 1980, p. 114 footnote). A table's identity, for example, is bound up essentially with the exact wood that is used to make it. With Parfit, it seems if the genetic information is exactly the same, the result is the same individual.

Parfit asserts that TDP is only contingently true. Since we can imagine science fiction scenarios dealing with random swamp gases combining in special ways or other causally unlikely factors that would result in the same person being conceived, then in some possible world or worlds, this can and does happen. Therefore, TDP is not a causal necessity, but the facts of our world correspond to TDP's proposition a sufficient number of times to make it reliable.

Although each person's complete identity is not wholly dependent on when she was conceived, that event does have great bearing on who the person becomes. First, the individual's genetic make-up is fundamentally significant to who he is and who he will be. In order to be X and Y's biological son, for example, he must have a combination of X and Y's DNA. Thus each person's identity is in part the result of a particular maternal ovum and a particular paternal spermatozoon meeting in a particular way at a particular time (Parfit 1992, p. 352). Parfit goes on to state that if it had been a different ovum and the same spermatozoon or a different spermatozoon and the same ovum, then whether or not the same person would have existed would be indeterminate. The alteration in genetic material entails that we cannot tell if there is sufficient DNA similarity to be the same or a different entity. However, if both the ovum and spermatozoon were different, then it clearly would not have been the same entity as that X and Y would have produced with their particular ovum and spermatozoon at the original time. For Parfit, there is insufficient resemblance in this case even though ova and spermatozoa from the same sources would have produced an entity with a high degree of genetic similarity.

There are other factors that determine an individual's identity. The characteristics the DNA causes an entity to exhibit are part of who the person is. A person deaf from birth due to a genetic condition will turn out to be a much different individual from one born with auditory capabilities. That is, if the two are virtually identical in every way except for the fact that one has deafness as a genetic trait while the other does not, they will be essentially different. The former's conception of life, communication, experiences, etc., allow him to lead an equally good but significantly altered existence from someone who will have incomparable auditory experiences. This result will also be true for anyone who grows up in a sufficiently different set of circumstances, regardless of whether the circumstances are in part caused by a genetic trait. Besides nature, the nurture element or environment in which one is raised or finds herself helps craft who the person is. Although there are other elements to personal identity over time that could be considered here, these three examined by Parfit – time, nature, and nurture – are all that is required for the argument.

Besides his personal identity view, Parfit bases his arguments on a moral principle that is familiar to many. Parfit is a utilitarian who focuses on lives worth living. He believes that:

> An act benefits someone if its consequence is that someone is benefited more. An act harms someone if its consequence is that someone is harmed more. The act that benefits people most is the act whose consequence is that people are benefited most. (Parfit 1992, p. 69)

As Parfit argues, this notion of benefit is not the one most used in ordinary discourse. Generally, when people are talking about an act benefiting another, what they mean is that the action directly produces the benefit or that it is the benefit's primary contributor.

Parfit's definition has two features not found in the standard usage. First, his allows for remote causations as long as the action is necessary to the benefit being received. That is, if the action did not occur, then the benefit would not have occurred (Parfit 1992, p. 69). Second, it provides a better way to measure benefits. Suppose that I am feeling in a generous mood and have a $20 and two children who want it. I can give it to either one. If I hand it over to the first one, then the second receives nothing. If I give it to the second, then the first one's mother will give her child $20. According to Parfit's definition of benefit, I do not benefit the first child by giving him money because regardless to which child I hand the money, he will receive $20. However, I do benefit the second only if I give him the bill because he would not have received any money except for the fact that I gave it to him. Therefore, benefit should be considered against what else would have occurred had not the action been performed.

The identity and benefit issues are especially relevant to future generations of experiencing creatures; such as if we have obligations to them to guarantee they have the best lives or merely lives worth living. The question that will be tied to transgenic organisms' impact on future generations is whether a person who is or would be actual can be benefited merely by being brought into existence as a result of genetic engineering. Parfit states the answer to the question is an affirmative if the relevant context is provided.

> Causing someone to exist is a special case because the alternative would not have been worse for this person. . .for this reason, causing someone to exist cannot be better for this person. But it may be good for this person. (Parfit 1992, p. 489)

Although what Parfit claims seems confusing, understanding it hinges upon realizing that he is comparing different states of affairs in two situations. First, consider whether or not causing someone to exist is better or worse for the individual on the grounds that he exists in one situation and does not exist in the other. Given that if a person does not subsist, then he is no worse off because he does not exist to be able to be worse off; therefore, it cannot be better for him to exist than not to exist. On the other hand, if the person has a life worth living, then causing him to exist is a benefit to or good for him provided that he subsists or would subsist. It is morally better for him to exist than not to exist (Parfit 1992, p. 391). On the other hand, if his life is not worth living, then causing him to exist would be bad for him. Moreover, the person's life can be compared to other people's lives or an ideal life for him. Some people have lives that are barely worth living, while others have lives that far exceed the minimal level (Parfit 1992, p. 489). If a person is brought into existence with a life barely worth living, then he has been benefited, but a person with a life well worth living is benefited much more than the former. However, it is good for both individuals to exist.

The questions are whether what we do now, including following risky policies, depleting resources, and so on will harm future generations, or if we have a duty not to bring into existence creatures whose lives will not be the best all things considered, but are better than lives not worth living (Parfit 1992, Chapter 16). To find an answer, we must recognize that if we act in one way, then one group of future people will exist, and if we perform an alternative action, then we will cause a different set of people to exist. Assume that both sets of people have good lives, but in the first instance the lives are barely worth living and in the second, they are well worth living. Since not causing an individual to exist does not harm it and causing someone to exist with a good life is a benefit, then it follows that no one is injured in either alternative. This result is counter-intuitive. If one group is far better off than the other, then it would seem obligatory to strive for that outcome, even in risky policy cases.

Our intuitions, however, are mistaken. In Parfit's Risky Policy case, due to a decision we make about social policy right now, future people will die at the age of 40 from radiation poisoning (Parfit 1992, p. 372). However, even though they have shorter lives than they and others would desire, their lives are still overall worth living. On the other hand, if we had taken a Safe Policy, then a group of people would exist who would be able to live much longer. Their lives are far better than those produced through the risky policy. Even though it might appear obvious that our duty is to take the safe policy, we cannot argue for that conclusion based on the injuries caused to those who would exist in the two alternative worlds. Given that no one is harmed by either choice because if they do not exist, then they cannot be injured and if they do, then they have lives worth living, it follows that nothing is done wrong merely by picking the risky policy over the safe one (Ibid.). There might be other reasons for the action to be unethical, but referring to the future generations' life values will not help establish that case.

Bringing together the various concepts about identity, benefit, and the utilitarian normative principle, if a moral agent brings a person into existence that has a life worth living, *ceteris paribus*, then the agent has done something morally right. Alternatively, if the life is not worth living, then the agent has acted unethically. Moreover, for those with lives worth living, depending on the person's life value, the agent has done something better or worse by bringing into existence the person. That is, the better the person's life, then the greater the good derived from creating him. The opposite is true for those who have brought into existence persons who will have lives not worth living.

The question now arises as to whether agents injure future generations by creating technology that had catastrophic consequences. The answer depends upon the life that the individuals will lead. If it is worth living, i.e., one that has more positives than negatives, then no harm has been done. Injury is committed to the entity only if its life turns out to be not worth living. In this latter case, further arguments in favor of the technology would have to be made to justify the harm. For example, the disvalue caused to them by their lives is outweighed by the good that will come from them, or human beings have a right to use the technology now so that the former can survive, could be the foci of additional reasoning lines. The point is that mere injury or risk is insufficient on its own to require a precautionary approach or principle.

Is it possible to have a practical Precautionary Principle? Perhaps Cass Sunstein's work on the Precautionary Principle should be mandatory reading for anyone who is struggling to understand and apply the principle to real life situations. First, Sunstein understands that the general idea behind the Precautionary Principle is one used by reasonable people. We should normally avoid hazards even when there is a lack of decisive evidence that such hazards will occur (Sunstein 2005, pp. 18, 23–4). Second, unlike many in the field (Whiteside 2006, p. ix), he recognizes that general claims about the attitudes of US and European Union citizens toward precaution are often mistaken (Sunstein 2005, pp. 14–15). People on both sides of the Atlantic Ocean tend to take a precautionary approach when psychologically inclined to do so. For example, if people fear a certain technology, then they are more likely to take precaution than are those who do not have the same negative reaction.

Third, Sunstein's explanation of how people evaluate risk is extremely useful in deciding how to comprehend precaution. People's risk evaluation and cost/benefit analyses are based not on pure reason alone; rather, they are heavily influenced by the information each person considers most pertinent to her and the social situation she is in at the time the decision is made. More specifically, there are cultural and social dimensions of fear and risk perception, as well as the predispositions each individual brings into his or her evaluation process (Sunstein 2005, pp. 92–3). Other variables affecting an agent's decisions include the "availability heuristic" (Sunstein 2002, p. 33) which makes a person believe that a risk is more probable if the person can recall an incidence of its occurrence in the past from news reports or other sources. The risk perception becomes stronger if the incidence is closely associated with personal experience. Another factor influencing decisions is probability neglect, which causes people to focus on the worst case no matter how improbable it is (Sunstein 2005, p. 35). In addition, if the public sees technology as having low benefit, then it is more likely to associate the technology with high risk. If the technology has great benefit, then the public's assumption is that the risk is relatively low (Sunstein 2002, pp. 40–1). Social cascades and group polarization can also affect cost/benefit analyses as well as risk perceptions. For the former, if availability and salience are sufficiently high for the population, one person's fear cascades through the rest of the group, building as it goes (Sunstein 2005, p. 94). In group polarization, on the other hand, like-minded people will move from moderate to more extreme positions as they discuss an issue (Sunstein 2005, p. 98). Although it would seem counterintuitive – they should have a better understanding after public discourse – the force of persuasive arguments, the over confidence caused by having one's views echoed and magnified, the emotional contagions of groups, and the value of seeing oneself as part of the group work together to make the group less able to evaluate risks thoughtfully than the individual herself would have been. Other psychological factors include how the evaluator views the badness of the outcome and whether that outcome is vivid for him. As the perceived badness and vividness of the outcome increases the less accurate the risk evaluation becomes (Sunstein 2002, p. 45). People also tend to favor saving a higher proportion of people in situations in which people are endangered than they are in saving more people overall (Ibid., p. 47). That is, if 100 people are at risk and 10% of them can be saved, and 100,000 people are at risk and 1% of them can be saved, then many people will select the former over the latter even though in the first case only 10 are saved, while in the latter 1,000 people are helped. Finally, when adequate information is unavailable, people's attention is selective not only for the above reasons, but also because of loss aversion which causes a negative change from the status quo, the non-rational belief in nature's alleged benevolence over the artificial, and system neglect, which is the inability to realize that risk is inherent in all systems and intervening in said systems also creates risks (Sunstein 2005, pp. 36–46).

It is now that we should begin to understand why the Precautionary Principle has been criticized by so many individuals, and should be viewed with skepticism. If people in general have these psychological impediments to an adequate evaluation of controversial technology, then why should they be the arbiters of public policy, especially since policy based on these features rather than reason can adversely affect a great number of businesses and their stakeholders? There has to be a more objective standard that can be used to fairly make decisions about precaution on which most can agree. Perhaps examining similar principles developed by critics will lend some insight.

Indur Golkany's version of Precautionary Principle is useful to the reasonable person in times of uncertainty and technological progress. First, the public health criterion – a combination of the human mortality and human morbidity criteria – places the flourishing of human beings over that of members of other species and the environment (Golkany 2001, p. 9). Although this seems too restricted a view as will be shown in Chapter 4, it can be a reasonable position for a reasonable person to take.[34] At the very least, it demonstrates a need for a hierarchical system for evaluating competing claims and resolving conflicts. The immediacy standard states that more immediate threats are to be given priority over those that will happen later. This makes good sense since there is less time to prepare for the latter than for the former, and the longer term threats might be able to resolve themselves given that other actions will be taken. The uncertainty principle states that we should address threats with higher probability before those with lower, all things being otherwise equal. The expectation-value criterion requires us to take the alternative with higher expectation value, *ceteris paribus*. The adaptation measure allows the negative impact of technology to be discounted to the extent that there is available technology to nullify the adverse consequences. Finally, the irreversibility criterion is clearly part of almost all Precautionary Principles. It states that priority should be given the outcomes that are irreversible or likely to become more persistent over those that can be reversed (Golkany 2001, pp. 9–10). The last standard should be interpreted based on the negative value of the irreversible or persistent outcome. Some irreversible outcomes are preferable at times depending on what is being done. For example, if

[34]Whiteside rejects Golkany's position as a Precautionary Principle on the grounds that Precautionary Principles entail "no presumption in favor of either the environment or human mortality" (Whiteside 2006, p. 45). I think Whiteside's attack requires an unnecessarily narrow criteria set for Precautionary Principles. If the principle deals with precaution and technology, then it should receive the Precautionary Principle label regardless of whether or not we agree with it.

we were able to eliminate a harmful parasite species by making it go extinct, then the outcome is irreversible, but one we should pursue.

Sunstein's Anti-Catastrophe principle is an attempt to provide a more practical, objective alternative to the Precautionary Principle. Precaution of the type required by a formal principle is limited strictly. "Unless the harm would be truly catastrophic, a huge investment makes no sense for a harm that has a one to one billion chance of occurring" (Sunstein 2002, p. 103). Basically, Sunstein's principle states that in cases of uncertainty and potential harm, we are generally obligated to identify the worst case scenarios and choose an approach that will reduce the worst potential or catastrophic outcomes (Sunstein 2005, p. 109). Moreover, this principle should be followed even thought it might be costly and has individuals objecting to such expenditures (Ibid., pp. 112–13). Limitations avoiding catastrophes includes preventing social risks that might lead us to create another catastrophe in order to avert the original one, using the least costly means to achieve safety, distributing the cost burden to those best able to bear them, and not separating the extent and expense of precaution (Ibid., pp.114–15). All of these are common sense rules that people apply in everyday life. For example, children should not have access to toxic cleaning chemicals. If there is a choice between moving the chemicals out of their reach and watching them every single moment of the day, the obvious solution is the former. The other rules are just as obvious to a reasonable person.

Since, governments and other powerful groups – not individuals in general – more often have the ability to prevent large catastrophes, Sunstein's principle is especially applicable to them. Among the tools that can be used by the government to reduce risk and avoid costly regulation are requiring companies to disclose their accurate risk information, provide companies economic incentives that punish them for bad behavior and allow them to trade rights, e.g. emission rights, use risk reduction contracts to produce social goals, and promote free market environmentalism (Sunstein 2002, p. 251). Each of these devices is consistent with autonomy, democracy, capitalism, and utilitarianism. Social utility is achieved in a permissible manner. If avoidance of an epic disaster places too great a burden on society's weakest members, and they do not receive compensating benefits for what they suffer, then there may be no obligation to prevent the catastrophe from occurring. Unlike the Precautionary Principle that takes away people's autonomy by making paternalistic decisions about acceptable risk for them, Sunstein's generally lets people make their own decisions. Every stakeholder, including businesses and consumers, gets to make the informed decisions to which she is entitled, without too much government interference. Of course, paternalistic actions are not always illicit, but usually, people need to be respected enough to be allowed to make their own decisions about what risks they are willing to take in their lives.

The reason that Sunstein's principle appears more practical than the Precautionary Principle is that it is not used for all technology about which people have safety concerns but rather is limited only to that technology which might cause catastrophes. The result of the narrower focus is less expenditure, regulation, enforcement, and so on, with their particular costs to individuals and the community, in establishing technology's moral legitimacy, which will permit the technology and its beneficial consequences into the free market at a much faster pace. The efficiency created will in turn lead to capital being placed to its best advantage in the market, instead of being squandered on regulatory costs.

Although Sunstein's principle is appealing, it faces two serious challenges. First, as Sunstein acknowledges, the Anti-Catastrophe Principle leads to impractical conclusions in too many situations. Suppose that there are a number of potential catastrophic harms that might arise from a particular technology and the information to assign probabilities to these hazards is lacking. Avoiding the harms can only be achieved through great cost to the society. Furthermore, assume that there are a number of non-catastrophic harms that are well enough known for probabilities to be assigned to them. While preventing these harms are costly, no greater harms are generated (Sunstein 2005, pp. 113–14). The odd result produced by the Anti-Catastrophe Principle is that the society should address the potential catastrophes even though there are more convenient avenues to pursue. Although Sunstein acknowledges this serious problem, he merely states that the Anti-Catastrophe Principle has a place in law and life (Ibid., p. 114).

Sunstein's optimism is somewhat puzzling. If the principle does not provide practical guidance in all cases in which it applies, then it is not a better alternative to the Precautionary Principle. The latter principle at least tells us what to do in every situation. What would make matters worse is if all or the vast majority of situations turn out to be those in which the Anti-Catastrophe Principle tells us to prevent catastrophes based upon our ignorance of probabilities over preventing smaller harms with high probability of occurring. Here is where the problem caused by ignorance always arises for Precautionary and Anti-Catastrophe positions. Why is it better to act out of ignorance than it is to address situations that we know more about and can address more readily? It does not seem to mean much to state that there is a potential catastrophe about which we know nothing about, and then assume that this should have more evidentiary weight in our decision making processes than does the less potentially damaging but better known outcome. Given that we do not currently know everything there is to know about our action's impact on others and the environment, it is possible that everything we do might be catastrophic. Since we do not know the probabilities, there is no way to separate the farfetched possibilities that anyone can become upset over from the more probable ones. Therefore, although Sunstein's principle does not state that all actions are unethical in the manner of the too strong Precautionary Principles, it does allow for technology and freedom to be severely curtailed on the flimsiest evidence while we are busily trying to avoid catastrophes that might or might not be likely to happen.

Perhaps this result could be eliminated if the terms employed in Sunstein's principle were fully explicated. The meaning of "catastrophe" is the most important for analyzing and using this principle, but it is unclear what it actually is. Obviously, the term must be *relative* to the context of the technology and the environment into which the technology will be introduced. For example, the deaths of 20 people in general would be catastrophic for a small community of 100 but not so for a community with 1 million citizens. However, if those 20 people were evil individuals keeping the remaining 80 in the small community in thrall, then it might actually be beneficial for those evil people to die. It is from the term's relativity that we should begin to see the glimmer of how a reasonable person would take precaution in new technology situations. If the person's goal is the flourishing of intrinsically valuable individuals, then he would use that ultimate end to decide which precautions, if any, are necessary given the situation. In the case of non-catastrophic harms with known probabilities and possible catastrophic harms with unknown probabilities, the reasonable person would address the most serious harms with high probability first since these are most likely to affect flourishing, as Golkany's principle would require us to do. As the probability and potential harm levels fall, then it becomes more open as to what the reasonable person should do. Sunstein's approach becomes more helpful here with its democratic approach, tools, and ways for minimizing harms that distributes burdens and benefits in a socially just manner.[35]

Although government agencies are often faulted for inefficient bureaucratic regulations and a lack of common sense, they can implement reasonable precaution measures. The US government agencies in charge of overseeing TOs, for example, use rules that are consistent with both Sunstein and Golkany's principles. If a transgenic is sufficiently similar to the safe, original organism, then the former is Generally Regarded As Safe or GRAS. Since it is GRAS, no precaution needs to be taken beyond that normally performed for the corresponding non-TO product. On the other hand, if the product is not GRAS or if there are risks, then it undergoes a rigorous decision procedure by the appropriate federal agency using the petitioning company's data and any other information the agency might require. The United States Department of Agriculture (USDA) protects agriculture from pests and diseases, the Environmental Protection Agency (EPA) regulates pesticides and potential toxic substances, and the Food and Drug Administration (FDA) oversees the safety and proper labeling of all plant-derived food and feeds (US 51 FR 23302). Although these safety processes do not satisfy those who hold the Precautionary Principle, all the agencies and the businesses they regulate take risk seriously and attempt to minimize it because it protects society, is good business practice, and is the right thing to do.[36,37]

To summarize this long explanation of every reasonable person's necessary features, in general, a reasonable person is someone who recognizes the true nature of morality. When figuring out what is the best alternative in a situation, the reasonable person correctly and consistently analyzes the value of the data available to him for the situation in the time available and incorporates all relevant facts about external world society's rules, practices, and customs; rules and responsibilities associated with specific roles the agent is playing at the time; claims that others

---

[35]The distributive justice principle for accomplishing this goal will be discussed in a later section of this chapter.

[36]The US Regulations will be revisited in Chapter 5's free market and trade barriers argument.

[37]There are those who attempt to manipulate the system or commit some type of deception or fraud. These people are of serious concern, but their bad behavior is not localized to transgenics and biotechnology. Whatever works to correct the condition for other businesses should work here as well.

have on the agent and the agent has on others; value of consequences; and all relevant factors into their decision process. If new relevant information becomes available, the reasonable person re-analyzes the situation to see if his position should be altered accordingly. Finally, in the pursuit of the ethical, he desires and works toward achievable good goals and the means to reach them in order to make himself and other intrinsically valuable entities better, provided doing so does not surrender something of greater moral value.

2.4 RPU's Benefits

Reasonable Person Utilitarianism should be accepted for at least two reasons. First, its simplicity makes it a much easier principle to use than standard act-utilitarianism. Mill and other utilitarians recognize there are times in which it is implausible to classify an action as wrong, according to utilitarianism, and classify the agent as blameworthy (Mill 1988b, pp. 20–1). For instance, if the agent could not have foreseen her actions would end in unmitigated tragedy, and she had the best of intentions, motives, and other mental states, then it would be, at least, counter-intuitive to say she is blameworthy of wrongdoing even though she failed to maximize utility.

What makes matters worse for standard act-utilitarianism are cases in which the agent is a victim of another's unprovoked, vicious attack. Suppose, for instance, that Jane promptly leaves her job in a hospital at a very late hour and walks to her car unaccompanied. This is exactly the opportunity Jane's stalker has been awaiting. He attacks and rapes her. Since Jane would have done better waiting for her friend rather than going alone to her car, according to standard act-utilitarianism, she has done the wrong thing, even though she could not have known her action's outcome. But this classification is an outrageous result of standard act-utilitarianism for many reasonable people. Our common moral sense correctly tells us that no one should blame the victims of attacks for what has happened to them. They are not at fault; the criminal perpetrator bears that responsibility himself.

Reasonable Person Utilitarianism is a simpler, more intuitively appealing, and useful theory than standard act-utilitarianism in these cases. First, RPU helps tie an action's morality into the praiseworthiness or blameworthiness of the agent performing it based on probable rather than actual utility. If an agent does not choose the alternative that probably will produce the greatest utility according to a reasonable person, then she has done the wrong thing and is blameworthy. If she has performed the correct action, she is automatically praiseworthy. The contraries hold as well. Hence, the principles for praiseworthiness and blameworthiness are:

1. S is praiseworthy for doing A only if S performed A and A is morally permissible, according to RPU.
2. S is blameworthy for doing A only if S performed A and A is morally impermissible, according to RPU.

Since a reasonable person would have concluded that Jane's action is likely to produce the greatest utility, then leaving from work at the time and manner she did is morally permissible for her. The fact that she is raped is irrelevant to the utility calculations given that no reasonable person weighing the information available would have considered that to be a sufficient risk. Hence, RPU states that Jane did the right thing and the principles incorporating RPU say she is not blameworthy. Thus the paradox created by standard act utilitarians can be eliminated without losing the basic idea of doing the best we can.[38]

Moreover, RPU is a principle, which is useful, in part, because it does not incorporate an ideal observer.[39] Much like Hare's Archangels, ideal observers are godlike creatures, who can perceive much more than mere mortals can and make their decisions based upon reason alone. Hence, we really have no reference as to what an ideal observer would be like or how it would actually come to the conclusions it reaches. We would have as much chance of figuring out this entity's views as we would a bat's.

On the other hand, the reasonable person is a more comprehensible, objective standard for any thoughtful person to use. One of ethics' presuppositions is that agents are reasonable people or are capable of behaving in this way; otherwise, they are not agents in the first place. If the evaluator is a reasonable person, he will have the ability to understand what is required of him given the information available to him at that time, even if he does not use the ability. Without necessarily agreeing to pursue the same course of action for ourselves, every reasonable person can still understand why another comes to this decision because she is also a reasonable person and can evaluate if the other acted reasonably in the situation. Moreover, since RPU is based upon the evaluation of the situation by a reasonable person reasonably believing what would maximize utility, it does not incorporate the undisciplined, biased views of merely anyone. Racists and others who base some of their judgments on unjustified and impermissible beliefs can be reasonable people overall, but are not reasonable people by RPU standards if they are evaluating a situation involving their biases. RPU is an objective standard, rather than one based on the subjectivity of the individual and his idiosyncrasies.

The second reason for adopting RPU is that it can be used by both rule and act utilitarians.[40] First, the former theory is specific enough to tell an individual what to do in every circumstance the person can encounter. Hence, we will not have the problem plaguing most rule utilitarians of when, if ever, it is permissible to violate one of the rules of conduct that is supposed to make the society as good as it can be in order to perform an action that greatly maximizes utility, but is proscribed

---

[38] A lingering problem for RPU and all consequentialist theories is if Jane would have done the wrong thing by leaving if she had good reason to believe a rapist was on the loose in the area. I think the solution lies in whether or not she treats herself as a mere means. Hence, the Kantian principle coming next, QCI, when conjoined to RPU, will classify these situations correctly.

[39] This is the same problem Kant's purely rational agent has.

[40] Attempts to formulate a rule utilitarian principle that is not merely extensionally equivalent to act-utilitarianism or based on implausible assumptions about ideal societies are unhelpful here.

by a rule, or that is extremely disvaluable but is required by a rule. There may be generally useful rules-of-thumb to maximize social utility, but if following the rules would likely be less useful than doing an alternative action, RPU says to perform the alternative. For example, rules against causing death and pain can be abrogated by exigent circumstances such as self-defense.

Finally, the utility achieved if everyone adopted RPU would be arguably much higher than other social rules, even when the second part of my overall distributive justice theory – QCI – is lacking.[41,42,43] RPU is the best candidate for use in all exceptional circumstances because it addresses every possible situation. Whatever state of affairs the agent finds herself in, she is to do whatever a reasonable person would reasonably believe would probably maximize utility. For example, all reasonable people reasonably believe it is better in general to keep one's promises than to break them. However, they also correctly understand there are times in which it would be better to break the promise than to honor it, e.g., when one can save another from unnecessary pain and suffering with little moral cost. Since the details of each situation are available to the reasonable person, given the existing condition, she will make her decisions based upon them rather than being restricted to more general rules, which are not sufficiently specific. The flexibility of RPU entails utility will probably be maximized to a greater degree than if the more general rules are adopted and followed.

Furthermore, Reasonable Person Utilitarianism avoids irresolvable moral dilemmas. With more than one rule of thumb, there are bound to be conflicts between principles when conflicting moral factors come into play. Moral dilemmas in which an agent is torn between two or more incompatible alternatives exist.[44] Since there is a conflict, there must be some way of settling it; otherwise, we encounter the morally repugnant conclusion that the agent must act unethically no matter what she does. But building in rules that allow for the resolution of dilemmas eliminates one problem while creating an equally serious one. If there is an overarching moral rule or a condition in the rules-of-thumb to solve conflict situations, then the process of moral evaluation is even more complex than at first thought. Of course, the

---

[41] Of course, the rule to always maximize utility will have the highest overall utility, but it is not a practical rule to use, so it is rejected.

[42] Although if the two parts of the theory were universally adopted, the best state of affairs would be most likely brought about. People would be trying to maximize utility in the consequences of their actions at the same time they are treating people with respect in both thought and deed. The world in which both of these goals are met has to have greater value than a world lacking either.

[43] Since there are a large number of exceptions to just about any simple rule, such as promise-keeping, the rule becomes too complex and open to interpretation the moral community's members. After all, if there are many exceptions focusing on this or that matter, people will reasonably start expecting that unforeseeable exceptions will arise. This fact leads to a further conclusion: no matter how diligent we may be in formulating a rule, we are bound to be unable to account for all circumstances that might arise in the future. The best we can do is to make the rule specific enough to deal with most exceptions, but it has to be general enough to be useful overall.

[44] Though they might not be of the technical sense in which all alternatives open to the agent are prohibited.

more complex the evaluation processes, then the more difficult it is to learn and use; thereby making answers to dilemmas more difficult to find and to defend to others with opposing viewpoints. By being a single, simple rule, RPU avoids conflict situations entirely, while also being very easy and practical. On utilitarian grounds alone, RPU is the correct rule to adopt.

Returning to the sample case, suppose that when Mary rejected her boss's request to create cross-pollinating transgenic wheat, he acted out of character and fired her. After that, he had another researcher engineer the wheat, which, as predicted, destroyed the livelihoods of the organic and conventional farmers. Mary's action, hence, fails to satisfy the rule to do the best one can. According to standard utilitarianism, Mary should have chosen to originate the wheat because the outcome for the farmers is the same in either case, but her employment being terminated for taking a stand increased the disvalue of rejecting her supervisor's request. Honoring the request, on the other hand, would have been better for her because it lacks that disvalue. However, we know that Mary still did the right thing, even though the effort to stop her boss from carrying out his decision failed. Under RPU, we are morally obligated to try to bring about what appears to us to be the better alternative; it is irrelevant if it truly is not the best we could do. If circumstances beyond our control prevent us in our duty, then no fault can be laid at our door. We tried and failed, but the morally relevant fact is we did try.

2.5 Respecting Persons

The second part of the moral principle governing the permissibility of transgenics, biotechnology, and other technology is Kantian in nature. As Kant instructs us to do in the *Groundwork* and elsewhere, when acting, all rational agents must always respect all the people affected by their actions as ends in themselves (Kant 1956, pp. 62–3). Rational people are:

> Capable of following rules, drawing inferences, generalizing, and making free choices. [They] are capable of altering [their] conduct because [they] recognize the truth of some propositions and their importance of certain interconnections among them. (Holmes 2003, p. 114)

Of course, all rational people are also autonomous, but rationality is a higher standard than mere autonomy. In addition to being able to act intentionally, with an understanding of relevant consequences of action, and with sufficient freedom of will, each rational person must be consistent in her thinking processes (Holmes 2003, p. 115). As stated in the Introduction, most people are imperfectly rational persons because they do not always use pure reason to determine their duties as a fully rational person would. However, even though we are imperfect, in part because we use desire and emotions to make decisions, we can use our reasoning ability to determine what a rational and reasonable person would do. If we act as such a creature would act, then we must be doing what is morally right. At least, morally right according to this pseudo-Kantian theory and principle.

Each person, to be a person, must have autonomy. I will assume each person acts autonomously only if the person acts:

1) Intentionally,
2) With sufficient understanding to be able to determine some of the relevant consequences and potential risks of the action, and
3) Without compelling influences beyond the person's control.

While my formulation is based on Faden and Beauchamp's model, my conditions 2 and 3 are different (Faden and Beauchamp 1986, p. 238).[45] The second condition is, however, consistent with the reasonable person standard used in RPU. Not only must the person act according to what a reasonable person thinks is best, she must be able to think as a reasonable person would think. In other words, if a person cannot appreciate the fact that her actions have consequences, then it is unlikely she is a reasonable person, or even a moral agent. Prime examples of these types of beings are toddlers and the severely mentally ill. Society does not hold them responsible for what they bring about because they were unaware their physical activity/event – which is not an action – would result in evil consequences. In a very narrow sense, what these humans do is analogous to a tree falling on a person. The results are tragic and evil, but the wind blowing over the tree is not at fault for the physical event because it cannot understand the consequences of the event.

The third condition is connected to freedom and free will. Agents unable to make decisions because they are under overwhelming duress cannot be considered to be moral agents acting autonomously, although they retain their status as moral agents. Suppose someone has been reading too much work by BF Skinner and decides to confirm some of the results. Mary is brainwashed by a serious of psychotropic drugs and pain conditioning to fear the sight of white rats. Her flinching when she sees a white rat is the result of the conditioning and not a conscious choice she makes. Hence, her reaction is not autonomous, and we cannot classify the action in regard to what she does as morally right or wrong nor her as praiseworthy or blameworthy. We would be able to classify the action as morally wrong on the logical behaviorist's part.

Other factors can reduce if not eliminate free will and an agent's freedom to choose. Pain, pleasure and other extreme emotional states overwhelm the rational component necessary to choose as a moral agent. For example, a person who is in extreme terror sometimes behaves in ways she would not if she were in a calmer state. She might lash out in an attempt to protect herself from a danger existing only in her panicked mind. If she were not in that state, then she would have been able to evaluate the situation rationally, and then conclude there was no imminent harm. In

[45] Of course there must be an informational component to the deal. The fact that the consequences are not clearly known should be imparted to the intended recipient of the product. Information is a vital necessity for making autonomous decisions. (See Judge Robinson's decision in "Opinion in *Canterbury* v. Spence" and the articles by Ingelfinger and Bok.)

these extreme circumstances, the agent still exists, but she is not acting in the moral sense.

The qualifying clause of condition three eliminates some excuses agents might make to minimize, or altogether eliminate, their responsibility for an action and its consequences. Without the clause, drunk drivers and others could claim they did nothing wrong because they were not able to make autonomous decisions at the time they decided to drive drunk or commit normally unethical behavior. According to them, being drunk was the compelling interest which renders them unable to act in the moral sense. Drug and other addicts could also claim their behavior was not autonomous based upon the compulsion of the addiction.

However, the clause states that the compulsion has to be beyond the agent's control. This can work out in two different ways. First, we hold people responsible for their addictions because they made the decision to use substances or become drunk in the first place. That is, they autonomously choose to engage in known risky behavior; hence, they cannot legitimately mitigate their responsibility for their behavior after they chose to render themselves non-autonomous. A drunk driver is guilty for driving in that condition by putting himself into a position in which he robbed himself of his autonomy so that he drove inebriated. Second, many of the compelling forces people cite as mitigating reasons for their lack of autonomy turn out not to be all that compelling. Given drug and alcohol treatment program successes, it is not clear that drug addicts and others are unable to stop their addictive behavior. Even in the face of addiction, many thoughtful people believe addicts have the ability to choose to be a recovering addict, although that path is extremely difficult. Some of those who are afraid of new technology might have a similar treatable condition that influences their decision making processes.

However, the duty to respect people includes the duty to respect people's autonomous decisions regarding technology use, even if they lack some information about the possible consequences of such use. Since autonomy is a necessary condition to personhood – and being a person is the reason why we have the duty to respect people – it can be set aside only when that autonomy illicitly interferes with another agent's autonomy or personal integrity.[46] The formal version of PMC's second normative principle is QCI.

Quasi-Categorical Imperative

An act is morally right only if in doing the act, the agent does not treat any person or intrinsically valuable thing as a mere means.

To begin to understand how to use QCI in the moral evaluations of actions, the Law of Reversibility – if you do not want it done to you, then do not do it to anyone else – and the Law of Universalizability – if it is wrong/right, respectively, for everyone to do it, then it is wrong/right, respectively, for you to do it – are helpful. Both

---

[46] Some may argue that the only rational goal for a person is what is in the person's best interest (Derek Parfit calls this the "Self-interest Theory". (Parfit 1992, p. 4)). The problem with this view is that it classifies altruistic acts, especially supererogatory ones, as irrational without any sort of argument.

laws incorporate a respect for intrinsically valuable things as ends in themselves by requiring moral agents to treat them commensurately to their actual worth. For example, if you are a valuable entity because you are a person and you do not wish for someone to harm you without just cause, then because others have the same value as you, you are not allowed to harm them without just cause. The Law of Universalizability, on the other hand, demands equality; no one has special privileges without some moral justification for them.[47]

Although my version of the Categorical Imperative is not limited to respect of persons, it does seem to capture the spirit of Kant's thinking. The reason moral agents must respect all persons affected by their actions as ends in themselves is *because* they are ends in themselves.[48] Being an end in itself is synonymous with being intrinsically valuable. By broadening the class of what is intrinsically valuable based on the belief that it is better to be more inclusive and be wrong than to exclude and be wrong, the principle is not limited to persons only. It might be an ontological truth that persons are the only things intrinsically valuable, but in Chapter 4, a plausible case will be made for treating other things, such as animals and nature, with the respect Kant reserves for people.

Although Kant maintains that the ideal person we are to emulate is any perfectly rational being acting rationally, his standard is unrealistic and contrary to what morality requires of us. As was seen in RPU's reasonable person requirements, the problem with using the rational person standard is a person can act rationally yet do the wrong thing. For example, an Egoist can be purely rational, yet not be a good moral agent. Moreover, since we must use emotion to make decisions, how a purely rational person would act is too obscure for us to be able to discern and use to evaluate alternatives.[49] To eliminate the ontological problem, the reasonable person

[47] Of course, this particular law is flawed. Many actions would be wrong merely because not everyone should perform them. For example, the devastation caused by lack of necessary services from everyone being brain surgeons would destroy society; thereby, making it wrong for everyone to become a brain surgeon. However, common sense tells us that it is permissible for some to follow this career path, especially if they are really suited for it. This problem is eliminated if the Law is tempered with the proper respect for persons.

[48] Unfortunately, this is one of the places where hand waving is appropriate. It is impossible to prove that intrinsic value creates duties of respect for whatever has the value. This could be one of the axioms of ethics that cannot be proved but allow us to have the ethical system we actually do. However, anyone who rejects the linkage between intrinsic value and respect to me seems to be an unreasonable person unlikely to understand morality.

[49] Kant tries to eliminate this problem by formulating a linked theory of virtues. According to Kant, rational persons are virtuous persons; hence, a fully rational person would never act in a way that disrespects the intrinsic value of any person affected by his action. However, the connection between a fully rational person who uses pure reason without much emotion and a virtuous person is dubious. Many of the virtues demand a caring attitude to others, including but not limited to charity, care, compassion, empathy, and so on. It is difficult at best to see how a pure sense of duty allows people to have the virtues. Being charitable out of a pure sense of duty is cold and does not seem to treat people as they ought to be treated. Being charitable out of duty and because one cares for the plight of those less fortunate than oneself seems to capture better what it means to respect individuals because of the care one has for them as persons. Of course, a footnote cannot

standard from RPU will be used here as well. Instead of acting as a rational person would act, we should act as a reasonable person acting reasonably would act.

Now that the necessary feeling/emotion/desire underlying all human-person ethics has been sketched out, attention can be turned toward what it means to treat someone as a mere means. Since Kant leaves the definition of treating a person as a mere means unclear in the *Groundwork*,[50] there is a need to develop it further for practical application. Merely stipulating that treating someone as a mere means is equivalent to treating that person as a mere tool to achieve the agent's ends is too vague to be useful for making and defending decisions. However, the explanatory process can be started by stating that part of this obligation includes having the appropriate attitude toward those affected by the act, which may include, but not be limited to, good intentions and motives on the part of the agent performing the action.[51] I will first examine one definition, which equates degrading an individual with treating the person as a mere means, and then reject it as inadequate. I contend a version of Thomas Hill's definition captures the letter and spirit of what it means to treat someone as a mere object; hence, it will be adopted.

Although it is often used this way, degradation is not equivalent to treating someone as a mere means. Degradation only comes about when the agent has decided she has been demeaned; the mere attempt by another agent to degrade is insufficient to devalue who she is as a person. The difference between the attempt to degrade and being degraded can be seen in one of Marcus Aurelius's *Meditations*.

> Today I will be meeting with interference, ingratitude, insolence, disloyalty, ill-will and selfishness...But for my part I have long perceived the nature of good and its nobility, the nature of evil and its meanness...therefore none of those things can injure me, for nobody can implicate me in what is degrading. (Marcus Aurelius 1996, p. 45)

Thus, a person who is called a vicious name need not be demeaned, though the name-caller attempted to do so. If the victim decides she has been degraded, then in fact she has because she is the only one who can devalue herself by accepting the event as degrading. On the other hand, if she decides the contrary, then she has not been degraded.[52]

---

contain a full discourse on Kant's complex moral theory; so I will have to leave it at this state of development and move on.

[50] Kant (1956, p. 96, 104, and 105).

[51] Unfortunately, the definition for treating someone as a mere means is very vague.

[52] Some may argue that treating someone as a mere means is equivalent to exploiting them by forcing the person to unfairly shoulder the burdens of an action without reaping an adequate amount of the benefits. (This counter argument is a modification of the positions found in the Crouch and Arras, Grady, Glantz et al. articles.) The argument, however, only works if the person does not autonomously choose to have this happen to her. If relevant information is withheld from her during the decision procedure, then she cannot autonomously select the alternative in which she is required to receive more burdens or fewer benefits than she deserves because this was not one of the options from which she had to choose initially. Since an agent must act intentionally, which requires making plans to bring the action about, she could not have acted intentionally to be exploited. She did not have the information needed to form the proper plan or intention. However, if she did have the relevant information, then she could have agreed to accept the additional burdens or fewer

Treating a person as a mere means is distinct from degrading the person. For the former, the treatment is under the control of the agent attempting to degrade the person, while the person to be degraded is in control in the latter situation. The result is that one agent can treat another as a mere means even if the latter refuses to be demeaned by the action, but the agent can never degrade her. A nun raped by a thug need not be demeaned by the action – she might even forgive the rapist – but the rapist acted wrongly by treating the nun as a mere means. On the other hand, a person can be degraded even when the person was not treated as a mere means. In ordinary business relations, interactions are often impersonal but ethically appropriate. If a client, for example, feels degraded by the impersonal actions of another because she desired a more caring approach, she is degraded because she has chosen to be degraded, but she has not been treated unethically by anyone other than herself. Hence, degradation and treating a person as a mere means are not equivalent.

Thomas Hill formulates a more plausible definition of treating a person as a mere means. In regards to the respect for persons which is merely the respect for moral law, Hill claims:

> [E]veryone should approximate, to the extent that she can, the ideal of a person who fully adopts the moral point of view...One must, so to speak, take up the spirit of morality as meet the letter of its requirements. (Hill 1997, p. 545)

For Hill, not only do people have to abide by the maxims governing the situation, they must also have a respectful attitude toward the moral law. "A respectful attitude toward a system of rights and duties . . . involves holding the system in esteem, being unwilling to ridicule it, and being reluctant to give up one's place in it" (Ibid.). On these grounds then, respecting someone requires two conditions be met. First, the agent is obligated to follow the rules governing the situation. Second, the agent must have the proper attitude of respect toward all the agents affected by the act as well.

Although Hill's definition may be theoretically illustrative, it is not very helpful in the practical realm as it could be. It does not identify what is entailed by the proper attitude of respect. Based on the reasonable person definition and Kant and Hill's work, following from the definition I will trace out some of the practical requirements for agents.

First, Hill discusses morally commendable motives and desires as justifying some actions while bad motives or desires making some actions wrong (Hill 1997, p. 543). Hill's Uncle Tom, for example, may permissibly act servile to white men as long as it is for a commendable goal, such as saving his children. On the other hand, if Uncle Tom acts deferentially for bad reasons, such as a desire for a minor advantage, then he has treated himself as a mere means (Ibid.). Hence, for the agent to fulfill his obligations, he must have primarily "pure" intentions and motives. The person failing in this duty may follow the letter of the law but not the spirit. I will take intentions to mean the goals of the act, and motives to be the emotions that

---

benefits. In this particular instance, though the act is imprudent, it is autonomous, and should be respected as such.

motivate the agent to try to achieve the goal of the action.[53] Each of the concepts requires explication to give it a practical understanding. I will begin with intentions.

Intentions are one type of mental state affecting moral classifications. Intentions can be defined as a rational part of an action; they are what the act is intended to achieve based on what the individual has set as his goal and the methods for achieving it. For example, merely intending to pump water into the house helps make the behavior of *pumping water into the house* into the act of *pumping water into the house*. Furthermore, although the behavior is exactly the same as in the preceding case, *knowingly pumping poisoned water into the house to kill the family living inside* is an act of murder. However, if one pumps the water into the house without knowing or having reason to believe the water is poisoned, the action is not one of murder or even negligence. It remains *merely pumping water into the house* (Anscombe 1963, Section 23).

Although there can be many different goals, the one that is not only the purest but necessary for every moral action is to respect the persons involved as ends in themselves. This means the agent will adequately evaluate the goals and desires of all those affected by the act to determine which she can assist in, if that is necessary, or avoid interfering in, if that is possible. When conflict situations arise, then the goals that are obligatory or permissible to obtain, in that order, will be the ones that the agent will attempt to bring about, while rejecting the impermissible ends.

Motives are the second type of mental states having an affect upon the classification of an action's morality. Motives, which stimulate us to perform particular actions, are non-rational parts of acts. A completely rational person may arrive at a conclusion he should act in a particular manner given the facts of the situations, moral principles he holds, moral reasoning processes he has, and so on, but until he has motivation to act, he will do nothing. Motives are part of each action as can be seen by the way in which we classify acts. Some actions are greedy, others altruistic, and so on. The various actions have these labels because of the motivations underlying them.

Motives are also of great importance in regards to the moral status of actions involving two or more people. The primary desire or combination of desires to bring about a particular goal must not be evil, such as greed, hate, revenge, etc., otherwise the act is wrong. A good intention to treat others as ends in themselves is inadequate to overcome the evil of, e.g., the desires to merely draw praise and acclamation onto oneself for respecting others. We ought to be tolerant of ethically permissible differences, but acting tolerant is unethical if it is done primarily to garner compliments for oneself.[54] To help make the action right, one must do it because one primarily desires to do what is right or from some other good desire.

---

[53]GEM Anscombe rejects this distinction in Section 12 of *Intention* on the grounds that the two terms have more overlap than can be accounted for by the distinction.

[54]Some utilitarians such as Mill claim that the agent's intentions and motives are irrelevant to the act's morality. Others such as GEM Anscombe hold that the intentions are parts of the act. I will adopt the latter position and also include motives.

Besides motives and intentions, the attitude agents have toward performing an action is also a factor in the action's rightness or wrongness, as well as being indicative of the person's goodness or badness. More specifically, each agent's attitude toward her performance of an action and to those affected by the action has an effect on her action's morality. In order for an action to be morally permissible, the agent acting must have a good/positive attitude toward her action. If the agent has a bad/negative attitude towards it, then the action is morally wrong. Suppose Mary has promised a colleague to help in his research. Unfortunately, the work can only occur after normal quitting time, which is when Mary would much rather pursue her leisure activities. However, Mary does not break her sincerely given promise. Mary helps, she intends to fulfill her promise, which makes her behavior the action of fulfilling the promise to help with research, and Mary performs the action with the good motive of honoring her word.

However, although the behavior, intention, and motives remain the same, there can be a significant difference between fulfilling the promise with a primarily positive attitude versus fulfilling the promise with a primarily negative attitude. Actions for the most part performed with a good attitude are done with the proper respect for person affected by the action. On the other hand, if she performs her action angrily, grudgingly, resentfully, or with some other negative attitude that causes her to be unable to treat intrinsically valuable objects with respect, then she does not value them as they should be. If Mary keeps her promise with resentment, the co-worker is being viewed as an imposition rather than as someone who is an end in himself to whom she owes a duty. Furthermore, Mary, by not controlling her feelings, is not respecting herself because she is not acting as a moral agent, but rather as a petulant, immature person. Her attitude demeans her in a situation under her control.

The proper attitude toward a situation is also the proper attitude toward morality. Recall from the definition of reasonable people that they recognize and incorporate in their decision processes that morality is a enterprise in which we try to "realize in ourselves and others nurturing goods such as caring, considerateness, compassion, sympathy, and love" (Holmes 2003, p. 217). Respecting those affected by our actions requires us to promote or instantiate nurturing feelings/goods in us and them. In order to promote or instantiate them, then we must be able to feel them, otherwise they would be empty terms. Therefore, all ethical actions must be performed with a nurturing/positive attitude.

Fourth and finally, in order to do what is moral, not only must an agent possess primarily good intentions, motives, and attitudes, she must also have an actual feeling of respect that all people deserve *qua* people for each individual affected by her actions. Since it is an emotion, respect cannot be defined descriptively, but let it suffice to point out those things which we tend to respect, such as parents and distinguished comrades, to elicit the feeling accompanying contemplation of these people, and say the resultant feeling approximates the meaning. The only difference is the feeling is not *quite* the same. The examples mentioned here are people with whom we have had considerable contact or are familiar with in some way. The feeling of respect we should have for all people affected by our actions is a milder, more general version since they have a more distant relationship to us than do those for whom we care most.[55] That is the best I can do to define ostensively how one should feel. Since most people have already experienced this state, they will know the relevant referent.

Hume's discussion of sentiment or what is commonly called empathy helps us in explaining not only how to do this, but why each agent needs to have respect for all persons affected by his actions. No one would disagree each person has some sort of ethics by which he judges his actions, thoughts, emotions, and so on, and then acts accordingly. The reason human persons have these ethical systems is human nature gives all human persons the same ability. Everyone has the natural sentiment of sympathy or empathy, unless it has been destroyed by severe psychological conditioning (Hume 1948a, p. 239). Those who have been habituated in such a way might no longer qualify to be human persons because of their inability to bond or feel anything for any other person or thing. However, as Hume recognized, even pirates and robbers have a system of ethics which is in general identical to all ethical systems, although the details of each system might be different in its particulars and customs (Hume 1948a, p. 205).

Treating others as mere means is a very complex and difficult topic to discuss, but it might be simpler to understand how to do it without having to explain it. It is easier for people to use ethical systems as they do than it is to explain what the systems are. Perhaps "a hundred volumes of laws and a thousand volumes of commentators" will not be found sufficient to explain how ethics actually works and justify every claim made (Hume 1948a, p. 200). At best, the most that might be achieved in philosophical discourse of treating people with respect they deserve is to give approximations of what people actually do, and then hope the listener sufficiently understands what is trying to be conveyed. Perhaps telling people the two Kantian rules of "if you do not want it done to you, then do not do it to others", and "if it is wrong for everyone to do it, then it is wrong for you to do it", along with describing examples of when they classify incorrectly, will give listeners the ability to modify the ethical systems they are using to become better people. This issue will be indirectly explored in the sections on distributive justice and the Moral Paradigm Test.

2.6 Working Toward a Plausible Distributive/Social Justice Principle

According to USDA survey respondents and the professional moral codes in Chapter 1 and common morality, not only are utilitarian and Kantian ideas necessary components of an adequate moral code, but for justice as well. Distributive

[55]It might be possible to insert a care ethic along the lines of Kheel if someone has an objection to respect (See Kheel 2008, Chapter 7). Although empathy and care are important to understanding my tenuous concept of respect, the combination is not equivalent to it. Respect carries an esteem or regard connotation that will make it more likely that people will act in a certain positive way that is missing from empathy and care. A person has empathy and care for a child, but will not take a child's advice in the same way that she will for a person she respects.

justice seems to be of particular importance to most people discussing technology, transgenics, and their impact on the developing and developed worlds. In this section, a theory of justice, the Distributive Justice Principle, incorporating intuitions underlying the other justice theories/principles will be formulated. Although the new theory focuses on distributive justice, it can be used for all types of giving people their just deserts, including retributive, compensatory and reward justices.

No one would disagree that distributive justice can be plausibly defined in a variety of ways. In the interests of time, however, I will stipulate the most plausible definition is a fair distribution of a society's benefits and burdens to each member of the society. Distributive justice focuses on two different values: benefits and burdens. Benefits are owed to each person by her society merely because she is a member of society. Each society has benefits or goods that its citizens can share, all of which an individual herself generally cannot procure on her own, e.g., security from unjustified attack, companionship, and political representation. Burdens are what each person owes to her society merely because she is a societal member. In order to be able to receive benefits, each member must bear some of the burden of maintaining the society, such as paying taxes, serving the government, and obeying laws to prevent each from unfairly harming others. Exactly what is considered to be fair in the allotment of burdens and benefits to each of the society's members is determined by what type of system is adopted.

Of the distributive justice theories, capitalism and socialism are the most relevant to technology markets. Capitalism states that a free market will produce the fairest distribution of society's benefits and burdens. Capital moves to where it is most efficiently used based upon the demand of buyers and supply of sellers. Furthermore, both buyers and sellers are able to freely enter or leave the market, or make contracts with each other for goods and services provided they have not already contracted away the same products. One of capitalism's central tenets is if competition is not illicitly interfered with, then society as a whole will be much better off than it otherwise would have been.

Socialism, on the other hand, is oriented toward helping those who cannot adequately help themselves, while demanding all to operate at the peak of their potential. Under socialism, society's benefits are distributed according to need, while social burdens are distributed according to ability. Hence, those who need more will receive more, as long as it is available, while those who need less will receive less. Furthermore, citizens able to produce more will be obligated to do more than those people who can produce less.

Although both capitalism and socialism are interesting, plausible theories and have had an enormous impact on the lives of most people, I will focus on the distribution theory expounded by John Rawls in *A Theory of Justice*. Rawls' theory incorporates many of the good ideas of the former while avoiding some of the most significant problems at the same time.

One of the keys to grasping Rawls' justice is to understand his pure procedural justice method for a fair distribution of society's benefits and burdens. Unlike imperfect and perfect procedural justice, pure procedural justice focuses on developing a fair procedure without first stating what the fair end is. That is, if we use a fair procedure, then the result of the distribution will be fair, although we will not know what the final distribution will be prior to using the procedure.

Imperfect and perfect procedural justice, on the other hand, begins by stating what a fair end is, and then developing principles to produce that result. With imperfect procedural justice, we do not have a fool-proof method to achieve the desired results as we do with perfect procedural justice (Rawls 1971, pp. 85–90). However, these two justice types suffer a flaw that pure procedural does not. Although the end is fair for the former, the means employed to achieve it might not be, while pure procedural justice by definition has ethical means to its end.

Given Rawls' focus on fair procedure to yield an ethical result no matter what it actually turns out to be, it is necessary to establish a procedure that actually is fair. The conditions of Rawls' thought experiment in which a social contract is created and signed – the original position – includes agreement to the contract must be unanimous, the contract is permanently binding, and the contract can never be altered. The veil of ignorance is used so that no one will know any personal identifying information so that the contract will not be written to favor people with those features. Finally, each person creating the contract is a rational, self-interested individual which entails she will be doing everything possible to maximize her long term self-interests, even though she knows nothing about her personal identity or anything allowing her to distinguish herself from anyone else.

The imposed conditions will create a fair process. Since people do not know what position they will occupy in the society and each wants to maximize her self-interest, no one will rationally sign a social contract including slavery or other disadvantageous social positions because she does not want the veil to be lifted only to find herself in the worst off class. Each person recognizes the possibility she could occupy the undesirable position; thus, in self-interest, she will make sure no one has too low a quality of life while she is contracting. On the other hand, there will be very few people who are extremely well off. No-one will see it is in his self-interest to make a few people much more powerful and wealthy than everyone else. Each person will realize in the original position that he probably will not occupy the very best social class; so, he will make sure no-one else will have a considerable amount of power over him.

According to Rawls, his thought experiment will yield two principles that every rational or reasonable person will accept.

1. Each personwill agree to the most extensive set of liberties for themselves that is consistent with the same set of liberties for all.
2. The benefits and burdens of society will be distributed so that:
   a. they are to the greatest benefit of the worst off, and
   b. the offices with greatest benefit will be open to fair competition for all. (Rawls 1971, p. 302)

If and when principle 1 conflicts with 2, the first principle always overrides the second. Equal, extensive sets of liberties are always more important than irregularities in distribution of benefits and burdens. The latter imbalances can eventually be eliminated if rational, self-interested people pursue their liberties, but principle 2 is unlikely to alleviate any problems involving the limitation of liberties.

Elements of the distributive justice theories of egalitarianism, libertarianism, capitalism, and socialism can be seen in the two principles. The first ensures each person has equal liberties with all other individuals – a form of egalitarianism – and making liberty the primary principle and guaranteeing as extensive of a set of liberties as possible are Libertarian tenets. Capitalism is part of the second principle with its second condition allowing benefit disparities if they are attached to offices open to fair competition from all qualified. Finally, the second principle's first condition of benefits and burdens being distributed to the benefit of the worst off in society is socialistic. Socialism requires that each person receive according to his needs and the worst off have the greatest needs. Although there are other points at which parts of each of the four justice theories can be discerned in Rawls' two principles, it is time to turn to the theory's problems, and then to a more pragmatic theory of justice.

2.6.1 Problems with Rawls' Theory

There are a number of practical puzzles raised by Rawls' theory. First, many people are not rational, self-interested people. They make decisions based upon factors other than their long term best interests, as happens in the case of smokers, those who waste money instead of saving enough for retirement, and so on. Second, the fact each person knows nothing about herself other than she is a rational self-interested individual entails her personal identity is lost. If a person does not know anything of her unique personal information, then it is not she who is signing the permanently, binding social contract. Hence, she cannot be held to a contract she did not sign. There is a way to overcome these problems without resorting to the veil of ignorance, but a lack of space prevents taking up this task.[56]

Rawls' two general principles face an even more severe problem than those listed above. By making the principles general enough so that everyone will unanimously agree to them, Rawls has stripped them of the details required to make them practical. Everyone is in favor of guaranteeing the most extensive set of liberties possible to every person, helping the worst off, and allowing greater benefits be attached to offices requiring the greatest work is open to fair competition from all. However, it is not at all obvious what these two principles mean for people in the real world. In

---

[56]First, the contracting group is divided into two groups of equal or almost equal number. One group will write the social contract, all the time knowing that the second group will be allowed to pick their positions in society first. Whatever is left over will go to the members of the first group. The result is the same as that of Rawls' thought experiment. The first groups will try to make the distribution of benefits and burdens as equal as possible because they know that the second group will take the very best positions.

order to be able to do what they should be doing, people first need to know what the principles entail.

What proves Rawls' theory impractical as developed is the fact that the principles can be interpreted legitimately in many different ways. A capitalist, for example, would say totally free markets allow individuals equal freedom, benefit the worst off by distributing benefits and burdens according to fair competition, and permit buyers and sellers to have the offices of power through fair competition. A socialist, on the other hand, would advocate a different set of liberties, a distribution of benefits and burdens based on needs and abilities rather than on market forces, and competition based on the notion of good work and what allows each individual to fully realize his potential. Rawls' theory is unable to choose between the two or any other alternative way of satisfying the two principles. The result is that every egalitarian, socialist, capitalist, and libertarian is right when he interprets the principles in his own way, which makes it difficult at best to find solutions to moral dilemmas that can be respected by all reasonable people.

To get things done, it would have been better merely to accept some form of socialism or libertarianism that tells people what to do, although not everyone needs to agree with it.

2.6.2 Seven Conditions of Distributive Justice

What is to be taken away from Chapters 1 and 2 to this point are seven conditions which must be incorporated into any adequate distributive justice principle. First, we need to take care of those who cannot care for themselves, if we are able to do so. Second, we want each person to have as much liberty as possible, as long as everyone has an equal amount. Third, the worst-off in society should be as well off as society can make them. Fourth, no one should have more than she needs, if there are insufficient resources to meet everyone's needs. Fifth, as long as everyone's needs are met, it is permissible to have more than one requires, but it should be achieved by fair competition open to all who can compete. Sixth, competition is good for society in some circumstances, as long as everyone is respected in the proper way. Finally, each person who can do so must contribute to the social good by making himself and others better as understood along the lines of flourishing lives. The Distributive Justice Principle incorporates all seven conditions, as will be demonstrated after its two sub-principles are examined.

2.6.3 Distributive Justice Principle

Brutally stated, distributive justice does not make sense without utilitarianism and Kantianism. We can talk about what one is owed, but social desert is meaningless unless we discuss how benefits and burdens should be distributed overall and how the re-distribution from current allotments should be accomplished. Each discussion requires utilitarian and Kantian theories and principles. First, we have to consider what distribution would be best for society. The distribution should be geared toward making society as well off as is practically possible, with the ultimate goal of creating a good society. But we also have to take into account the fact that we cannot treat people as mere means when we are distributing society's benefits and burdens, otherwise we might have the best society according to utilitarianism, but it is an unethical community that treats its people as objects. So the best distribution cannot be one that fails to respect all people affected by the distribution as intrinsically valuable.[57]

Furthermore, the mere fact that no one is treated as a mere means in a particular society does not entail the society is a good one. There are many different ways of satisfying the Kantian imperative, but in order to give people as good a life as they can have, we should also be concerned with bringing about the best society within this limitation. Hence, the combination of the two theories is required in any adequate distributive justice principle.

Combining RPU and QCI gives us the best of both worlds: utilitarianism's prescription to maximize utility and the Categorical Imperative's respect directive. According to justice, we must try to make society and its members as well off as we practically can, while at the same time treating everyone affected by the actions as ends in themselves. Put more formally, the theory is:

Distributive Justice Principle (DJP):

X is a fair distribution of society's benefits and burdens if and only if

1. X is a state of affairs that a reasonable person would reasonably believe to probably maximize utility, and
2. neither the means to achieve X nor the maintenance of X treats anyone as mere means.

According to DJP, if a distribution of society's benefits and burdens maximizes utility, but at the same time disrespects the value of an individual in society, then the distribution is unjust. Furthermore, even though there may be many different distributions and ways of maintaining them that do not treat anyone as a mere means, a morally correct dispersal is one that will probably maximize utility according to a reasonable person.

It is helpful to see how the principle works in practice. Since she is interested in flourishing, the first thing a reasonable person would do if she were using the distributive justice principle is to examine society's overall resources to discover what types of lives are available for the society's citizens. If there are abundant resources, then each person's life could be very good. However, if very few means exist, then the citizens' lives would necessarily be worse than those in a more privi-

---

[57]There can be disparities in wealth to foster the best utility, but they must not be implemented or maintained if it would treat people as mere means.

leged society. The reasonable person would want for every citizen to have the very best possible life given the circumstances of the society.

For the reasonable person, there are three different types of lives that would make a society good overall.[58] *Subsistence* is the first and lowest level. This is a life level at which only a person's physical needs are met so that she does not die in the very near future from starvation, dehydration, or from lack of resources to fill any of her other physical needs. Anyone who has this type of life has a life barely worth living because it is hardly more valuable than disvaluable. If her resources are adversely affected, her life would become disvaluable at that moment. Basically, an agent living at this level is on the cusp of a bad physical life.

Second, the *minimal utility level*, which is similar to the subsistence level, has every need the person has being met, but if she loses one unit of value, her life is not worth living. The difference between these and subsistence lives is that the latter considers only physical needs, while the former incorporates both physical and non-physical requirements. For example, most people need community of some sort to make their lives worthwhile. Among other things, they require feeling part of a relationship of care. Additional psychological needs, such as emotional security, are also met in the minimal utility life, but not in the subsistence life.

Since each person is different, their non-physical needs will be sometimes dissimilar from each other, but they still fall within a specific group of requirements for the set of human persons. This result should not be surprising to anyone. If different people have different physical needs, then they will most likely also have different psychological requirements. Some people require more physical goods than others because of their body type, for example, but the needs are from identical groups, including but not limited to hydration, nutrition, and shelter (Khatchadovrian 2001, p. 298).

Non-physical needs vary in the same way as physical requirements, except for the fact that they incorporate different types of necessities which an individual can have, in addition to the ranges of a particular need. However, regardless of who the particular person is, she will have some minimal non-physical needs that must be met. Even hermits, for example, psychologically require spirituality or study in their lives to make them worth living on the minimal utility level. Regardless of what they are, if any of the individual's physical or psychological needs are not met, then the person's life will fail to meet minimal utility and not be worth living due to that physical or non-physical reason and the precarious state of the other elements of the person's life.

The final level of life is the one to which all reasonable people aspire, which I shall call the *maximal life level*. The maximal life is one in which someone is provided the goods to fully realize herself as a happy, moral agent, whose life is

[58] I will not consider the problem if there are insufficient resources to give everyone at least a subsistence level life. In those cases the value of the society cannot be positive because at least someone is dying from her needs not being met. In societies in which there is little available, the reasonable person will take the least evil alternative.

flourishing fully, rather than being on the cusp of either physical or non-physical needs deprivation. What distinguishes this level from the others is that all the person's needs are met, and yet she can lose some resources and still have a secure happy life. There is in effect a buffer providing security in its broadest sense to the individual from falling into a minimal utility life with its precarious grasp on fulfillment or the subsistence life that has to be an unhappy life given that the person's psychological needs are unmet.

Since the highest level of life is ultimately the one a reasonable person would try to promote with resource distribution, if he can, it is useful to consider some of the life's necessary features. Each of the seven characteristics provides pleasure or other intrinsic value and enables the individual living the life to grow into a better, happier person. First, a good life requires shelter and other physical goods, which provide for not only the bare minimum requirements of physical need, but are pleasing to the person. These might be aesthetically appealing or have some other feature the person enjoys. Second, caring relationships are necessary for a number of reasons including but not limited to create the nurturing goods of considerateness, compassion, sympathy, and love in the person. The goods provide benefit not only to the individual but to others as well. A person with them is more likely to given them to others in the same way they were given to her. Third, satisfying work makes the person feel useful, besides being creative and challenging for the individual. Fourth, entertainment and relaxation are necessary for happiness. Though work, exercise or other pursuits are useful and interesting, but if they are imbibed in excessively, they become wasteful and even irritating. A person needs to have a little time to do something else in order to be able to appreciate the other valuable things in life. Fifth, exercise, medical service, dental care, and related goods required for both the mind and the body's health. Sixth, security is needed in more ways than one. Not only do people desire to be safe in their person and possessions, but they require psychological and financial security as well. Security allows them to make plans for the future that will help them to realize their potential rather than being a mere method for preventing them from losing all of their valuables. Seventh, an education that facilitates pragmatic critical reasoning is essential for a few reasons. The most important of which is that it enables the person to know what evidence is and which values to assign to it, and how to make reasonable decisions, to become a reasonable person and a good person who does the right thing if she chooses, and to pursue and maintain a flourishing life for herself and others.

If all seven conditions are fulfilled, then the person is capable of living a happy life and is more likely to do so than for any other type of life, but is not guaranteed to do so. After all, to flourish, the agent must be a good person, which entails that she has control over her life. With control, however, comes the ability to choose poorly. It is possible for any agent to waste her opportunities, which will result in her having a substandard and bad life through her own fault.

From the vast difference in the types of lives, a reasonable person would attempt to procure the best possible life each person could have given the society's resources and the desire to promote sustainability for each individual, the society, and the world. The uppermost limit on resource consumption and requirements for any society is the amount needed by the society if everyone in the society is living a maximal life for her entire lifespan. Anything beyond that level would be waste. Waste becomes unethical provided there is another society or group of intrinsically valuable entities which could use the excess to improve its members' lives from a lower to a higher level.

A reasonable person would want a resource distribution that appears likely to maximize utility, as DJP's first condition requires. In order to have utility maximized, along the line of Rawls, it is reasonable to believe that achieving a certain sustainable standard of life for each member of society will be better than having a large economic gap in which some people fall below the good life level, while permitting others to far exceed consumption of resources required to obtain it. The pain and suffering caused and created by those who have less than they need would probably be equal to or overwhelm the value generated by those who have a great deal more than they need for a good life. Furthermore, allowing some to have substandard lives while others have excess does not respect the value of the former; thereby violating DJP's second condition.

According to DJP, before the developed world countries take their surplus resources and redistribute them to other nations, each of the former has to fulfill its own citizen's maximal life level needs. Since there are people, even in the developed world, who cannot lead this level of life due to inadequate access to the products required for it, then they should be benefited first. It is most efficient not to have to transport goods and respects people since they rely on their fellow community citizens to alleviate their plight before their fellow citizens take on non-citizens' problems. After everyone in the particular society has achieved the minimal level for the maximal life, any additional surplus must be transferred to other parts of the world to improve their existence in such a way that they are able to obtain and maintain the maximal life level for their populations.

DJP does not require the supererogatory when it comes to resource distribution. According to DJP, the developed world has no duty to drop below the maximal level to that of minimal utility or subsistence merely to help other nations. Giving up a good life is a charitable action, which is permissible but cannot be required (Singer 2006, pp. 257–8). Moreover, RPU supports the conclusion that maximal life societies provide a firm foundation for creating positive change in other areas of the world that minimal utility level societies cannot. In a world of minimal utility, there will be more people with lives worth living than in a world with inadequate resources and some people living maximal lives. However, the former world has people living on a precipice. If they lose one unit of value in their resources, they have fallen below minimal utility, and their lives no longer have positive worth. At this level, it will be very difficult for them to help others because they have nothing to give that will not cause them to fall to a subsistence or worse level. In addition, they will be unable to maintain their lives in the face of adverse events, such as drought. On the other hand, the world with maximal lives has people who by definition can spare resources to help others. They will be able to assist their world for long term benefit, while the minimal utility citizens are living hand to mouth[59] (Parfit 1992, p. 387).

DJP also helps discover a process by which to reach the best possible world in a practical, sustainable manner. Although such a reality would be one in which everyone has adequate access to the materials required for living the maximal life, it will not be possible to achieve this goal right away.[60] Provided that there already is a group of people with maximal lives, a reasonable person would start by achieving the subsistence level for everyone else in each nation, and then work her way up the scale. As societies create better lives for more of their citizens, these individuals in turn can assist their societies in raising other citizens yet higher. Once a society has achieved the maximal life pinnacle, it is then obligated to help other nations with their upward climb.

Furthermore, unlike Garrett Hardin, I believe that if we help people to achieve maximal lives, we will not need to spend as much as we do now on our military or policing agencies (Hardin 2002, p. 199). With the DJP distribution, there will be more people than there would have been otherwise – provided that achieving the maximal life does not actually decrease birth rates and people do not foresee that such large number of inhabitants will reduce the likelihood of everyone having a maximal life. However, although there are more people, they will be more content with their lives because they are maximal lives and no one has excess before everyone has her best opportunity to create a flourishing existence for herself. Therefore, the extra people with strength given to them from more abundant food and other resources will not be the dangerous, rebellious hordes Hardin makes them out to be (Hardin 2002, p. 200). Although starving people might be "completely selfish," civil disorder and revolution is not a necessary result of the wealthy helping the poor, especially if the assistance takes the form of life improvement (Hardin 2002, p. 200).

DJP might sound foolishly idealistic, but there are many relatively simple practical actions that can be taken to redress some of the unnecessary life level gaps. Even if it is impossible due to inadequate resources to have a world in which everyone has a maximal life, there is a great amount of misspent wealth that could be transferred to those who need more to achieve at least the subsistence level of life. First, the developed world can distribute much of what it spends on elective surgery and other unnecessary medical procedures to its own citizens who do not have maximal lives and other parts of the world needing basic health care resources. In 2000, for example, the American Society of Plastic Surgeons reported that approximately $7.5 billion was expended on unneeded elective procedures such as breast augmentation, facelifts, and chin augmentation in the United States alone (American Society of

[59]This will be as much as will be said about theoretical examples such as these. Since ethics is practical, counter-examples that are possible but extremely improbable will not be considered to be legitimate.

[60]I am assuming that Earth has the resources required to give everyone on it a good life. Of course, this would require a major attitude change from consumerism to sustainability.

Plastic Surgeons). Since these procedures are unnecessary for a maximal life, the money would be better used by giving it to people whose countries cannot afford health care, food, clean water, or other necessities.

Second, many developed world citizens needlessly bring a number of their health problems upon themselves. Instead of the denizens of the developed world indulging themselves with bad products, such as cigarettes, or an excess of good products, such as overeating chocolate, they can transfer the money that would have been spent on easily preventable health problems to those who need it (Finkelstein et al. 2004; Colditz et al. 2002). The cost of obesity in the United States alone is estimated at $75 billion dollars for 2003 (Finkelstein et al. 2004, p. 18).

Third, the amount of money the developed world spends on its pets is morally disturbing once we realize that many human beings cannot attain the subsistence level of life or beyond due to a lack of resources. In 2005, $35.9 billion (estimated) will be expended in the United States on pets (American Pet Products Manufacturing Association 2006). Of that amount, the American Pet Products Manufacturing Association National Pet Owners Survey states that in the basic annual expenditures for each dog or cat are $34–$68 for treats and $29–$45 dollars for toys (Ibid.). Granted having a pet is very good for a person because of the caring relationship and relaxation, it cannot, however, justify the amount of resources wasted on pet greeting cards, holiday costumes, monogrammed beds from designer catalogues, designer food, gadgets, and vet bills for diseases resulting from over-breeding, among other avoidable problems.

Even though a substantial number of wasted resources have been mentioned, it is merely a short list of goods that could be put to better use elsewhere. If we included many other factors, such as the fact that the developed world overuses energy resources, food, water, and so on, above and beyond that required by a maximal life, there would be a huge amount of capital freed to uplift others, especially in biotechnology and food production. Given all of these squandered resources, DJP correctly classifies the present distribution within developed world countries and the world as a whole as morally wrong, while pointing to a morally justifiable redistribution.

To return to the issue beginning this discussion, DJP also can classify the creation and marketing of transgenic organisms and other technology. If the developed world expends resources on biofood research and marketing unnecessary for its citizens to achieve or preserve a maximal lifestyle, then it acts wrongly.[61] It would have been

[61] There is a difference between the duties of governments and those of corporations. Although the social good is a vital consideration for companies, in order to survive and increase overall utility, they must ethically compete in fair markets. One of their primary foci, therefore, needs to be upon their own probable agent-utility. Agent-utility is defined as the result of subtracting all the evil produced for the agent of an action by the action from all the good produced for the agent by the action. Governments, on the other hand, should focus on the needs of their citizens, which entails the pursuit of the society's "agent utility." Once social agent-utility seems to be maximized, the governments should help other countries' to sustainable achieve and maintain the subsistence, minimal utility, and maximal utility, in that order.

better, in all likelihood, to have transferred capital to provide greater food production for the citizens of the developing world, either by conducting research that will provide affordable production or improving the farming resources Third World nations already have. Furthermore, by devoting resources on needless research and other activities, the developed world does not treat the people of the developing nations as ends in themselves. As we have seen before, it is impossible to respect the value of others when spending more on oneself than required to achieve a good, flourishing life. Hence, DJP is a useful theory for examining research and marketing performed in both the developed and developing worlds and the technology developed from it.

So how does DJP tie into bio and other forms of technology? New technology creation and use will be judged as to whether or not they satisfy PMC's two normative principles. If the technology does not treat any person as a mere means and it is reasonable, according to a reasonable person, to believe that it is likely to maximize utility, as measured in terms of sustainable, flourishing lives possible within the situation's constraints, then it is morally permissible to generate and utilize innovations. Technology that would easily fulfill these requirements would be transgenic organisms designed for developing world conditions, especially if it was to elevate the very poorest of the poor's lives. On the other hand, if the new product will serve few with little effect, and there are avenues much more likely to produce better results, then DJP would not classify its development as morally permissible. New plasma televisions, for example, are beneficial for those with the means to afford them, but there is no real necessity for them when subsistence needs in the world are not being met.

Existing technology would undergo the same analysis as new technology, although because it is already in the market, it will have different considerations. We need transportation to get to where we work, shop, and carry out the everyday activities that make life worth living. However, there is no need for SUV's that use far too much gas for too little positive result, which is one reason that Ford Motor Company and other car manufacturers have decided to refocus production on more fuel efficient vehicles. SUV's and large trucks make no sense under DJP for the average driver because they are highly unlikely to maximize utility, while their expense and overuse of limited resources disregards those people who struggle to buy gas for their minimal travel needs. Of course, other essential or unnecessary products might be too ingrained in the social fabric to replace so easily or at all, but every piece of technology needs to be considered in a world with limited resources.

Although DJP would clearly force the developed world to rethink its actions and lose many of their materialist products, the fact is that DJP's social justice demands it. It will be difficult for many of us to give up our technological luxuries, but when we truly understand how they hurt others and that they prevent us from doing to pursue our own flourishing lives, then acceptance becomes easier.

2.7 The Moral Paradigm Test

Although the Practical Moral Code's normative principles are sufficient to classify at least one morally right action in every case, they can prove to be too theoretical to capture how people make or should make moral decisions. Calculating utilities as reasonable people would, or trying to figure out which agents are affected by an action so that they can be respected properly is difficult intellectual work that often shifts the evaluator to using too much abstract reasoning from an appropriate balance of both reason and emotion/feeling.

One way of testing if a person has found the correct solution – or has decided upon a morally permissible action – is to apply the Moral Paradigm Test. It is a very useful tool, which helps us to incorporate emotions correctly into any moral decision process, rather than merely relying on pure reason alone. This test might not always capture moral reality, but it works in the vast majority of situations that most people encounter.

The Moral Paradigm Test is a five to six step process, depending on whether or not the person using the test has selected the correct answer with PMC before she begins. First, each agent is instructed to think of an actual person whom she believes to be a paradigm of morality, and would like to emulate. The person must be someone the moral agent personally knows, such as a relative or teacher, with whom the moral agent has established the right emotional connection. The best person to choose for this test is someone who the agent knows will see through any attempt at deception or rationalization on her part and hold the agent to her moral principle, without letting the agent off the hook. The importance in selecting an appropriate person as a moral paradigm is he or she will be making a pronouncement the agent cares about enough so that it will cause either to affirm her conclusion or to re-do the decision procedure. If the agent does not sufficiently care about what the paradigm thinks about her conclusion, then the test will be more likely to fail.

The moral paradigm chosen must be someone those using the Moral Paradigm Test care to please. He or she is someone the agent loves, cares for, or respects in a positive way. These relationship types are more positive than that of fear. The agents wish to become like the paradigms, which entails that the paradigms have qualities the agents believe to be good. After all, who would want to become worse than they already are, unless they somehow think doing so is a good. A moral paradigm for these people motivates them to be good because they inspire them. Hence, when they select a paradigm, they should pick one who will motivate them to emulate the person based on a desire to please the person out of positive emotion, rather than out of fear of being punished.

The second step in the test is for the moral agent to carefully evaluate the paradigm she has chosen to find any moral flaws the person has. Most paradigms have some evil traits or habits, which will need to be eliminated to get more accurate answers with the test. For example, does the person have racist tendencies? Does the person sometime act callously toward others? Does the person rush to judgment at times? And so on, until the obvious flaws have been identified.

Now it should not be thought the Moral Paradigm has to be a Perfect Paradigm. That is, the Paradigm is not an Angel or divine entity because those creatures are beyond our understanding.[62] In order to be humanly understandable to us and create the requisite feelings of likeability, respect, awe, and so on which makes us want to emulate the paradigm, the potential paradigm has to be human. We have never encountered a human Perfect Paradigm; so trying to make an emotional connection to such a creature is as impossible as trying to perceive as a bat perceives. Since we have no reference points by which to make comparisons, we cannot hope to understand what they would do and why they would do it so that we can feel more strongly that we have found and can defend correct answers to our moral dilemmas.

The third step in the Moral Paradigm Test is once each person finds all of the relevant moral flaws in the paradigm's character, she then replaces them with the contrary virtues, such as making the person being egalitarian instead of be an racist. The result is a virtuous ideal, potentially real person, with whom the moral agent has an emotional connection that makes the agent want to emulate the individual, i.e., the Moral Paradigm, who will be used in the test.

In the test's fourth step, the moral agent is asked to perform a thought experiment. In her mind, she is to tell her Moral Paradigm her solution to the moral dilemma or what action she would select under the circumstances. Furthermore, the moral agent is to inform her Moral Paradigm of all her intentions, motives, attitudes, and feelings, as well as the relevant information she used in arriving at her conclusion.[63] For example, if she says Mary should conduct research on transgenic wheat, then she will explain her reasons for why Mary should conduct the research, including how utility will probably be maximized, according to the reasonable person, and establish that no person is treated as a mere means by the action. In certain cases she might even explain how the alternative situation is likely to be worse or that it treats someone in a disrespectful manner.

The moral agent, in the fifth step, is then asked to decide what the Moral Paradigm would say about her decision. If her paradigm would inform her that her conclusion is correct or not lose some respect for her for reasoning, then the moral agent knows she probably has selected the correct solution. The agent can stop at this point with some justification for believing she has found a moral truth.

On the other hand, if her paradigm would lose some respect for her or tell her she has erred in her evaluation, then the moral agent knows she probably has made a mistake. She has to go back to her decision process and find the moral factors she

[62] Recall that we are trying to avoid needless controversy and work. If we assert that angels or divine entities exist, then we automatically become responsible for proving our contention is true or at least prima facie plausible. In order to make the practical moral code as practical as possible for all, then it is best not to introduce elements that are not needed to make it work and would cause some to abandon it because it does not represent their beliefs.

[63] The information that a reasonable person would need to evaluate the outcomes of actions would be used here as well.

missed or re-evaluate those given an improper evidentiary weight. Reconsidering or discovering evidence is the sixth step of the process.

The Moral Paradigm Test's greatest advantage is that it makes moral agents realize ethics do matter and it is not a mere abstract thought experiment thought up by professors who have never had much contact with the real world. The test provides evidence for at least three important claims. First, the rationalization and illicit biases of moral agents can be stripped away. This directly attacks the position that people cannot find objective moral truths. Second, moral agents generally know what the right thing to do is provided that they are given a proper framework to help them make judgments. Third, moral agents can be motivated to do the right thing if they are emotionally connected to the situation in the proper way. Any one of these benefits would recommend the Moral Paradigm Test, but the three together are sufficient to justify its adoption.

2.8 Conclusion

Developing this part of the Practical Moral Code is really only the first step in building a moral decision procedure. At the moment, all that exists are the principles, how to test results from using PMC, and a little bit of information on how to discover evidence for the decision process. What is missing is an axiology or theory of value to know what is intrinsically valuable, and how those things with intrinsic value should be compared together. In Chapters 3, I will eliminate the natural and evolution intrinsic value candidates and arguments based upon them. In Chapter 4, a practical hierarchical axiology will be developed that will help any reasonable person know what good to maximize and what good to minimize in any situation confronting him.

Chapter 3
Are Transgenic Organisms, Biotechnology, and Technology Unnatural?

3.1 Introduction

One of the first negative responses to any new technology is to claim that it is unnatural, and therefore, morally wrong or bad. Airplanes, wind turbines, cars, telephones and a host of now socially acceptable technological achievements have faced this objection. They eventually overcame the resistance barrier as a result of people adopting and becoming comfortable with them, and are now part of the *status quo* in many societies. Transgenic organisms, which have been around since the 1990s, are still undergoing the same objection, although it is still too early to determine if they will also gain general acceptance.

In order to avoid the time and effort spent by so many in continuing to respond to this type of attack, I will attempt to show its philosophical uselessness here once and for all.[1,2] Arguments which rely upon all transgenics' alleged unnaturalness are inherently defective for at least one of two reasons, both of which are tied to the difficulty of plausibly defining "naturalness." First, some of the individual definitions in conjunction with the Unnatural Is Unethical (UIU) argument result in an absurd conclusion, such as all actions are morally right or all man made objects are morally bad. Second, even if the definition avoids the absurdity problem, it entails conclusions that the opponent of transgenic organisms cannot or will not accept.

To evaluate UIU fully and fairly, it is necessary to consider as many initially plausible or commonly used essentialist definitions as possible. First, Burton Lieser's

I would like to thank Gary Comstock for giving me the idea for this chapter.

[1]Some people would reject this philosophical analysis of the unnatural based on the claim that the "natural/unnatural distinction is one of which few practising scientists can make much sense" (Nuffield Council on Bioethics 1999, p. 15).

[2]Jan Deckers tries to defend the natural/unnatural argument by showing that there are still concerns that both adherents and critics have (Deckers, 2005). The problem is Deckers never establishes whether these concerns are rational ones to have. The mere fact that people feel a particular way, no matter how knowledgeable they are in a specific research area, is insufficient to establish a claim other than that they feel that way. What would have to be done is to argue that the feelings are justified, which requires a clearer definition of the natural.

D.R. Cooley, *Technology, Transgenics and a Practical Moral Code*, The International Library of Ethics, Law and Technology 4, DOI 10.1007/978-90-481-3021-4_3,

five definitions of inherent unnaturalness will be analyzed in detail and rejected for at least one of two reasons listed above. In addition, three more complex meanings will be critically examined. In *Vexing Nature? On the Ethical Case Against Agricultural Biotechnology*, Gary L. Comstock evaluates fourteen intrinsic arguments against TOs – three of which are especially relevant to the Unnatural Is Unethical Argument.[3] Opponents of transgenic organisms have claimed that the existence and creation of transgenic organisms and agricultural biotechnology are unnatural because either they "transfer the essence of one living being into another," "change the telos, or end, of an individual," or "illegitimately ... cross species boundaries" (Comstock 2000b, pp. 189, 191, and 193). Although Comstock's rejection of each argument and definition covers many plausible interpretations, there are still several versions and alternate criticisms which further illustrate defects in UIU.

In addition, it can be proved that transgenic creation is morally identical to some non-artificial phenomenon or human intervention which is not inherently wrong, therefore it follows that both creating transgenic organisms is not inherently wrong and the organisms themselves are not inherently bad, at least on these grounds. Hence, the same method used in rejecting Leiser's general definitions can be utilized again for the subtler ones Comstock examines. As a result, the Unnatural Is Unethical Argument is irrelevant to transgenic organisms, biotechnology, and other forms of biotechnology, and should be abandoned for more promising avenues.

3.2 The Unnatural Is Unethical Argument

The Unnatural Is Unethical Argument has only one form but comes in two varieties: one for actions and the other for objects. Basically the argument is an Aristotelian syllogism of mood AAA and Fig. 1. Put formally, the two versions of the argument are:

Unnatural Is Unethical Argument:

Version 1:

P1. All unnatural actions are morally wrong.
P2. *All X actions are unnatural actions.*
C. All X actions are morally wrong.

Version 2:

P1. All unnatural objects are morally bad.
P2. *All X objects are unnatural objects.*
C. All X objects are morally bad.

---

[3]The reduction-of-life-to-its-chemical-components-is morally-wrong argument is a variation of this theme. Bernard Rollin does an excellent job formulating the strongest case for it, and then showing why it fails (Rollin 2006, pp. 138–41).

Since both versions share mood and figure, they are valid. A valid syllogism, of course, is one in which it is impossible for the conclusion to be false while the premises are true.

However, the mere fact an argument is valid does not entail it is sound as well. Soundness includes validity in its definition, but also requires that the argument's premises are all true at the same time. Of course, it is the soundness of UIU's two versions in the transgenic debate with which we are interested. The only way to determine if either version is sound is to substitute "creations of transgenic organisms" and "transgenic organisms" in place of each respective variable, and then determine if both premises correspond to reality.

3.3 The Weak Definitions

Burton Leiser's five definitions of unnaturalness in his argument against employing UIU in the debate over homosexuality are useful in the examination of transgenic organisms' naturalness. The definitions are general enough so that they can be applied to any object or actions someone classifies as unnatural. Leiser's definitions are:

> O1. X is an unnatural object = df. X's existence violates the descriptive laws of nature.
> O2. X is an unnatural object = df. X is an artificial or man-made object.
> O3. X is an unnatural object = df. X is an uncommon or abnormal object.
> O4. X is an unnatural object = df. X is an object that results from using an organ or instrument contrary to its principal purpose or function.
> O5. X is an unnatural object = df. X's existence is morally bad. (Leiser 2007, pp. 127–33).

If we adapt Leiser's meanings for actions, then an unnatural action is one that has at least one of the following five characteristics.

A1. X is an unnatural action = df. X is an action which violates the descriptive laws of nature.
A2. X is an unnatural action = df. X is an artificial action.
A3. X is an unnatural action = df. X is an uncommon or abnormal action.
A4. X is an unnatural action = df. X is an action which uses an organ or instrument contrary to its principal purpose or function.
A5. X is an unnatural action = df. X is morally wrong action.

To most efficiently examine UIU and the various meanings of the unnatural, the corresponding definitions under objects and actions will be paired together, such that O1 and A1 make definition pair one, O2 and A2 make definition pair two, and so on.

3.3.1 Rejecting the Definition Pairs

Even though UIU is frequently used and seems to be a powerful argument in the minds of those opposing TOs and other technology types, it cannot adequately establish either that any technology is morally bad or the creation of it is morally wrong. UIU fails to be convincing rationally because each definition pair, when combined with it, either entails an absurd or unjustified conclusion. I will consider each pair in turn.

The first definition pair's broadness renders it useless to those who would try to prove that transgenic organisms are unnatural. If this definition set is correct, it follows that all human activity and the products of such activity are natural merely because all human endeavor is done in accordance with all the laws of nature. Everyone would agree that the scientific activity of creating TO and the TO themselves neither are miraculous nor do they violate nature's descriptive laws. Therefore, creating transgenics is not morally wrong. Moreover, since transgenic organisms are able to exist within the system governed by the laws of nature, they must be natural entities and not morally bad. Hence, UIU cannot do the practical work transgenic opponents desire it to do.[4] Definition pair one should be rejected on these grounds alone, but it has another fatal flaw that makes it impossible to adopt.

The only things that are unnatural, according to definition pair one, are those actions or things, whose mere existence violate the descriptive laws of nature; in other words supernatural activities and objects, such as witchcraft, astrology, or even positive supernatural actions and objects. Consider miracles, for instance. Miracles are supernatural by definition; hence, because they are states of affairs that could not have arisen without some force beyond that found in nature, they are morally wrong or bad. Moreover, God or any supernatural creature, according to UIU and the first definition pair would be a morally bad thing as well. Since UIU classifies the divine as morally wrong or bad, which is a result that most people who believe in the goodness of miracles and God cannot countenance given the latter's widely ascribed inherent goodness, they must reject definition pair one.

Definition pair two based upon the artificial fares better than the first because it, combined with UIU, does the work opponents want it to do. First, UIU and the definitions do not classify all human actions and non-supernatural states of affairs as morally permissible or good by definition. Second, and most importantly, the artificial production of TO and the organisms themselves are morally wrong and bad, respectively. If an opponent of transgenic organisms wanted to prove that transgenics are morally bad, then definition pair two would certainly do the job.

---

[4]If UIU actually was the Natural-is-Ethical Argument, one of whose premises is that natural objects/actions are morally good or right, the result of definition one would be that all human activity and the products of such activity would be morally right or good. In fact, it would be impossible for humans to ever do the wrong or evil thing because humans do not have the ability to violate the descriptive laws of nature. The Nazi's eugenics programs on these grounds lose their status as moral atrocities and become morally good and right states of affairs. That cannot be correct.

However, it is unlikely that any thoughtful person would resort to this pair due to it entailing an absurdity. By these definitions, all human actions and human made products are unethical. The result is that there is nothing any human being can do that is ever morally permissible and every object caused by human intervention is morally bad. This means that opposing TOs is morally wrong, for example, at the same time and in the same way that supporting TOs is unethical. Furthermore, working to prevent transgenic organisms from being incorporated into agriculture and the food supply is just as immoral as working to incorporate them. Although definition pair two is helpful for opponents of transgenic organisms in the short term, when the focus is solely upon creating transgenics or their existence, its universal condemnation of all human activities and the products of such reveals it to be implausible in the long run.

Moreover, this definition pair destroys any reason to be an ethical person or to act morally. Under definition two, provided that there is nothing that one can do that is right or permissible, there is no incentive or reason to do one thing rather than another. Opponents of transgenics, on these grounds, would lose the moral impetus they want to achieve by labeling transgenics and the human activity of creating them as unnatural, and therefore unethical. If creating TOs is morally wrong, for example, while doing something else is not, then there is an ethical reason for doing the something else and an ethical reason not to do the former that any reasonable person would understand and make part of her decision process. Yet, if every action that an agent can do is morally wrong, then there is no ethical justification to choose or do one over the other. Furthermore, if every object that will result from every alternative will be equally morally bad, then there is no morally significant reason to prefer one object over another. There is no more justification to support the desires of the opponents to prevent or eliminate transgenics than there is for those of the transgenics proponents who want them in the marketplace. Definition pair two is as useless in the transgenic ethics debate as the first pair.

Definition pair three – the unnatural is the uncommon – seems to avoid the mistakes of the first two pairs by being able to classify some objects of human endeavors as morally good, while others as morally bad. In addition, the third pair allows for some human actions to be morally right, while others are morally wrong. By being able to classify human activities and the results of them in different ways rather than necessarily right/wrong or good/bad, definition pair three captures our intuition that there are instances when we should and can act rightly and produce good things, while avoiding doing what is forbidden or making bad objects.

Although it is initially more promising, this definition pair's vagueness requires some development of what it means for something to be abnormal or uncommon. When someone claims that an object or action is abnormal, he is actually stating that the object is uncommon or abnormal relative to some set of objects, which is either explicitly or implicitly expressed. Car ownership is common in some areas of the United States, for example, but unusual in others, such as in very large cities. Therefore, someone remarking "It is unnatural to own a car" has to be understood in context of the set of objects to which she is referring in her statement. Furthermore, the claim is true if and only if the object truly is uncommon in the group of objects.

For instance, "It is natural/normal to be a woman" is false when referring to world leaders because there have been few women in the set of world leaders. But the same proposition is true if referring to the set of US citizens because there are more women than men in America. Since the truth value of the proposition is relative to the group that is being referred to by the statement, we must be careful to identify identify the set that is being referenced to understand and evaluate the assertion.

When referring to the set of all organisms that have existed, exist, or will exist, the proposition "Transgenic organisms are unnatural" is obviously true. Using UIU, it follows that TOs are morally bad since they really are unusual in the group of living objects. On the other hand, common organisms, such as those developed through evolution alone, e.g., *E coli*, Mycobacterium tuberculosis, given they are in the majority of organisms, are morally good.

Furthermore, it is safe to say that the creation of such transgenics remains relatively rare in the set of purely scientific endeavors and other activities as a whole. Besides the multitude of scientific activities that have nothing to do with biology, most of those involving biology in some way, e.g., plant and animal breeding and medical research, do not incorporate transgenic development into their work. The result is that creating transgenics is morally wrong and their existence is morally bad, according to UIU and definition pair three. On the other hand, doing almost anything else is morally right for an agent to do because doing anything other than creating TOs will always be in the majority of the set of human actions.

There are three devastating problems with definition pair three, any one of which renders it rationally impossible to accept. First, the definition pair and UIU entail that the rare is morally evil or wrong merely because of its being unusual. This conclusion is a mistake because many rare things are positively good. For example, the supererogatory action of saving another person's life at the cost of one's own is clearly morally right, although it is rare when considering the full set of human actions. Moreover, being a moral saint is a very good thing to be even though it is extremely unusual for people to be able to rise to that level.

Second, pair three and UIU incorrectly entail that the mere fact something is common makes it morally good or right. For example, if the implied set is the set of acts by despotic rulers acting as despotic rulers, then Hitler's actions of eliminating perceived enemies of the state are morally permissible because they are common relative to the set. An inherently evil thing, such as pain, is morally good if the implied group is the set of states of affairs identical to pain.

The third problem with the third definition pair and UIU is their misclassifications of morally neutral characteristics or products. For example, the morally irrelevant characteristic of being a human male is morally bad under UIU as interpreted with this definition pair because being male is in the minority relative to the set of all human beings. Moreover, any neutral form of human activity or existence can be classified as morally wrong or bad as long as the implied reference set has a greater number of members not having the quality than members sharing the feature with the object. Merely because there is only one individual that is identical to who we each uniquely are entails, according to the definition pair three and UIU, that it is morally bad to be who we are, and that each of our individual actions, because there are so few of them relative to all other human actions performed, are necessarily morally wrong. Such absurd results entail that definition three is at best incomplete, if not outright incorrect.

The last two problems show that this definition pair is based too much upon the arbitrariness and subjectivity of the individual making claims about naturalness and morality. Instead of a more universal morality, if he is a clever person, he can alter morality based upon the group he chooses as his set. Suppose an ardent racist enjoys torturing animals. His morally repulsive actions are common and natural if he limits his set to that of actions in which animals are tortured. On the other hand, he can legitimately state that interracial marriage is morally bad given that it is unusual in the set of all marriages. But, as we have seen in the preceding two chapters, morality has at least some objective standards, thereby making this subjective, idiosyncratic principle much too arbitrary. Moreover, if everyone is allowed to choose their own referent sets, then opponents and proponents of a technology cannot convince anyone else of the truth of their claims unless they share a common referent set. The result is that people will talk past each other instead of addressing moral controversies in ways that can find practical and reasonable solutions.

The fourth definition pair is most commonly used in the debate against the naturalness of TOs. Unnatural objects and actions are, respectively, the products of a misuse of another object or their creation does not employ an object for its primary function.[5] The Catholic Church, for instance, uses these definitions on a variety of issues and technology, including but not limited to abortion, birth control, homosexuality, and in-vitro fertilization. According to Catholic dogma and Aristotle, when a person uses an organ naturally, she, by definition, uses the organ for its primary purpose (Aristotle 1941c, Book II). Furthermore, if organisms/organs are goal driven systems with a primary purpose, then all such objects have a function that must be the result of some sort of design (Aquinas 1989, pp. 12–14). All human actions, which by definition are designed, have an intermediate sort of end, which is identical to the intention of the agent performing the action, and a final end, which is the goal of true happiness. Those things which are not the result of human design are assumed to have had some other sort of other designer, e.g., God, nature, or evolution.[6,7] Acting or being in accord with the designer's function entails the naturalness of the action or object, and hence, its moral rightness or goodness.

Design in this argument also has a hierarchy. A human being might desire to use an object in a way not intended by one of the more powerful designers of the object. In these situations, the most powerful designer's plan trumps the designs of lesser powers. Hence, if God designs an object for a particular purpose, which is contrary to an end for which a human desires to use the object, then God's design

[5]In Section 3.5.2, Comstock's definition incorporating the idea of a telos is different from the one encountered here. Comstock's definition deals with altering or changing an object's telos, while Leiser's leaves the telos unchanged and the object is used contrary to its telos.

[6]Henk Verhoog argues that each animal has a telos based on its needs as a result of evolutionary processes on its species (Verhoog 1992, pp. 274–6).

[7]The function of objects will be discussed in much greater depth in Chapter 4.

and function triumphs over the human's. The human's design is morally bad, while God's is morally good. On these same grounds, if there are other beings with less power than human beings, they would be also morally bound to use artificial objects in the manner determined by their higher creators.

Critics claim that transgenic organisms are unnatural according to definition pair four because humans are interfering with the function or design of the organism. If God, for example, had wanted tomato DNA to incorporate flounder DNA to create tomatoes less likely to freeze in certain conditions, then He would have already created tomatoes with the requisite genetic material. But He did not. Humans interfered with God's design by injecting flounder genetic material into the DNA of tomatoes. Therefore, the creation of this and other transgenics is morally wrong and their existence is morally bad.

Although not everyone believes in a Creator, it is still the case that functions derived from evolution and nature will lead to almost identical conclusions as those of the Divine Designer position. The forces of nature and evolution did not produce a tomato with a gene identical to that of the flounder which helps to prevent it from freezing in certain conditions. Moreover, nature and evolution cannot now create the transgenic entity no matter how many tomatoes are placed in the tanks of fertile flounders interested in sexual reproduction. Hence, it is wrong for mere mortals to create transgenic tomatoes and other organisms and transgenic organisms are morally bad.[8]

There are many problems with the fourth definition of unnatural. First, it is not obvious to many, unless one already sees some sort of design, that organisms have a primary purpose or function as a designer would give them. Some strangers might see the rock in my yard and believe that it was placed there to fulfill a function, while others assume that the rock was merely left there by glacial activity. The reason why neither conclusion is better than the other is that there is no evidence to support either claim. The same holds true for organisms: it is as non-rational to believe that organisms exist without an outside purpose to serve as it is to believe they have a purpose. Since TOs opponents would have this argument prove that we should not create or allow transgenics to exist, they have the burden of establishing their case. If they wish their arguments to appeal to the reason of others, then what the former must avoid at all costs is resorting to evidence which is merely the product of faith rather than rational justification.

Second, an object's function may only be relative to the person using the object at a particular time for a particular purpose (Teitel 2002, p. 24). A hammer's primary function is to hammer one thing into another - usually nails into some type of solid surface. However, if I want to use the hammer to hold the door open to let in a breeze, then employing the hammer contrarily to its primary purpose is neither morally wrong nor bad. Furthermore, the different functions of hammers may be limited only by the imagination of the person using it and natural laws governing all matter. In fact, all physical objects share this feature of mutability to what a designer using

[8]The evolution position will be fully addressed later in this chapter.

them wants them to be. Hence, transgenic organisms could have whatever function the person creating them assigns to them at a given time. If the person gives them the functions that they currently serve, then transgenics are morally good and creating them is morally right on the sole grounds that is what their creators want them to be and do.

Furthermore, as was seen in the first definition, people are a part of the natural process and not some supernatural entities standing distinct from it. If all parts of the natural process are in turn natural, it follows that whatever human beings do is a component of the natural process; therefore, their works or the products of their works must be natural. According to UIU and definition pair four, nothing humans do and nothing humans produce can be morally wrong or bad, respectively, which is the same problem that forced us to reject definition pair one.

Moreover, if God is the master designer, and He created humans to interact with the environment as we do, then what we make in turn must be a part of His overall plan. If God did not want us to do something, then He would have made it the case that we could not do it. Hence, the function that God gives to an organism is identical to the one that humans give to it, since God is responsible for the overall design which includes all human designs.[9]

Definition-pair-five guarantees that UIU will be valid and sound. In fact UIU is a tautology, given that the first premise, after the proper substitution of definitions, is identical to the conclusion of the argument. It necessarily follows from the premise that X is morally bad in UIU, for example, that X is morally bad. If this definition pair and UIU were sound, any noun or verb that someone substitutes for X would be morally bad or morally wrong, respectively.

The problem here is one of circularity. The fact something was unnatural was supposed to lead us to the discovery, through other premises and justified reasoning, that the thing was either morally bad or morally wrong. If "unnatural's" definition is merely being morally bad or wrong, then there is no reason to talk about something being unnatural, when we already know that it is bad or wrong. UIU would be a waste of time and resources to even consider.

The fifth definition pair does raise the issue of the "yuck" factor and the moral taboo argument raised by Leon Kass and others. Basically the idea is that there are certain intuitive or physical responses people have to various situations that indicate whether or not something is moral/good or immoral/bad. In Kass' argument, the feeling of repugnance or "yuckiness" people feel helps establish that some taboo has been violated by the action itself or whatever entity the action produced (Kass 1997, p. 20). Robert Strieffer goes so far as to claim that in some cases, "we know that an action is wrong merely on the basis of our reaction to it, even if we cannot satisfactorily justify that reaction" (Strieffer 2003, p. 38). In other words, the feeling is a sufficient condition for indicating the ethical status of an object or action. It follows that in the case of transgenic organisms and other technology to which people

[9] If we adopt this view, then God becomes responsible for all evil actions.

have an adverse reaction of this type, either the creation of the technology is wrong or the technology itself is bad.

The moral repugnance argument and the negative reaction have both been attacked on legitimate and illegitimate grounds. The legitimate grounds generally focus on why the supposedly intuitive repulsion is sufficient to indicate anything about the morality of an action or entity. Granted people have this reaction to new technology and other things, it does not follow that they should have such a reaction. Consider the early years when interracial marriage was recognized by the state as legal. Just because many people were morally repulsed by people of different races being wed and having the lifestyle that comes with it did not prove that their marriages were morally wrong or bad in any way. In fact, having this reaction said something negative about the person experiencing it, namely that he is unjustifiably prejudiced. The same argument can apply to new technology. The mere fact that a few to the vast majority of people are repulsed by technology such as transgenic organisms is inadequate on its own to establish that the latter are bad or their creation is unethical.

Even with this considerable problem, I do not want to dismiss the "yuck" factor from all moral consideration, especially since I have asserted that morality requires a blend of rationality, emotions, and feelings. Yuckiness does have a role to play in morality and decision making, but it should never be thought to be sufficient to establish knowledge of an action or object's morality. Instead, repugnance can provide some guidance as to what we should do in the situation, the least of which is the necessity to take greater care in our moral evaluation than we otherwise would. Although I cannot prove a nuanced version of emotivism is true, it seems that for ethics to exist in the way that it does, certain conditions must first obtain. One of them is for us to be drawn in some psychological way toward what is good and right and to be repelled by what is bad and wrong. How else could ethics be action guiding? Furthermore, we would not have moral terms such as "bad" if we did not first have these types of reactions. Repugnance, then, is part of the foundation of ethics as well as being a standard appropriate reaction to certain conditions, once people have been taught what moral terms mean to the worst types of people and actions. Other emotions and feelings have roles to play as well.

There is, however, a limit to the use of emotions and feelings in making moral evaluations. As stated before, just because one has a certain reaction does not entail the reaction is justified or tells us anything about morality. I propose that PMC be used to decide if a particular feeling has merit in cases in which decisions can affect people in morally significant ways. If the feeling passes the test of respecting all persons and is one that is likely to maximize utility according to at least one reasonable person acting reasonably, then it is a justified feeling that can provide incentive to act ethically. Consider the racist and an interracial marriage. By having a feeling of repugnance, the racist does not respect all intrinsically valuable beings in the way he should. At the same time, it is unlikely this feeling makes the overall situation better than it would have been had the racist had a neutral or positive feeling about races or such marriages. PMC can be used for transgenic organisms as well. If a person is repulsed by a particular transgenic and the feeling clearly passes both RPU and QCI, then the feeling is justified. The most obvious case of this occurring would be the creation of chimpanzee-human chimeras that have the mental functioning of human beings but will be used for lab testing in the same way that chimpanzees are used. This type of being should create a great deal of repugnance in anyone familiar with animal research. If it is unclear whether having the feeling passes both RPU and QCI, then the feeling does not provide any credible indication of morality. Better evidence must be sought.

3.4 A "Playing God" Argument

An emotionally charged criticism of new technology, especially biotechnology, is to accuse researchers and intellectual property owners of "playing God" with DNA, species, ecosystems, or some other "natural" thing, which will make the object unnatural. The mere assertion alone implies hubris fit for a Dr. Frankenstein who attempted to steal the secret of life from God, and was punished, along with his community, in a spectacular but appropriate fashion. In many cases, there is no argument here other than one designed to appeal to the listener's fear of new technology and inherent dislike of those humans who would place themselves on equal footing with the divine.

Rejecting such poorly constructed objections out of hand is de rigueur (Sherlock 2002, p. 149), but finding a more thoughtful approach is difficult. Fortunately, Gordon Graham has developed just such a position in his handy book on genes. According to Graham, creating designer babies is unwarranted on the grounds that it violates three boundaries.

> Anyone who believes that he or she can engineer an improvement in the sorts of human being who are likely to arise from more normal processes must believe, first, that they (sic) can predictably secure a certain outcome, second that this outcome is demonstrably superior, and third that their (sic) judgement of its superiority transcends or overrides the first-person judgement of the alternative, non-designed person. (Graham 2002, pp. 180–1)

The first might be impossible to do given the lack of knowledge about human biology which is likely to be with us for as long as we exist. The lack of agreement of what counts as valuable makes the second unobtainable (Graham 2002, p. 179). Finally, the third is what introduces hubris into the equation (Graham 2002, p. 181). For one person to decide the value of another is to engage in overwhelming arrogance. Since the only person who can evaluate a person's life is the person herself, it follows that designing new individuals, by its very nature, implies that non-designed individuals' lives are defective in some way that the designed individuals' lives are not.

Gordon's position is the best of the lot of those who develop a playing God argument; however, there are several false assumptions which ultimately undermine it. First, Parfit's ontological argument about lives worth living and existence adequately refutes Gordon's third condition. Since the designed babies would not exist save for the fact they are designed, then we cannot compare the designed lives to those that are un-designed. They are two different, incomparable entities. Furthermore, if they do not exist at the time of their creation, then their viewpoint of their lives' value is irrelevant to morality if it turns out that they will have lives worth living. Second, if one is selecting from the group of excellences, then it does not matter if there is a lack of agreement as to which one is superior. Intelligence, strength, longevity, beauty, and so on help people lead flourishing lives; therefore it is unimportant which particular ones they have in which quantities as long as they believe their lives were worth living. Finally, lacking absolute knowledge of procreative outcomes should be no hindrance to "normal" or technological procreation. After all, people have children for all sorts of reasons; some are meritorious, while others are purely selfish. There are parents who are attracted to each other on the mere grounds of their sensual appeal to each other, and want to create children that are similarly endowed. If scientists can make children with desired traits through artificial methods, then we should not be concerned unless we are worried about and willing to act against the "normal" method as well. The main moral focus in both cases should be whether the offspring have to opportunity to have flourishing lives. If the answer is no, then there is at least good reason to reject genetic engineering and normal procreative methods used. If the answer is yes, then there does not seem to be a problem with what the researchers have done. The result is that we will have to place to the side this definition of playing God to seek more fecund possibilities.

3.5 More Complex Meanings of the Unnatural

Although Leiser's five definitions of the unnatural capture the vast majority of the uses in the transgenics debate, there are three others he does not address. I believe that these three are more complex in part because they require a greater depth understanding, but their intricacy also stems from the incredible vagueness and ambiguity of terms such as "essence."

In this section, each of the three definitions of unnaturalness will be considered in detail and rejected for at least one of two reasons. If it can be shown that the creation of transgenics is morally identical to some non-artificial phenomenon or human intervention which is not inherently wrong, then it follows that neither creating transgenic organisms is inherently wrong nor the organisms themselves are inherently bad at least on these grounds. Hence, the same method employed to reject Leiser's general definitions can be useful for the subtler ones Comstock examines. As a result, even with the most plausible definitions, UIU provides no advantage in debates about the morality of transgenics or other technologies.

3.5.1 The Transferring Essence Definition

Comstock's seventh intrinsic argument against agricultural biotechnology will provide the first of three complex definitions of "unnaturalness." According to some transgenic opponents, "To engage in ag biotech is unnatural because it is to transfer the essence of one living being into another" (Comstock 2000b, p. 189). If essence is understood to mean the definition of a thing or the necessary characteristics a thing must have in order to be the thing it is, then this version of unnaturalness assumes *the* essence of organisms is indivisibly bound up in their DNA. After all, the only material transferred between the donor and recipient is genetic material from the former; so, the DNA must either be the essence or the essence is necessarily expressed by the material in some manner. Perhaps, it supervenes on the genetic strand. It might also be the case that it is the expressed traits of the individual. Furthermore, this definition of unnatural clearly implies that when a transgenic organism is created, the *entire* essence of the donor organism is transferred to the recipient organism.

The Unnatural Is Unethical Argument using the essence definition is relatively simple. Because genetic material containing the essence of one individual is transferred from one individual organism to another, the resulting organism is unnatural. All unnatural actions are morally wrong. Hence, the creation of TOs is immoral.

Moreover, if the actions which created the unnatural entity are morally wrong, then it is plausible, although not necessary, to conclude that the entity itself is an inherently bad thing. The recipient organism incorporates in its very being that which made the action creating it unnatural. The entity's essence is now unnatural as the result of unnatural actions that corrupted it. Since I want to provide the strongest case for those who oppose transgenic organisms, I will stipulate for the purpose of this discussion that if it is morally wrong to create TOs because doing so is unnatural, then TOs are morally bad due to the fact an unnatural origin creates unnatural entities with corrupted essences.[10]

Before being able to evaluate fully the soundness of UIU and the essence definition, using the Principle of Charity it is necessary to broaden the definition to include all the different plausible ways it can be utilized. First, the essence definition of unnatural is too strict and needs modification. After all, only a small portion of the donor organism's genetic material causing a few characteristics is transferred into the recipient organism, not the entire genome. Unless it is implausibly assumed that the donor organism's entire essence is somehow contained in the relatively small section of transferred DNA, then at most, only part of the organism's essence is shifted to the recipient. In the interests of presenting the most inclusive version of this type of unnaturalness, Comstock's definition will be revised to read "transfer some of the essence of one living being into another" from the narrower "transfer the essence of one living being into another." The former is broad enough to capture the range of partial to complete essence transfer between organisms.

Second, to include all variations of UIU using the essence definition, we must incorporate more entities than individual organisms in the meaning. Comstock discusses whether or not it is possible to change an organism's fundamental nature, when it seems clear that some of the critics of transgenics are actually referring to

[10]It could also be the case that possessing the DNA essence of another thing is sufficient on its own grounds to make the object morally bad.

changing the essence of a species (Comstock 2000b, p. 190). And critics are correct when they assert that transgenic researchers are creating new variations in species. Although natural selection would favor certain character traits for some species, it is impossible for a species to acquire such traits if it is not within its possible variations. For example, golden rice, which has two genes from daffodils and one from a bacterium, could not have been generated from conventional breeding techniques. Since the rice species did not have a variation capable of providing the gene needed to produce Vitamin A, neither the genetic material for production nor the property of producing the vitamin were in the rice species' fundamental nature. Hence, by engineering golden rice, researchers gave rice a new variation/essence that could not have existed except for the researchers' actions.

UIU and the expanded essence definition is much the same for species as it is for individuals. By introducing genes from one species into another, the fundamental nature of the recipient organism's species has necessarily been altered; thereby, creating a hybrid with the essence of two different species. Since changing a species' essence is unnatural, and the unnatural is morally wrong, according to this argument, it follows that creating new transgenic species is morally wrong and the resulting species is morally bad.

Even though the definition has been broadened to be as inclusive as possible, and seems to be the one that many opponents of transgenics use, there are still questions of meaning to be answered before evaluating UIU's soundness. It is therefore imperative to define terms, mostly to try to clarify what "essence" means and how it will work for this anti-transgenic objection.

The location of a thing's "essence" as used in this criticism can be understood in one of three ways. First, essence could be equivalent to the genetic material transferred from donor to recipient organism. Second, rather than essence being the genetic material itself, it might be that the word refers to the properties exhibited by an organism as a result of its genetic material. For example, the essence of being red-haired is the property of being red-haired rather than having the genetic material that causes the organism to be red-haired. Bald people can have the genetic material in their cells that will determine them to have red hair, but do not have the property of being red-haired. Finally, essence might be a combination of the two. Essence, in this third alternative, is both the organism's genetic material and the properties the genetic material causes the organism to have.

The second alternative for essence is the most plausible candidate for reasons other than the obvious fact that, in general, an individual is not identical to its DNA.[11] Comstock states that if things have essences, then the essence of a particular thing is the set of "intrinsic and indispensable conceptual characteristics" of the thing (Comstock 2000b, p. 190). In other words, each member of this particular characteristics' set is a necessary feature of the entity. "When we think of a property as essential to an object we usually mean it is true of that object in any case where it would have existed" (Kripke 1980, p. 48). If at least one of the object's intrinsic

[11] Environment, at the very least, matters as well.

and indispensable properties is altered, the object is no longer the same thing and is significantly different. For example, if corn loses the essential property of being corn by becoming a tomato through genetic manipulation of its genome, then the entity is no longer corn.

On the other hand, objects can have what is called accidental or contingent properties, which can be changed without the object being altered into an essentially different thing. For example, if corn's color is changed from yellow to red, the corn is essentially the same object as it was before the alteration occurred.

Since unnaturalness' definition has been expanded to include individual essential properties of entities, as well as the complete set of properties essential to the entity, it is incumbent upon us to show how the greater inclusiveness will affect the Unnatural Is Unethical Argument. It is clear that critics using "essence" think there is some essential set of properties, but it is vague as to which properties they believe to be part of the thing's fundamental nature.

A radical position is to assume that each property an individual entity has is one of its essential characteristics. This position expands Kripke's claim that origin, substantial makeup, and type membership are essential properties (Kripke 1980, p. 57 footnote). That is, the set of all properties an object actually has is identical to the set of all essential properties of the object. If one property is changed, then it follows from this definition that the object is necessarily no longer the same entity. For example, if any property of an ear of corn where altered in any way, such as one kernel being moved one millimeter to the left of its current position, then the new object is not the ear of corn that existed before the alteration. An ear of red corn is not the same thing as an ear of yellow corn even though the origin of the red corn is the yellow corn.

The benefit to critics of adopting a radical essence definition is that any transfer of genetic material from one organism to another is unnatural; hence, morally wrong according to the Unnatural Is Unethical Argument. Furthermore, the created object is unnatural, which means it is morally bad. Since the fabrication of a transgenic organism necessarily is the transfer of genetic material from one individual to another, then all organisms created in this manner are morally bad and their creation unethical. Hence, it is unnecessary to consider the morality of transgenic organisms on a case by case basis because all TOs are morally bad.

The insurmountable problem with stipulating that every bit of genetic material is essential to an object is the fact it leads to obvious moral misclassifications. On the same grounds as transgenics being unnatural, the creation of life through sexual or asexual reproduction would also be immoral. As everyone knows, in sexual reproduction, genetic material from the male is transferred to the female's egg. What everyone might not agree with is the claim that the resulting fertilized egg is morally bad because it contains the transferred fundamental nature of the male. Hence the creation was morally wrong, according to the Unnatural Is Unethical Argument and the radical essence definition. The result not only is that procreation of non-transgenic animals is unethical, but the creation of crops through conventional and organic breeding techniques is always morally wrong. Anything transferring DNA would be impermissible and its products morally bad.

It is more plausible to assume that the essence definition's proponents mean something less extreme when they talk about essence transference when transgenic organisms are developed. Comstock focuses on the narrower meaning there is some property or properties "which all and only the members of [a] species possess ... some characteristic unique to and shared by all members of the [species] which explains why they are the way they are" (Comstock 2000b, p. 190). According to this definition, in order to be a species' member, the individual must have the same individuating species' characteristic as all other individuals of that class. But Comstock is correct in arguing that no species has a "single essence identifying all [members of that species]" (Comstock 2000b, p. 191).[12] Hence, this version of the necessary-characteristics-for-species-membership position can and should be readily cast aside.

Unlike Comstock, however, I will suppose that there are combinations or clusters of characteristics that are sufficient to being a member of a species.[13] In order to be able to classify an organism as belonging to one species rather than another, there has to be a set of characteristic combination sets that allow us to distinguish between species. In other words, there could be many different sufficient sets of characteristics for a species, at the same time there is no necessary feature of a species other than satisfying one of the sufficient sets. In order to be a species member, an individual must satisfy at least one set of possible characteristics that establishes the bare minimum required to be a member of that species. For example, poodles and Pomeranians are canines because each is a variety of dog, while a Siamese cat shares some identical characteristics with dogs, but does not posses any of the sufficient combinations of characteristics to be a canine.

In the transgenic organism debate, the claim might very well be that the genetic material being transferred between species supports a set of characteristics sufficient for membership in the donor's species. The result is an organism that is now a member of two different species: that of the donor organism and that of the recipient. For example, it could be claimed that Roundup Ready wheat is also a variety of bacteria provided that it satisfies one of the set of sufficient conditions to be a bacterium of that type.

Another way of interpreting the essence definition is to state that although membership in a species merely entails satisfying one of the species' sufficient sets of characteristics, each characteristic in each set is essential to that set. If it is possible to transfer some of an entity's fundamental nature, then it must be that some, if not

---

[12]Comstock rejects this version of the Unnaturalness Argument on the grounds that genes can be transferred without essence transfer, it is not proven that there are such things as essences, and "it is impossible to identify the essence of a thing simply by describing its genome without describing its environment" (Comstock 2000b, p. 191).

[13]This definition is sometimes called the "homeostatic property cluster view" (Robert and Baylis 2003, p. 3). I take it that John Searle's cluster or family description of proper names is very similar to this view (See J.R. Searle's "Proper Names" *Mind* 1958, 67:166–73).

all, of the characteristics in the set of intrinsic and indispensable characteristics are also intrinsic and indispensable to the particular entity's identity. For example, a red ball must be red. If the color of the ball is altered, say to blue, the red ball has lost one of its essential characteristics. When transgenic organisms are engineered, at least one of the necessary characteristics for the sufficient set is transferred into the recipient organism, thereby creating a new variation in both species. Although it is controversial to assert that there are such simple essential characteristics of transgenic organisms,[14] in order to fully develop the argument for this definition, it will be assumed that there are such characteristics and they can be transferred to different organisms and species when creating TOs.[15]

Although the change-of-essence-is-unnatural position can be strengthened using either of the latter two interpretations of essence, an insurmountable problem is immediately encountered. The essence definition makes the illicit assumption that all essences or essential characteristics are morally good, and not altering them is morally right. Instead what should be kept in mind is that all essences and essential characteristics are, at best, intrinsically neutral and the relativistic circumstances in which the species occurs generally determines if the essence or characteristic is good or bad. Consider the HIV virus. In human beings, it is a devastating disease killing many people, especially in sub-Sahara Africa which currently has 29.4 million infected people (UNAIDS/WHO 2002, pp. 17–20). However, for primates other than human beings and chimpanzees there is little replication of HIV-1 (Watanabe,www.the-scientist.com/yr2003/jun/research1_030603.html, p. 1). In other words, this virus is not inherently evil in all situations although it produces devastating consequences in other circumstances. Moreover, if researchers were able to alter the HIV virus' fundamental nature using transgenic means so that it could no longer negatively affect people, then it would be implausible to argue that researchers acted unethically, although they changed the essence of the particular virus or the species. Hence, the transgenic organism's context plays a vital role in the evaluation of the morality of the organism and its creation. It is good or bad because of what it *does* rather than what it *is*. As will be seen again, most of the arguments for TOs being inherently bad are surreptitiously based on what they will allegedly do in certain circumstances, such as destroying a particular environment or species, rather than on their alleged inherent badness.

In addition, those who hold the essence definition and others like it seem to be drawing conclusions based upon some sort of antiquated Platonic world view ideology that does not represent what actually occurs. Part of the definitional problem of what a species is stems from the fact that people often assume that kinds such as species are natural with essentialist rather than conventionalist, definition. That is,

---

[14]Even the characteristic of being resistant to glysophate in bacillus thurigensis is complex.

[15]If it was maintained that it was wrong to transfer the essential characteristics of a sufficient set of species characteristics to another organism, then the same problem that arose for the radical interpretation would arise here. Creating new organisms with those characteristics would be morally wrong or bad, even if it was not artificial.

they believe that each kind of thing has an essence that is absolute and universal in the same way that elements on the periodic table are absolute and universal.[16]

However, there are no eternal and unchanging Platonic Forms that are essential to what a thing is, as can be seen through the problems posed by relations and the Third Man Argument. Species and many other kinds are true most of the time generalizations rather than being natural concepts. This is why there are at least twenty-concepts of species and the "consensus among biologists is that no single species concept will be sufficient for all situations" (Baylis and Robert 2006, p. 1). The most common definition in use vaguely defines a species as a "group of individuals. . .that share certain morphological criteria that render them distinct from other 'species'" (Reiss and Straughan 2001, p. 61). The question, of course, arises as to what this overly broad definition actually entails other than what the user wants it to mean is hard to determine. Given the lack of consensus of the brightest minds on such a vital issue, if there was a natural definition of species, then it would have been found by now and be as controversial as the definition of hydrogen.

There is at least one plausible competitor to naturalistic definitions of species. In his explication of a C.I. Lewis inspired Conceptualistic Pragmatism, Richard B. Miller argues that the definitions of conventional kinds such as species should be evaluated on pragmatic grounds.[17] According to Miller, "Human beings are tool-making animals, and concepts are intellectual tools" (Miller 2009, in press, p. 12). Therefore, the value of a concept in a particular situation is determined by how useful it is in those circumstances. Moreover, concepts are neither true nor false as propositions are but rather are "wise or unwise" choices (Miller 2009, in press, p. 11). That is, the wisdom of selecting a concept is based in part upon the purpose the individual or group has in wanting to make a kind distinction in the first place. If the concept better fulfills that purpose than does a competitor, then the former is a wiser selection than the latter. The definition of what it is to be a species, on these grounds, would be determined by the situation and what needs to be done. The concept that should be picked is the one that will work best in the situation, although it might not be the wisest choice for all situations.

Although some might argue that adopting a version of Miller's alternative method for defining kind terms invites a relativism which will undermine my fundamental assumption that an adequate ethics is practical, nothing can be further from the truth. Admittedly it would be simpler for moral decision making to have natural definitions that a person could reject only if he was willing to have his position labeled irrational, and therefore justifiably ignored. However, as has been stated in the Introduction, for controversial issues it is always better to use what is actual and practical rather than creating a perfect non-existing world solution for a real world problem.

---

[16] I am grateful to Richard B. Miller for his examples and helping me explicate his work in this area.

[17] Bernhard Glaeser argues that nature is a cultural concept whose meaning changes according the culture and situation. Nature is real, but its reality is one that has been molded by human feeling, perception and thought (Glaeser 1995, p. 146).

Once the situation has been stipulated, reasonable people can discuss how concepts should be defined given the purpose each has to want a differentiation in the first place. Ad hoc definitions that serve the sole purpose of attacking or supporting technology are less useful and wise a choice than those that produce better results, i.e., generally, a definition that works in more ways for more people in the same situation. Consider the definition of species. Those who endorse transgenic technology could eliminate the definition of species entirely or make it so vague that it does no work. On the other hand, those oppose transgenic organisms could stipulate very restrictive definitions to support the arguments about species' integrity. Neither of these is as practical a definition as one based on the best scientific and other relevant evidence that both sides and everyone else can use to classify entities and discuss the debate over transgenics. The concept of species, for example, has a conventional definition that contains natural elements. The natural elements are to be found in the term's deep scientific roots, such as the inability to interbreed.[18]

To determine which definitions are the wisest in a particular situation, I will once again put forward PMC. If adopting a definition does not treat anything with intrinsic value as a mere means (QCI) and it is reasonable for a reasonable person to believe that adopting and using the definition is likely to result in the best outcome at this time (RPU), then choosing the definition is both wise and ethical. Ad hoc definitions, on these grounds, would not be wise because they generally are intended to be used to stifle other reasonable people's views from being expressed and they rarely are useful beyond supporting an unyielding position. Although the combination of PMC and Conceptualistic Pragmatism are not a definitive answer for people wanting universal, absolute answers to problems, it is practical and will be used throughout the rest of this work.

3.5.2 Changing the End or Telos Definition

Comstock's eighth argument provides the second complex definition of the unnatural. "To engage in ag biotech is unnatural because it is to *change the telos*, or end, of an individual" (Comstock 2000b, p. 191).[19] According to this meaning, certain species and their members have teloi, which are inconsistent with their original, non-artificial ends, forced upon them by researchers importing in new genetic material. In other words, transgenic organisms are unnatural not because they are not being used according to their actual ends but because they have corrupted ends. For example, wheat, corn, and soybeans do not have the telos of being resistant to glysophate, and are supposed to die when exposed to herbicide's containing it. By giving these crops the Round-Up Ready ability to live in glysophate's presence, researchers have illicitly altered the original crops' end rather than merely interfering with it but

[18]I take Verhoog to hold a similar position on the definition of species (See Verhoog 1992).

[19]Michael W. Fox is a proponent of this argument and adds that changing the telos of a natural object is "playing God" (Fox 1999, p. 4).

leaving it unchanged. Hence, creating TOs is morally wrong, and transgenics themselves are morally bad.

Although UIU with the telos definition may appear initially plausible, it is impossible to evaluate the combination without examining in more detail how "telos" is being used. First, in order to determine if altering or replacing a telos is unethical or even possible, it is necessary to know what end an individual or species *qua* individual or species, respectively, has. Altering something's telos has to be broken into components which examine at least four types of objects: humans – potential and actual persons; animals – capable and incapable of feeling pain; plants; and inanimate objects, such as minerals. There is a further classification of telos of each individual itself – either as particular individual teloi or as a species being – and the end of the individual in a biosystem or relative to its surroundings

Second, once the telos of an individual or species *qua* individual or species is discovered, then there must be yet another classification of function of the object itself and the end of the object in context of its surrounding circumstances, be it a biosystem, environment, or something similar. A plant, for example, can have both its plant end which does not depend upon the environment it is in, and a telos relative to its environment. If a weed is merely a plant out of place, then some plants have the end of being flowers in one biosystem, and weeds in another, while simultaneously having the plant telos of survival and reproduction.

Third, in order to evaluate the unnatural as a change of telos definition and the UIU's soundness, it is necessary to consider the sources of each individual's end. Otherwise it will be impossible to determine if there is a telos being interfered with and whether the interference is unethical. For example, human persons might be able to give humans, animals, plants, or inanimate objects a telos. It might also be the case that humans can naturally alter the end of any or all of the four types of objects. Another possible creator source of a telos is the biosystem, which could make the flourishing of the biosystem or nature as a whole, the end of all entities within it. I will begin with the third issue, and then move to the first two.

There are four possible states of affairs for teloi and their sources.[20] First, if no telos exists, then there would not be a source. Second, a telos source might be internal to the object with the telos, which means the individual's end is self-determined in some way. Third, an individual's telos might come from outside of the individual. Fourth, the source of an end could arise from a combination of internal and external sources. I will consider each in turn.

If there is no function, then the telos version of the Unnatural Is Unethical Argument can be efficiently dismissed. No end entails that there is nothing wrong with interfering with an individual's telos because there is nothing with which to interfere. Hence, there is nothing unnatural about creating TOs or TOs existing.

[20] A thing cannot acquire an end through random generation. The design argument is based upon a global or local design argument, which implies a designer, which might be God, nature, human beings, or some intelligent entity. Unlike a designer, randomness cannot give an end to an object.

If an individual decides its own telos, then the source of the end is internal. This interpretation borrows heavily from Existentialism, which states that there is no worth in the universe until the individual chooses to create value for his universe (Sartre 1956, pp. 144–5). Individual choice is the only mechanism to confer worth, and that value is subjective to the individual. Some might choose to make their families the most valuable entities in their universe, while others decide that their careers have ultimate worth. If each telos is created in this manner, then there is no end for an individual until the individual chooses it freely for herself. Moreover, if she values consistency, then it would be unnatural for her to change or interfere with her self-imposed telos once it is established.

Internal sources of telos are not going to offer much assistance to those who oppose transgenic organisms for several reasons. First, non-thinking entities, such as hammers and plants, cannot choose functions for themselves. Rather, they either do not have a telos or some external thinking thing has to impose or assign an end to them. In either alternative, creating transgenic crops or other non-thinking organisms does not necessarily interfere with their end in any way. Once again, a telos' non-existence entails that there is no end with which unethically to interfere. If someone has already imposed a function on the entity, then using it in ways that are inconsistent with that original purpose would be unethical, according to this definition and UIU. However, a thinking thing imposing a telos upon an unthinking thing clearly is an external source of a telos, which will be considered later.

Furthermore, since it is at best difficult to believe rationally that animal life-forms, with the exception of humans, have the ability to make choices to create their own end, then it is not obvious that their telos is altered or interfered with in any way. To have an internally generated telos would seem to require some form of rationality. Once again, if a telos does not exist, then it is impossible to interfere with it.

Of course, someone might object that primates, and maybe other animals, do make rational choices based upon beliefs they hold. However, a telos, as it is used in arguments against transgenics seems to entail more than merely being the immediate goals for the entity. A hungry primate might have the end of becoming a fed primate, for example. A telos, on the other hand, is an object's purpose or function; it determines what the thing is. It is part of their identity. People, for example, can reasonably decide who they want to be as a type of person, e.g. scientist, parent, or married person, but an animal does not make choices about lifestyles affecting what they are as a member of that species. Hence, only a person can internally generate a telos for herself.

Second, if a telos is whatever the individual selects to impose upon herself, then thinking entities can permissibly change their internally generated ends at will. If they expressly decide to become a transgenic organism or it does not matter to them if someone makes them into a TO, then their original end has not been interfered with illicitly. In addition, if genetic material from a person is transferred to another, then the individual can change the telos of the transferred DNA as she chooses. The old function in both cases has been replaced with a new one.

In fact, under the Unnatural Is Unethical Argument with the telos definition of the unnatural, it would be morally wrong to prevent someone from fulfilling her internally generated end of becoming a transgenic organism or changing the telos of her genetic material, in the same way it is unethical to interfere with an internally generated telos of not becoming transgenic. If someone interferes with the agent's end, when the latter has selected a particular telos for himself, then the former acts unnaturally. Hence, the intervention is impermissible on the grounds of UIU and the telos definition.

Of course, the problem with adopting UIU, the second complex definition of the unnatural, and internally generated telos is it leads to serious ethical errors. The mere fact an agent selects a telos for herself does not entail that pursuit of the end is natural, much less good, nor does it follow that interference with the selected end is unnatural or morally wrong. If a person has chosen an evil function for her life, then it might be morally permissible, if not required objectively, for others to prevent her from achieving her end. For example, if someone decides to be a drug addict, then it is morally permissible, *ceteris paribus*, to stop her from being such a person. The point is that the combination of the three elements – UIU, the telos definition of the unnatural, and internally generated end – does not provide a mechanism to distinguish good teloi from bad teloi. Instead all ends are misclassified by definition as being good because they were internally selected by the thinking person. That fact alone is sufficient for rejecting UIU, definition two, and this origin for telos.

A third possibility for the origin of teloi is from a completely external cause. God, the natural environment, and even human beings are possible candidates for the source of an individual, group, or species' telos.[21] For example, God might create for a particular person his unique destiny, such as being the liberator of his people at a particular time. God might also give to members of certain groups their species' telos. An alternative non-divine source is evolution, which cannot provide an individual telos because it works on species, not individuals, but is a possible origin for each species' end. Evolution could "design" certain species to have a particular goal such as being a dog or more narrowly, a hunting dog.[22] Humans could create new plant varieties with a new telos using conventional breeding or give an individual entity a particular end, such as being a philosophy/ethics book about transgenic organisms.

Unfortunately, before the morality of altering a telos caused by these sources can be evaluated, there are two possibly unanswerable questions needing resolution. First, it is vital to discover who or what is the source of a telos. Without knowing its

[21] Aquinas argues that the design of the universe, which necessarily implies at least one end, is an adequate indication that God exists (Aquinas 1989, pp. 12–14).

[22] Evolution does not design any species for any end. It does have an influence on what species survive in a particular environment. This very weak form of influence I will call design for the sake of developing this argument.

creator we cannot truly know what the end is.[23] If the source is God, then the end can be very different from that created by the natural environment. God can ordain a particular telos for individuals as well as species' telos, which the more general natural principles cannot. After all, God as a person is able to address each individual rather than being limited to species or groups as a whole. Furthermore, a human person could give a weaker end to an individual or group than the other two possible sources, in part, because of the relative powerlessness of humans in comparison with God or nature. That is, a divine entity and nature have ability to perform more complete changes than does a human person. However, human persons can plan for the future, unlike nature, but not in the same manner an omniscient, omnipotent, and omnibenevolent entity would. With inadequate evidence of the telos and source, we cannot know what not to alter in order to act naturally; therefore, this ethical theory is not practical.

The second impossible question to answer is what kind of justification can be obtained to prove beyond a reasonable doubt whatever answer a person supplies for the first question is in fact true? That is, can we prove the alleged source exists? Second, if we can prove it subsists, then can we prove the alleged source actually is a source of teloi? Answers to both questions are probably more than anyone can conclusively establish to any neutral thoughtful person. The best that can be done is to try giving the telos definition the fairest hearing possible by carefully examining the evidence.

First, as philosophical history has made clear, it is impossible to prove that it is rational to believe in a God who gives anything a telos, much less prove the existence of such an Entity.[24] Of course, the fact that no one can prove a particular being exists does not mean that it does not exist. Epistemology is after all different from metaphysics. If humans, rational belief and knowledge had never existed, objects in the universe could and would have subsisted.

It is probably the case that faith, which gives no rational evidence to support or defeat a hypothesis, is the only way to address the issue. Many people have great faith in the existence of the Judeo-Christian God, but have no a priori or a posteriori evidence for the existence of such a being. However, if there is no more reason to believe there is a God or one that creates individuals and species with ends than to believe the opposite, then it will have to remain an open question as to whether teloi from this source exist. Therefore, due to the lack of evidence, to be rational, we must withhold judgment both on God's existence and Him/Her/It being the source of any telos.

Even if it could be established to a satisfactory degree that God exists and acts in this manner, there remains a vexing problem. If the particular individual or species' end does exist, then how do we discover what that end is? No one knows what

[23]The only way around this problem is if the end is self-evident. The fact that people disagree about what a thing's end is, *if* such a telos exists, is sufficient evidence to prove that the end is not self-evident.

[24]Although the problem of evil argument seems to be decisive to prove that a God with infinite goodness, power, and knowledge does not exist.

is in the mind of God or what His plans are for particular individuals, if any, so it would be impossible to believe rationally an individual has a particular telos. It would be simpler to discover an individual's species' telos because it would be general to the species population and easier to identify as a species characteristic from observing the members of the class. However, a divine end for the human race and other species is not clear, once again, because of the lack of reliable evidence. There are no a priori or a posteriori sources serving as adequate justification for proving that there is a telos for the human race. The same holds true for all species. The best many can do is to have devotion to the religious text they think is an adequate source of information, but of course, that is a matter of mere faith and not rational belief. Once again, we need rational justification for positions so that others in the debate can at least understand why we hold our beliefs and act as we do when it comes to controversial technology. Hence, given the position's impracticality, the possibility that God or any other supernatural entity is an external teloi source can be legitimately placed to the side, and then the focus turned to something for which we can provide at least some evidence.

Although it is clear that evolution exists, it is not obvious that general traits shared within a species are the telos of the species.[25] However, that is one way of interpreting the telos objection. Roughly, this UIU argument using the second complex definition and an evolutionary function claims that particular species have developed through evolutionary forces to have particular traits or variations; hence, those traits must be the purpose of the species. Wheat, for example, has characteristics allowing it to grow and reproduce, but not allowing its members to be resistant to glysophate. Therefore, creating variations of wheat with glysophate resistant traits is violating the telos evolution or nature has conferred upon the species.

There are three problems in assuming nature over time gives entities their teloi through evolution and other natural processes. First, there is the difficulty in knowing what the evolutionary end is for any individual organism. Some people point to environmental and genetic advantages to every feature a member of a species allegedly has. The reason humans walk upright, for example, is attributed to how much fitter for survival members of the ancestral species were than those continuing to perambulate on four legs. Although this is a plausible evolutionary explanation for a species' characteristic, descriptions of other features' causes begin to move into the realm of the incredible. At one paper I attended many years ago, a biologist asserted that American males in their late teens and twenties like a lot of bass in their music because evolution has designed them to prefer their groins being vibrated. Those ancestors whose reproductive organs were stimulated by deep noises were better fit to survive and reproduce for some reason the biologist did not explain. The heightened ability to reproduce increased the ability to pass on their traits to future generations as a result. Since males of homo sapien sapiens have the trait, then one

[25] Additional difficulties with using evolution to establish naturalness will be addressed in greater detail in Section 3.6 of this chapter.

of their goals is to use the characteristic in pursuit of the ultimate goals of survival and successful reproduction.

The explanation of the groin stimulation trait and its purpose seems to be an instantiation of the old saying that a man who has a hammer as his only tool starts thinking that everything looks like a nail. Basically, the speaker's theories craft the explanation of telos to fit the speaker's biases rather than providing any useful, objective evidence of particular ends. The fact that it is impossible to negate any of the explanations shows that it has the same defective nature as claiming everything that happens is God's will. Moreover, many of the explanations seem the result of biased speculation rather than anything supported by empirical or other acceptable evidence. We need objective evidence so that we can make decisions that reasonable people can appreciate.

There are two other difficulties with proving that nature through evolutionary and other processes has created a telos for a species. First, for very specialized characteristics, such as enjoying music with a lot of bass, it is virtually impossible to establish what role evolution played in the causal chain. Although possibly correct, the trait, if it is one, might have been accidentally acquired. Much like altruism and caring for human beings, which do not seem to have an evolutionary advantage, the characteristic might have hitchhiked on a strand of DNA containing the evolutionary beneficial trait which was passed on after successful procreation. In addition, causal sources, such as socialization in a particular society, could play a greater role in trait acquirement than that of evolution. It might be that loving one's groin vibrated by bass notes improved survival and reproduction as well as socialization of young males in a society or sub-society. Without knowing the actual causes, it is impossible to determine what the end actually is or from where it originates. Unless all human activities can be explained by evolution, which can only be done if there is no other causal source beyond those of evolution, it is an open question as to what an individual's species telos is.

Furthermore, a telos based upon a species' general characteristic might not be the end for every member of the species. Assuming reproduction is one of the ultimate teloi of homo sapien sapiens, then homosexual humans do not share their species' end with the heterosexual members. If an individual is homosexual as a partial result of natural processes, then being homosexual is a variant for the species. Hence, it is difficult to determine if there is a telos for all members of a species because there appears to be no one characteristic shared by all members of the species (Comstock 2000b, pp. 66–7). If every variant has its own end for the species, then for however many variants there are, there is at least the same number of teloi. One evolutionary explanation for an end within a species is therefore not going to apply to all members.

In addition to the other problems of externally created teloi, assuming biosystems are the sole or partial source of a telos causes yet further difficulties. In order to understand all the ends of a species or individual, then it is necessary to stipulate that the biosystem in which the species or individual exists. In any biosystem, an end of the individual or species is the role it plays within the particular system. For instance, in some schemes, corn is a crop with all that is entailed by being a crop.

However, if a field is planted with wheat, then volunteer corn plants growing from the former year's plantings are weeds, which mix with or contaminate the wheat. If an end is defined relative to a biosystem, then it is clear that altering or interfering with a telos means that the role the individual or species plays in the biosystem is altered or prevented from occurring.

One of the main drawbacks to defining the telos relative to a biosystem is the combination of UIU with this definition leads to a severe misclassification. Since this definition and argument has the same defect, as seen previously, of not being able to sort the good from the bad teloi, it follows that a bad telos in a biosystem should not be altered even if it destroys the system or morally valuable entities within it. Our interference is unnatural in this line of reasoning; so, obviously, something is wrong when an argument states it is unethical to protect oneself and others from undue harm merely because of the impact on a relative telos.

The fourth and final telos source possibility is a combination of internal and external origins. It might be the case that some outside source determines the partial end of an individual or the individual's species, while the individual itself has some ability to alter its own function.

The problem arising in this alternative is to identify the possible sources, prove their existence, and establish to what the extent they determine an individual's particular or species telos. How much of the end is determined by the outside source? What parts of the telos are under the control of the external source and what is under the control of the individual itself? These questions must be answered before we can take seriously the argument that altering or interfering with a telos is something wrong to do. But discovering the answers to these pressing queries seems as likely as proving that one moral theory is the one and only legitimate one. Much of any proponent's justification could be claimed to stem from their unconsidered ideologies. Hence, there appears to be insurmountable evidential problems rendering this definition of the unnatural too problematic for adoption, but perhaps the fault here lies with a lack of imagination.

The extreme difficulty of finding adequate proof for a fundamental assumption should not automatically bring to closure investigations on subsequent issues. The fairest course of action in moral controversies is to examine fully a position to determine if it has anything useful in it that will help to find and justify solutions. Granted the low plausibility of UIU and this definition given the above problems, to discover if they have any possible value, we should assume that someone might be able adequately to answer the source question, and then see if there is another insurmountable problem awaiting this argument and definition. If there is a source(s) of teloi, then the next query would be about the teloi for various classes of objects. More specifically, what are they? Without knowing the teloi, we cannot know if transgenic organisms or biotechnology violates them.[26]

As stated previously, one factor of the UIU's telos version requires much greater analysis to discover and prove the end for various entities in the environment. There

[26] A more in-depth development of teloi will be made in Chapter 4.

are at least four different categories of objects that could have teloi: non-living, plants and non-sentient animals,[27] non-human animals – which for expediency's sake I will merely call animals from now on to distinguish them from the last category – humans. It would be easiest to base both categories' telos on Aristotle's work.

First, since they are incapable of thinking in any manner, non-living entities, such as rocks, have external sources of teloi – provided they have a telos at all. Human beings, for instance, can assign a telos to a rock, but the rock cannot adopt a telos for itself.

The other three categories of entities, which are being currently assumed to have ends, are more plausible candidates for having species and individual telos, regardless of whether or not the teloi are internally or externally generated. Plants have the end of growing and reproducing, although it will not be decided here to what degree they are required to do either. Unlike plants, sentient animals not only have the telos given them by evolution and nature over time – to survive and reproduce – and their role in the biosystem, but also from the fact they can feel pain or pleasure. One part of the animal end is the avoidance of pain and the pursuit of pleasure, although the former seems to be more important than the latter. After all, it is possible for an animal to have a subsistence or minimal utility flourishing life without pleasure, but impossible for it to have such a life with constant pain or if the amount of pain outweighs the pleasure in the creature's life (Parfit 1992, pp. 357–8). Regardless, each telos would have to be consistent with the flourishing of the group's members.

Finally, humans have the telos of animals, in addition to that conferred upon them by the fact they have free will and rationality. Many people have believed each individual *qua* that individual or *qua* species being has an end or purpose in life to fulfill (Aristotle 1941b, 1097b 23-1098a 18; 1941c, 198b1-17; Aquinas 1989, pp. 13–14; Rollin 1996, pp. 29–30). For Aristotle, each person as a member of the human species has the ultimate goal of a good life (happiness/flourishing), which can only be achieved by developing and using his theoretical reasoning capabilities to a great extent (Aristotle 1941b, 1097a 35-1097b 22).[28] Let us call a general end of this type a species telos. From the fact there is a natural and good telos for each individual, it follows that if an individual acts to fulfill the purpose, then his action is natural and morally right (Aristotle 1941b, 1094a 1–3). After all, if the end is moral and an action is a true means to achieve the end, then the action must be ethical as well. In the case of human beings, developing and using their theoretical reasoning capabilities must be morally right. Moreover, the means to the end of humans is obligatory because it is the only way to achieve true happiness. Adopting a portion

[27]"Non-sentience" as used here means not able to feel pleasure or pain. Sentience is being able to feel pleasure or pain.

[28]There is a tension between the definition that Comstock uses and that of Aristotle. Comstock refers to the end of the individual, while Artistotle's refers to the end of groups. More precisely of groups of individuals, which seem to be divided along the lines of species boundaries, such as in the case of humans. In order to examine fully the issue of end, both types of end will be evaluated.

of Aristotle's position, the end of all humans seems plausibly to be a good life, which means a happy life or a flourishing life, which is already part of PMC.

In addition to a telos as a species being, there are other teloi types an individual can possess. First, a particular individual might have a unique or particular purpose, such as being the dutiful child of a particular person. Let us call these types of ends particular or individual teloi. Second, individuals might also have teloi because they are a member of some group other than a species. For example, if an individual is an aristocrat, then she might have the telos of *noblesse oblige*. Ends an individual has due to membership in a group other than species will be called group teloi.

One interesting feature of the human telos is that it is unclear how anyone can alter any person's end as a species being, including herself. As a species being, all humans have the ultimate end of happiness. Each person can achieve this telos or not, but they cannot change the end. Furthermore, one agent can prevent or interfere with another person achieving the end of happiness, but the telos does not change. In order to strengthen Comstock's definition, what should be included is definition pair four from above: one agent attempting to stymie or interfere with another from achieving the latter's species' telos is unnatural. The new definition will read, in part, "it is to *change the telos*, or end, or to attempt to prevent the achievement of the end of an individual," instead of the current "it is to *change the telos*, or end, of an individual."

Although changing the species telos for people does not seem possible, altering the individual's particular function is. Consider the Christian belief that Jesus' particular end of offering salvation through his death to those who believe he is the Son of God. When the devil tempts Jesus, he is attempting to alter Jesus' individual telos (*Bible*, Matthew 1974, Ch. 4). By trying to corrupt Jesus from his purpose, the devil hopes to change his end from salvation to something evil. If individual people have individual ends in addition to their species' telos, then these idiosyncratic teloi might be capable of being altered. If someone is a Great Person, such as Martin Luther King, Mohammed, or Confucius, who is the only one able to lead her society to great social change, then she could modify her function by not performing the actions necessary to enable her to achieve her ends. Moreover, others can change her teloi by destroying the characteristics she needs to fulfill her individual end as the Great Person.

The problem with all of the teloi mentioned above for the four object categories is they do not seem to be the proper type of telos to make the Unnatural Is Unethical Argument work. That is, a telos must be something which if violated or altered is clearly morally wrong at all times and in all situations. However, the ends above can be violated or altered and the violation is not clearly intrinsically wrong or right. Plants can be prevented from reproducing or growing, without doing anything wrong, even *prima facially*. Grinding wheat seed to make flour to feed starving people, for instance, is morally right, *ceteris paribus*. Placing animals in pain for medical testing in order to help humans eliminate diseases such as AIDS, tuberculoses, and so on is something few would find morally objectionable. Hence, altering or preventing an individual from realizing its end is not in and of itself morally wrong, especially when considering those objects without cognitive faculties of any sort.

Furthermore, UIU with the telos definition of "unnatural" is ineffectual in attacking the morality of transgenic organisms or their creation. It is a fact that TOs can satisfy the telos of their category, regardless of whether or not they are plants, animals, or even human beings.[29] First, transgenic crops have the designed end of surviving and reproducing, which they seem to do very well. Moreover, just as conventional or organic crops are, transgenics are either crops or weeds in the biosystem according to the relative biosystem in which they are located. In other words, they are no different on these grounds than conventional and organic crops. Transgenic animals, as do their conventional and organic counterparts, have a telos of reproduction, growth, avoidance of pain, and pursuit of pleasure. They also have identical teloi to conventional and organic. In fact, transgenics might be engineered to *better* fit their environment than those produced by other means, as in the case of enviropigs (Strieffer and Ortiz 2002, p. 2). This means they are superior at fulfilling their teloi to other creatures. Finally, transgenic human persons, if they still have the intellectual ability as non-transgenic humans currently in existence, would be able to achieve all the teloi in their species end, while creating ends on their own.[30]

The only way the telos argument is effective as an objection is if God or some other divine entity with appropriate power, or nature itself, gave the individuals a telos not to be a transgenic organism. That is, the recipient organism had a function not to be a transgenic, which is altered or interfered with when the individual is made into one. But this almost tautological claim would require a massive amount of evidence to overcome its ad hoc quality. If it is difficult to prove from observation that entities have a telos, such as growing and reproducing, in the first place, it would be next to impossible to establish if it had the very specific end not to be a transgenic organism, much less convince anyone other than another non-rational or irrational believer that this is the case.

If UIU with the telos definition is broadened to include the species *qua* species end rather than merely that of individuals *qua* individual or *qua* species being, it would still remain useless for rational discourse purposes. To date, no new species had been created by transgenic technology. Rather one species is modified by some genetic material from another species, but the new organism is still the same species as the former, albeit a new variation. Hence, the species' telos does not change. If there are teloi, then plants still have the end to survive and reproduce, animals have theirs, and so on. Furthermore, creating transgenic organisms in this manner

---

[29]There are no human transgenic organisms of the type of TO under discussion in this work. Of course, there are a large number of transgenic humans created through the old fashioned method of procreation.

[30]If human beings are created to be mindless servants or sources for organ donation, then there will be difficult ontological questions to answer, such as whether or not an engineered being has the end given to it by its engineer or if it has been harmed by not being allowed to have the end of non-engineered beings. The problem becomes more acute if the being is engineered to have a life worth living, according to how that is defined by Parfit, but never achieves the level of a full person. Since the almost human TO was never going to be a person in the first place and it has a life worth living, then is it morally wrong to create it?

is merely a mirror of microevolution. Microevolution produces the same results as transgenic engineering, although the former takes a much longer period of time to accomplish its results.

Even if a new species is created, it does not follow that the recipient's species telos has been altered or interfered with unethically. Rather the old species still has its old end, while the new species has the new end given to it by its creators, as well as that of the more general ones found for that type of thing: inanimate, plant or human. New plant species, for example, still have the universal plant telos of growing and reproducing, although their individual species' telos might be a little different from that of their ancestors. Hence, part of the new end is not altered or interfered with by transgenic engineering but created for a specific purpose. The telos is in part, then, whatever the researchers intend for it to be. Given the exhaustive examination of it and the failure to find one plausible interpretation, it is clear that "changing the telos" and UIU is not a practical objection to transgenics or biotechnology.

3.5.3 The Crossing Species Boundaries Definition

Comstock's ninth argument is based on the third complex definition of "unnatural." According to some opponents of transgenics, "[t]o engage in ag biotech is illegitimately to *cross species boundaries*" (Comstock 2000b, p. 193).[31] Putting aside for the moment the new way to define terms discussed in Section 3.5.1, I will assume the crossing of species boundaries refers to a species being fixed and immutable[32] as shown through the inability of members of different species to transfer DNA unique to their species through non-artificial processes, such as sexual reproduction. In fact, not being able to interbreed is generally one sign that a new species has evolved from its ancestral species (Curtis 1972, p. 520). The argument, hence, is one based upon evolution and evolutionary mechanisms. In order for this definition to be plausible it will have to avoid empirical mistakes about evolution's principles and processes, and not assume moral conclusions from evolutionary theory.[33]

The broad interpretation of the unnatural as DNA transfer across species boundaries can be attacked in two different ways. First, Comstock adequately addresses the scientific deficiency of the argument by showing that different species can interbreed to create a new species through sexual reproduction. Therefore, it is possible to have two species involved in DNA transfer, and people say little about its unethical nature.

Also, although unmentioned by Comstock, species are not genetically isolated because horizontal gene transfer exists (Reiss and Straughan 2001, p. 61), but to what extent the phenomenon occurs and its impact on evolutionary processes is a

[31] See Michael W. Fox, p. 7 for example.

[32] This definition can be found in Rollin (2006, p. 141).

[33] If illegitimately means unethically, then UIU with this definition becomes a tautology. Hence, I will only interpret the definition in ways that allow it to add something to the debate.

matter of debate.[34] Viruses, for example, transfer their DNA into that of their host cells, which can cause the host to become ill. The new genetic material or resilience to the illness could very well create a new variation in the species over time. Hence, if "illegitimately" in the third definition refers to evolutionary processes preventing gene transfer between species, then this UIU argument against transgenic organisms fails. Since this phenomenon happens in nature, regardless of reproduction or artificial intervention, then horizontal gene transfer is a natural process.[35] The mere fact that transgenic organisms are created through artificial genetic transfer rather than non-artificial horizontal transfer is an insufficient reason to judge TOs as inherently bad and their creation morally wrong for reasons stated earlier. After all, if merely being artificial is sufficient to make something morally bad or wrong, then the unjustifiable conclusion follows that all human activities and their products are unethical.

The use of the word "illegitimately" implies another interpretation. An illegitimate transgression could mean a violation of a non-artificial process, or a new species is created which could not have come about through evolutionary processes.[36] That is, certain gene transfer across species is possible without human assistance, while others are not. It is only when a TO that could not have occurred without artificial transgenic methods is brought about that the natural process is violated unethically. For example, Golden Rice had to be created from the rice genome being modified with one bacterial and two daffodil genes because the required variation could not have occurred through non-transgenic means (BIOTHAI et al. 2001, p. 1). Let us focus on the issue of generating from the DNA of two species a new organism and species that could not have existed through non-artificial horizontal gene transfer or interspecies breeding.

There are four problems encountered by defining the unnatural as the artificial transfer of genetic material between species that could not have otherwise occurred, which have already been stated. First, as to the concerns about a new species being created, the fact is no new species is fashioned in the process.[37] The second problem with this "unnatural" definition is if we are truly concerned about what is a natural action and product in the environment, then we cannot merely assert that sexual or asexual reproduction without gene splicing is natural and gene splicing is unnatural, and then stop at those two conclusions. The third difficulty with the definition is

---

[34]See the articles by Michael Syvanen for an excellent introduction to the controversy.

[35]Richard Sherlock rejects the isolationist argument on much the same grounds in his excellent article examining genetic trespass and subtle definitions of the natural/unnatural (Sherlock 2002, p. 154).

[36]The issue of evolutionary arguments for and against transgenics will be addressed in much greater detail 3.6

[37]If it turns out macroevolution is the process critics are concerned about violating, then as long as researchers do not do that, then their activities are not illegitimate under this argument. See Chapter 4 for more information.

even if we ignore the fact that homo sapiens acting within the environment is natural because it is a partial result of evolutionary processes, there is still the matter of whether unnatural actions and their products are inherently unethical. Fourth, the definition of species is relative to the circumstances and the purpose of the word. In one situation, species defined on the grounds of inability to sexually reproduce is the most useful definition, while in other circumstances, another meaning is more expedient. Hence, this definition also fails to establish the unnaturalness of transgenics or any biotechnology, for that matter.

3.6 Arguments from Evolution

As has been clearly demonstrated so far in this chapter, in bioethics and technology, and its subfields of biotechnology and transgenic organisms, there are a number of instances in which one side of an issue employs moral and scientific principles and arguments to support their conclusions without really understanding them.[38] The focus of unnaturalness and transgenics was on whether such technology was intrinsically unnatural based on a false Platonic ontology that assumed that all objects have functions and that unambiguous essences exist that allow us to delineate clearly. A species' boundary transgression, telos alteration, or essence transference were sufficient to make a transgenic organism inherently bad, at least in some people's minds. These arguments were shown to be baseless on the grounds of internal inconsistency, factual errors, and impracticality.

Evolutionary theory is another case in point of misused science. Critics of transgenics sometimes claim that the creation or existence of TOs violates natural evolutionary processes. From there, they argue that any action or object violating evolution in some way is unnatural and the unnatural is morally wrong or bad, respectively; hence, transgenic organisms are morally bad and their creation is morally wrong. On the other hand, some proponents of transgenics contend that science has naturally evolved to the point in which transgenic organisms can be created and sustained. Since TOs are part of the evolutionary process, they must be natural; hence, transgenics are morally good. Unfortunately for both, each argument makes one of two mistakes. Either the argument is empirically mistaken about the violation of evolution and its mechanisms, or evolutionary mechanisms actually are broken, but the transgression is irrelevant to morality.

Before examining the evolution arguments in favor of and against transgenics, it will be helpful to briefly described evolution, how it works, and some moral facts to note about it.

[38] Other types of transgenic organisms will prove more problematic than plants. Plants after all, do not have the capacity to feel pain or pleasure and certainly do not have rational thinking processes.

3.6.1 Evolution: A Primer

First, the theory of evolution holds that both the diversity and similarity of all living organisms result from the same process. Evolution occurs when there "is a change in the gene pool of a population over time," where the gene pool is the set of all genes of a species or population (Colby 1996, p. 1). Evolution tends to preserve traits conferring an advantage to a species, i.e., those enhancing survival and reproductive success. These traits are usually passed by the fitter members of a species to their progeny. Organisms more fit to survive in the environment are more likely to contribute to the gene pool of the succeeding generations than those less fit, where the fitness of an individual is determined by its relative contribution to the gene pool of the next generation. Hence, a species' gene pool will change over time as the genes from the fitter organisms predominate over less fit genes. Eventually the changes in gene pool, in part due to environment, competition, and predation pressures, will be sufficient to create a new species from the old, the latter of which will eventually become extinct (Curtis 1972, p. 532).

There are two types of evolution. First, microevolution occurs when there have been significant changes in the gene pool of the species, but not a sufficient amount to make the result a new species unable to reproduce with the old. For example, English moths underwent microevolution when darker members predominated over lighter ones as a result of pollution pressure from coal burning and soot on the trees on which the moths rested. Furthermore, when local coal use was reduced, the moths microevolved again. This time, white members were favored over darker ones because the tree bark was once again mostly white. Since microevolution occurs relatively quickly, especially in comparison to macroevolution, it is the only type of evolution observable by human beings.

Macroevolution, on the other hand, is much slower process and creates a greater alteration in species than does microevolution. Macroevolution forms a new species from a previously existing one, in the process of speciation. The two types of speciation are allopatric speciation, in which two or more groups of one species geographically split, and then evolve to a point in which they cannot interbreed, and sympatric speciation, in which two subpopulations become reproductively isolated without first becoming geographically inaccessible (Colby 1996, pp. 21–2). For our purposes here, we will use PMC to stipulate that, roughly, a new species is created from the old when members of the two groups can no longer interbreed (Curtis 1972, p. 520). Generally, since the DNA alterations required to generate a new species move at such a slow pace and in infinitesimal increments, researchers cannot observe macroevoluton from start to finish. However, using abductive reasoning and available empirical evidence, the best explanation for the current species and environment is macroevolution.

The five mechanisms of both types of evolution are gene flow, recombination, mutation, genetic drift, and natural selection. Of the five, gene flow, recombination, and mutation can increase variation, while natural selection and genetic drift decrease it. Gene flow occurs when members of a species migrate from their established geographic area to a new one, and then mate with members of the same species. Gene flow can also occur in horizontal gene transfer in which distantly related species from the same geographic areas reproduce (Colby 1996, p. 9). Recombination happens through "the process of meiosis ... [V]ariations arising from mutations are reshuffled, worked into the gene pool, and so brought into new combinations with other genes, eventually giving rise to new phenotypes" (Curtis 1972, p. 496). One method of recombination is sexual reproduction occurring between mothers and fathers from the same or different geographic areas. Finally, deleterious, beneficial, and neutral mutations arise when the cellular machinery copying the DNA makes a mistake and produces a novel genetic combination (Colby 1996, p. 10).

One argument against transgenics involves the reduction of diversity, but natural selection and genetic drift can decrease variations in a species as well. Genetic drift is a random event in which "certain alleles tend to get lost or are overrepresented" provided that the population is sufficiently small (Curtis 1972, p. 508). Natural selection, the only mechanism of adaptive evolution, generally either causes a beneficial mutation to fix in the species' DNA, which increases the population's members' because of that variation, or deletes deleterious mutations, so that the latter do not become fixed in the species' genome (Felsot 2002, pp. 5–7).[39] However, natural selection does not necessarily lead to a population with an optimal set of traits, although they might be almost as good as those of the optimal set (Colby 1996, p. 7).

Granted it would be beneficial to a species for natural selection to confer upon it every conceivable useful trait, the process can only distinguish between existing variants. Beneficial characteristics not part of the existing variants will not be selected because they cannot subsist given the particular species' genome. For example, it might be beneficial for deer to have skins impervious to bullets and other projectiles, but the species does not have the genetic variation that can be selected for the physical feature (Colby 1996, pp. 5–7). In other words, natural selection can only work with existing DNA combinations; it cannot create a new one based on how practical it would be.

Under natural selection, a common condition for all organisms is population growth tending to far exceed the available resources of the geographic area, which entails that more offspring will be produced than can survive. Furthermore, competition for resources increases as the amount of available resources decreases, *ceteris paribus*. The environment, hence, helps drive the origin of all species through the gradual accumulation of those traits steadily improving fitness over time for that particular environment. Of course, if this intense competition pressure was nonexistent, then new traits would not necessarily become universal in a species. There would be no genetic advantage to having that characteristic over another neutral one; so, there would be no reason for one to be selected and fixed in the genome over another.

[39] It should not be thought that natural selection is a force in the same manner as gravity. Rather natural selection is an effect.

Moreover, an organism's success in reproduction will affect the organism's contribution to future generations. The less successful the organism is, then the less input it makes in its offspring, to their offspring, and so on. Those organisms most successful in reproduction will have the greatest probability of contributing to future generations because of the simple fact that they can pass more of their genetic material into the gene pool than can those with less useful characteristics.

When evaluating moral arguments about technology, biotechnology, and transgenics, it is important to remember that evolution is a theory about a process populations undergo, and not about individual development (Curtis 1972, p. 498). Species or populations, not individuals, evolve. Furthermore, the only way a single organism of a species can have the essence of the entire species is if there is no variation within the population. That is, each species' member has to have the same genome without alteration. However, the only time members have identical overall DNA is in the case of identical twins or siblings. Moreover, there is generally a great deal of variation in the DNA of members of a species, with each member having its unique genome (Colby 1996, p. 3). Hence, moral conclusions about one member's particular genetic code cannot be extended to include all members of the species.

Another fact to keep in mind is that evolution is not a form of progress. Many people believe the new species have to be an improvement over those from which they descended because the new species is generally fitter for the environment in which it evolved than the former. However, in the moral sense, the belief is false for a number of reasons. First, "Populations simply adapt to their current surroundings. They do not necessarily become better in any absolute sense over time" (Colby 1996, p. 2). That is, new species are not morally better than their predecessors nor are they better physically than what came before. It should always be remembered that the species has to be judged according to the environment in which it exists. If the environment changes, then what was best fit in the former environment might now be less fit than a competitor.

Second, natural selection and evolution do not select qualities to maximize for upcoming utility. They cannot plan for a beneficial future of any species. Alterations in species result merely as responses to current conditions, not future ones. If natural selection could maximize utility, for example, then it would have modified people prior to being exposed to pollution created in the post-industrial age. People would have been able to withstand exposure to radiation and other current pollution problems than they currently can (Colby 1996, pp. 20–1). Hence, just because a species is the result of natural selection and evolution does not entail that it is morally superior to other species or has any special role to play now or yet to come.

Third, under evolution, all organisms are self-interested only in the sense that they do whatever is necessary to maximize their long term interest of inclusive fitness.[40] In other words, they try to survive and mate as often as possible. This is not

[40]Inclusive fitness is a combination of direct and indirect fitness. "Direct fitness is a measure of how many alleles, on average, a genotype contributes to the subsequent generation's gene pool by reproducing. Indirect fitness is a measure of how many alleles identical to its own it helps to enter the gene pool" (Colby 1996, p. 6).

a moral consideration unless the true ethical theory is egoism, which requires an agent to maximize his self-interest every time he acts. However, egoism is contrary to many legitimate moral intuitions, including but not limited to those underlying Kantianism and utilitarianism. For example, respecting the intrinsic value of others is right because it recognizes true value, rather than in the case of egoism in which the agent should respect another if and only if doing so maximizes the agent's utility. People should not be respected merely as a means to another agent's end.

Finally, organisms modify their environment in a number of different ways (Colby 1996, pp. 3, 28; Curtis 1972, pp. 511–13). The modifications are done without regard for morality, including the prudential egoism. For example, predators eat their prey, which affects the number of organisms available as food resources to them, and the reduction in prey numbers affects the prey species as well. The environmental modification can be very bad in regards to the continued success of the predator organism and its species. For example, if one species' members deplete all of their possible nutritional resources, then both species can become extinct, unless the predator can find alternative means of support. Hence, the lesson to remember is that evolution is not a theory incorporating moral aspects, but rather a scientific theory about the origin of species.

3.6.2 Arguments Against Transgenics

There are four main arguments against transgenic organisms based upon evolution: Usurping Nature, Genetic Instability, Abrupt Evolution, and Genetic Uniformity. Whatever the particular variation, all of the arguments maintain that TOs violate the evolutionary processes because their creation bypasses natural selection, gene flow, and the other mechanisms of evolution. According to the second version of the UIU, TOs are artificial or human made, and evolution is nature outside of human actions; hence, TOs and their creation are morally bad or wrong, respectively. In this section, I will consider each of the four arguments in turn. If the particular argument has a unique problem, then the criticism immediately follows its exposition. General problems faced by all four arguments end this section.

The evolution argument's first variation, Usurping Nature, claims that researchers are overtaking evolution's role when they splice the genetic material from one species into that of another in a manner evolution cannot duplicate.[41] Since transferring genes in a way evolution is unable to replicate is unnatural, the researchers' actions are morally wrong. Furthermore, the products of such activity must be morally bad. The result of an unethical transgression of evolution continues to violate evolution by its continued existence or the product of an unethical action is bad because it bears the taint of the wrongful deed.

In order to fully understand this argument, it must be evaluated in light of the two different types of evolutions: macro and micro. In order to use macroevolution

[41] The Playing God argument is a variant of this argument.

in the argument, researchers would have to be creating new species which cannot interbreed with others. There is some reason why people would believe the introduction of genetic material from a different species into another would create a new species because one species' DNA plus part of another species' DNA would add up to something entirely new. Furthermore, the new species' DNA code could not have occurred in nature, since the original species could never have interbred.

However, new species are not being created with TO technology. To date, the transgenic organisms reviewed by the various regulatory bodies, such as the USDA and FDA, have been substantially equivalent to their conventional counterparts (Harlander 2002, p. 164S).[42] Even though a genetic alteration has been performed to create a new variation, transgenic crops can still cross-breed with their conventional and organic counterparts, which is one of the risks often cited to prevent the formers introduction into agriculture. Since the new and old organisms can still reproduce with each other, then according to the tenets of evolution, they must still be the same species. Hence, researchers are not usurping macroevolution when they create transgenic organisms.

At best, transgenic organisms have undergone microevolution, much as the English moths did during and after the Industrial Revolution in England. The moral difference between the two examples of microevolution, some might cite, is that transgenic organisms were designed to undergo microevolution, while the moths were not. In the latter case, the moths' evolution was a mere byproduct of pollution in the area. No human being intended the moths go through the process of selecting for dark over white, and then again, white over dark moths. Hence, the change was an accidental, indirect consequence of the actions of human activity in the area, instead of being a deliberate attempt to bypass microevolutionary processes.[43] On the other hand, if researchers intended to create the evolved species from transgenic alteration, then they have usurped the role of evolution or one or more of its five mechanisms, in the process.

One of the strongest claims that evolutionary processes have been violated depends on natural selection and the features which can be selected for in nature. Recall that although a characteristic might be very beneficial for a species, it is not possible for it to occur in species' members if it is not already part of the species' variations. Being psychic might be useful for human beings, for instance, but that characteristic is an impossible variation for homo sapiens, regardless of what charlatans claim and the self-deceived believe.

According to this criticism, transgenic organisms illicitly create features that could not have occurred outside human intervention in the species' DNA, which cross the natural boundaries evolution has placed on each species' genome.

---

[42]The Rivermouth Action Group, Inc. claims that "corn (maize) which has been genetically modified to act as an insecticide ... is hardly 'substantially equivalent' to conventional corn" (RAG 2003, p. 2).

[43]This is not to assert that the pollution which stimulated the change in the moths is morally neutral or good; merely that it was not morally wrong because it violated evolution in some way.

> Natural agents [such as viruses] exist which can transfer genes horizontally between individuals...The natural agents are limited by species barriers, so that for example, pig viruses will infect pigs, but not human beings, and cauliflower viruses will not attack tomatoes. However, genetic engineers make artificial vectors (carriers of genes) by combining part so the most infectious natural agents, with their disease-causing functions removed or disabled, and designing them to overcome species barriers, so the same vector may now transfer, say, human genes, which are spliced into the vector, into the cells of all other mammals, or cells of plants. (Ho 1999, p. 2)

Not only does the process of engineering TOs violate natural selection and evolution by inventing entities that could not have otherwise existed, it is also asserted that there subsists a danger for those organisms consuming transgenic organisms. There has been evidence of horizontal gene transfer from bacteria in the human gut since the 1970s (Ibid., p. 4). More recently, British researchers have found that the marker genes used to identify transgenic organisms cells can be horizontally transferred to the bacteria in the human gut, which could adversely affect antibiotic resistance (Vidal 2002, p. 1).[44] This means not only have the intended species been altered, but other species have had their DNA unintentionally changed in an illicit usurpation of microevolution. Moreover, human beings and others might become susceptible to viruses and other diseases to which they are currently immune due to the violation of the natural genetic barriers already in place.

A second evolution argument against transgenic organisms, Genetic Instability, involves the alleged relative volatility of transgenic organisms' DNA in insufficiently controlled environments (Cummins 2002, p. 1; Perez 2000, p. 2; Lenski 1993, pp. 201–9).[45] Under the mechanisms of evolution, such instability is weeded out relatively early by natural selection without posing much of a hazard to the biosystem.[46]

> DNA transformation to make [TOs] is mainly by illegitimate recombination because homologous recombination is very weak except during meiosis. Higher plants seem to be more genetically promiscuous during their evolution and take more foreign DNA than do other kinds of organism. [TOs] are unstable in the first few generations because of the illegitimate recombination. (Cummins 2002, p. 1)

The volatility of splicing genes from one species into the DNA of another makes the resultant genetic material more likely to mutate than it otherwise would have. Since so few mutations are beneficial to the species or organism, the resulting

---

[44] Michael Antonio, a senior lecturer in molecular genetics at King's College Medical School, London, states that the findings are significant, although there are a number of inadequacies in the study (Vidal 2002, p. 1).

[45] Also see Jean-Claude Perez's *Planete Transgenique* (Ed. L'espace blue: Paris, France, 1997).

[46] Michael Syvanen argues in a number of articles that horizontal gene transfer played a much greater role in evolution than previously thought. He maintains that organisms that can incorporate DNA from other species are fitter than those that cannot. Hence natural selection favors the former over the latter. See Syvanen's "Conserved Regions in Mammalian ß-globins: Could They Arise by Cross Species Gene Exchange?", "Cross-species Gene Transfer: Implications for a New Theory of Evolution," "On the Occurrence of Horizontal Gene Transfer Among an Arbitrarily Chosen Group of 26 Genes," and "Recent emergence of the modern genetic code: a proposal"

organisms are harmed in ways they would not have otherwise been if evolutionary processes had been allowed to proceed naturally (Curtis 1972, pp. 146–7). The deleterious mutations are less fit; hence, if scientists had not created a situation likely to produce them, they would not have been able to pass their genes on to the succeeding generation and harm the individual and its species' gene pool.

Furthermore, the instability and bad mutations resulting from the genetic transfer could cause an otherwise impossible super-weed or pest. Outside artificial intervention, the variants for such a pest did not exist in the species' gene pool as evidenced by the fact that genes from another species had to be introduced into that of the recipient species to create the desired TO. Hence there is greater risk created by humans illicitly taking over evolution's role when they engineer their new species or modify the DNA of current species than would have otherwise existed if humans had let nature take its course or only produced novel crops through the tried and true methods of selective breeding.

Although there are the general objections to this evolution argument variation that will be addressed at this section's end, the second criticism has a particular response unique to it. While it is true that the results of splicing genetic material can be unstable, the volatility exists for a relatively short period of time. Seed companies and researchers do not want genetically unstable products to be released into the market because it would affect the marketability of the seed or other product. No producer desires a crop having a high incidence of mutations and related problems. Rather, producers and seed companies want a stable and dependable crop to raise and sell. To ensure that the transgenic organism is a reliable product capable of being marketed, the seeds are interbred until a stable line is produced. Hence, the genetic volatility of transgenic organisms can be reduced to be no more than that of conventionally created crop species, and if there is not a moral problem for conventional and organic products, then there cannot be one for transgenics.

Miguel Altieri and Peter Rosset posit one of the more sophisticated variations of the evolution argument, which is based on the acceleration of mutation fixation or Abrupt Evolution.[47] They claim that transgenic plants with Bt-pesticide will unethically alter the resistance of pests due to the rapid increase in gene fixation that could not have occurred outside of illicit human interference (Altieri and Rosset 1999b).

> In general, the greater the selection pressure across time and space, the quicker and more profound the pests (sic) evolutionary response. An obvious reason for adopting this principle is that it reduces pest exposure to pesticides, retarding the evolution of resistance. But when the product is engineered into the plant itself, pest exposure leaps from minimal and occasional to massive and continuous exposure, dramatically accelerating resistance. (Altieri and Rosset 1999a)

Instead of allowing microevolution to take its relatively slow time in the course of human history, transgenics are increasing the pace of species' microevolution, in

[47]Philip Davies uses a form of Abrupt Evolution argument to attempt to justify his claim that we should assume that each genetically engineered crop will have a significant impact on the environment until it is proven not to (Davies 2004, pp. 74–5).

this particular case, the resistance of pests to natural Bt-pesticide. Those members of the pest species with greater resistance to the pesticide are fitter to survive and pass on their beneficial characteristics to future generations, while those with lesser resistance are not as likely to survive or reproduce. Hence, the species' gene pool will include resistance to Bt-pesticide, and eventually those individuals without this genetic characteristic will be uncommon and unlikely to survive for very long in the bio-system. Furthermore, the methods in place to prevent this disastrous microevolutionary result from happening, i.e. refuge zones, are ineffectual (Altieri and Rosset 1999b). The overall result, it is claimed, is natural Bt-pesticide is rendered useless for organic farmers, and a pest has been created capable of overcoming the defenses of even the genetically altered crop.

In addition to insects, weed species have acquired the pesticide resistant gene as a result of microevolutionary mechanisms, and might do so again in the future (Holt 2002, p. 47). In one case in Argentina, it is alleged that "[b]ecause of the evolution of vicious new weeds, farmers have had to use two to three times more pesticides than previously" (Branford 2002, pp. 1–2). Hence, weeds developed pesticide resistance much more rapidly than they would have without the transgenic organisms passing on their transgene to their weedy relatives. The result is a violation of the slow speed which evolution generally moves in creating new varieties of species, as well as the breaching of natural species boundaries. It is only because of the gene from the bacterium that a super-weed capable of resisting Bt herbicides could have been created.

There is an exclusive objection to the Abrupt Evolution argument variation. As Martina McGloughlin points out, no biological solution to pest control is ever permanent. Pests will overcome any control method, regardless of whether or not it is used correctly, such as in the case of refuges and pesticide spraying (McGloughlin 1999, pp. 5–6). It is a biological fact that those pests fitter for the environment in which the control method is used will directly and indirectly contribute their genes to the next generation's gene pool. This pattern will continue until the control is no longer effective against the species because it has microevolved to be insusceptible to the pest control. Hence, the problem is not unique to transgenic organisms. It affects all biological control methods whether or not they are organic, conventional, or transgenic. It follows that the "directed evolution" creating TOs is not inherently bad *because* it is directed evolution (McGloughlin 1999, p. 6). Rather, the moral value of directed evolution's products must be determined in the same way that conventional breeding products are evaluated, i.e., on their impact on the market, environment, or things other than the crops themselves. If it can be shown conventional breeding is bad because it leads to control methods becoming ineffectual, then transgenic organisms causing the same effect are bad on the same grounds, and not because they have violated some evolutionary process, and vice versa.

A fourth variation on the evolution argument is the unethical creation of "genetic uniformity" from the introduction of transgenic crops into the bio-system (Altieri and Rosset 1999a). It is a fact that evolution has produced a great deal of diversity, which human beings have steadily decreased since the species originated. In fact, humans have become the "chief destroyer[s] of organisms" (Curtis 1972, p. 532).

The fear is that with very few seed and other technology companies, there will be a reduction in the number of variations evolution created in a species. The seed companies will generate as few varieties as possible to increase their profits; so the number of varieties available will be more limited than if transgenics were banned (Joly and Lemarie 1998, p. 2).

The evolution part of the argument is based upon the "genetic erosion" it is alleged will result (Altieri and Rosset 1999a). Once there is market and production uniformity – monoculture – if some sort of new pest or pathogen is introduced into the system, and the variety is unable to resist it, then the potential for destruction of the species is very high. If all wheat, for example, is vulnerable to rust, then rust's introduction into the production system will wipe out all the wheat. If there had been as many varieties of wheat as evolution created, then the fittest wheat would have survived and reproduced, while only the less fit would have succumbed to the rust. When humans limit the diversity evolution has given them, then they act in a manner contrary to evolution. Therefore, limiting variety is morally wrong and transgenic organisms are morally bad.

The Genetic Uniformity argument has several unique problems. First, there are those who believe that the available evidence is insufficient to convincingly prove claims that genetic diversity has been limited by the introduction of transgenics. In fact, the historical record shows diversity has not been lost due to human beings developing agricultural products (Bartsch et al. 2002, pp. 78–89; Harlander 2002, p. 161S). For example, the hybridization of corn in Mexico has not lead to reduced bio-diversity. According to Allan Felsot, even with "the tremendous amount of gene flow from non-local to locally adapted and selected cultivars, the [original] varieties survived intact as recognizable entities" (Felsot 2002, p. 5). Furthermore, the Royal Society of London, et al., recognize:

> The domestication of plants for agricultural use was a long-term process with profound evolutionary consequences for many species. One of its most valuable results was the creation of a diversity of plants serving human needs. Using this stock of genetic variability through selection and breeding, the "Green Revolution" produced many varieties that are used throughout the world. This work, carried out largely in publicly supported research institutions, has resulted in our present high-yielding crop varieties. (Royal Society of London et al., http://books.nap.edu/html/transgenic/need.html)

In other words, conventional breeding has increased the bio-diversity of crops able to serve human needs. There is no reason to think transgenic crops will not do likewise, and perhaps, genetic engineering will be able to create plants and animals that will help environments undergoing great environmental stress due to biotic, e.g., insects and other pests, or abiotic, e.g., drought and salinity, factors that turn functioning bio-systems into wastelands.

The second unique difficulty for the Genetic Uniformity argument is that even if genetic diversity has been reduced, the reduction problem is not solely limited to transgenic organisms. The problem is for any human activity limiting bio-diversity. Over the many years of conventional agriculture, a large number of species have been lost (Lenski 1993, p. 1). If it is morally wrong to limit diversity through transgenic research, then it must also be morally wrong to reduce it through the conventional methods of plant breeding used by the seed industry. Since few, if any, are complaining about conventional breeding techniques and effects, then on the grounds of consistency, they should not single out TOs. Hence, there is nothing unique about transgenic organisms and their development, regardless of whether or not it is in accord with evolution.

The Genetic Uniformity's third problem is its assumption that greater diversity is always good, when there are cases in which it is thought to be positively bad by those opposing transgenics. Friends of the Earth Europe (FoEE), for example, opposed Morphotek's plan to create new crop varieties because the group objected to the use of some of the genes identified as causing colonic cancer in humans to speed up evolutionary processes (FoEE 2002, p. 2). Roughly, Morphotek exposes the DNA from a number of crop species to the genes, which cause mutations in the species. Morphotek then selects the plants exhibiting the characteristics the company desires for new crop varieties (Cohen 2001, p. 12). This method is called morphogenics, which "switches off the [genetic] 'proof reading machinery' allowing for entire genomes of microbes, plants, and mammals to be genetically altered more frequently" (Ibid.). Radiation has been used in a different type of accelerated evolutionary process, called radiation mutagenesis, to improve varieties since the 1970s. To date over 1,800 different crop varieties have been developed using the technique (Harlander 2002, p. 161S).

Even though diversity is increased using morphogenics, FoEE believes Morphotek will not be able to remove the cancer causing genes from its new varieties, which the group is worried could produce cancer in human beings. Hence, diversity in itself is not necessarily a good to be pursued. Rather, what is morally important is the effect members of a species have on other species and individuals, especially on homo sapiens. If a species' members have a deleterious impact, then it is extrinsically bad regardless if its existence creates greater diversity. If the species' existence is beneficial to those species valuable to human beings or the bio-system, then the former is extrinsically good, despite whether or not diversity is increased.

A fourth problem for the Genetic Uniformity argument is based on a lack of harm for destroyed species. Even if human actions lead to the extinction of a species, it is not immediately clear that the extinct species has been harmed in any way (Russow 2002, p. 114). Although it makes sense to talk about harming organisms because they have individual interests, a species is not sufficiently similar to an organism's status as either a sentient or living entity. Rather a species is a group of individuals which is neither sentient nor alive. Bernard Rollin claims that "killing any ten Siberian tigers is no different than killing the *last* ten" (Rollin 1994, p. 78)[48] because it is the sentience of the members of the species that counts morally rather than the extinction of an "abstract entity" (Ibid.). Therefore, the moral focus should be on the harms and benefits to existing and likely members of a species instead of the species itself.

[48] See also Rollin (1996, p. 35).

A general reply to all four of the anti-transgenic evolutionary arguments – Usurping Nature, Genetic Instability, Abrupt Evolution, and Genetic Uniformity – is that if we must allow evolution to take its course without interference on our part, then it follows that all human made products contrary to evolution are morally bad and creating them is morally wrong, much as in the case of definition pair two. Since evolution has not made it the case that all homo sapiens can resist cancer, AIDS, Ebola, and other forms of diseases plaguing them, for instance, then it is morally impermissible to interfere with the evolutionary processes to save people afflicted by the diseases. If we are serious about letting evolution take its course, then we should stand back and allow those without disease resistant DNA to die off as soon as possible. It is beneficial to the species' fitness for them to die quickly, otherwise they might survive and reproduce, which will pass on the defective genes to their offspring. In the long run, homo sapiens will improve when the beneficial mutations become standard in the human gene pool. If we accept arguments against human intervention in evolutionary processes, then we must also accept the morally repugnant conclusion that any person who is not fit to survive in the evolutionary sense should not be helped in the interests of species development.

Of course, the argument that human interference in the evolutionary process lacks plausibility for a number of reasons besides that of the repulsiveness we feel toward it. First, even if we assume creating transgenic organisms is outside the realm of evolution, it is not necessarily wrong to do it.[49] Most people accept Peter Singer's principle that if we can help others without sacrificing anything of comparable moral worth, then we are obligated to do so (Singer 2006, p. 255). Given this widespread belief about moral duties, it follows that allowing morally valuable individuals to perish simply to assist morally neutral evolution is unethical. The sacrifice of an individual for the benefit of the whole might be justified in some cases, but to sacrifice so many merely because evolution made them too weak to survive and evolution will help to create a stronger species in the distant future cannot be true on its own.

Furthermore, even though the species will benefit from allowing the less fit members of the human race to fail to reproduce or take resources from those who could use them to better effect, there will be other diseases which will occur in the future for which evolution has not yet fitted the species. As humans acquire genes to resist cancer and other diseases, new ones will spring forth. For example, we might become immune to a certain variety of virus, but the virus will not cease to exist. Those members of the virus that are fitter will still inflict harm on the human race because the human race is not immune to them. The virus survives and passes its genetic material to its offspring, until the beneficial gene becomes a standard in the viral genome. There will also be new types of diseases to arise for the new human genome, which the current genome could be immune to or not. The point is there will still be problems for the species even if all of the sick individuals are prevented from passing on their genes.

---

[49] I will argue in the Pro-Transgenic section that creating transgenic organisms could be the result of evolutionary mechanisms; hence, it does not violate evolution in any way.

Another problem for the anti-transgenic evolution arguments is the fact that there have been very beneficial examples of genetic manipulation by human beings. First, some turkeys have been bred to reduce brooding behavior and increase egg production (Reiss and Straughan 2001, p. 183). Crops have also been vastly improved to the benefit of human beings and others (Harlander 2002, p. 161S). Finally, as stated above, humans have already had a significant impact on crops for many years through conventional methods of breeding (Royal Society of London et al. 2000). The crops developed in the Green Revolution via conventional means, for instance, enabled many more people to be fed than would have otherwise been the case (Ibid.). For most people, especially those helped directly, the reduction in starvation is a very good thing.

The issue of our obligations to those who are not as fit as others in the evolutionary scheme of things shows one very important moral truth. Just because something is the result of evolution does not mean it is a morally good thing. It is important to remember evolution is a natural process, but it has no moral value on its own nor does it confer moral value on actions or objects. Evolution is merely a process like gravity.[50] It is neither morally good nor bad that gravity exists. An object falling to earth is neither morally good nor bad in and of itself merely because gravity is operating. Gravity might even end up hurting someone, but it has no moral worth outside human valuation of it and its effects. Evolution and its processes are the same in this regard.

3.6.3 Arguments for Transgenics

The evolution argument has two variations supporting the morality of transgenic organisms. The first is common in many different venues and focuses on the evolution of technology or science. The second has not been explained or examined in any great depth, although it holds greater promise than the former. The latter draws its limited evidentiary strength from the fact that humans are organisms affecting their environment, which in turn, has a natural impact on evolution. I will begin with the first and weaker argument of the two.

The evolution of science argument is a poor one despite being used so many times. Some scientists and others attempt to justify TOs by saying that they are the natural products of naturally evolving science. In one instance, the term "evolution of the plant biotechnology industry" is employed to discuss the morality of TOs (Joly and Lemarie 1998, p. 2). Furthermore, Monsanto adopted the terminology to promote its products, which was found by Britain's Advertising Standard Authority to have misled the public in an ad by "[failing] to make . . . clear that scientific opinion was divided over whether or not genetic modification was an extension of the cross-breeding of plants which has gone on for centuries" (LRS, p. 1). The final positive example mentioned here comes from Peter Gregory, Director of Biotechnology

[50]Although gravity is a cause and natural selection is an effect.

International Programs at Cornell University, who stated, "agricultural biotechnology ... is not something completely new, but a natural evolution from traditional agricultural techniques such as cross-breeding" (GKCCB 2002, p. 1).[51]

Critics of transgenics have also noted this evolution argument variant. The Rivermouth Action Group wrote that, "Proponents of GMO food maintain that genetic modification of DNA is only an extension of genetic evolution through natural reproductive means" (RAG 2003, p. 1). Furthermore, Karen Charman claimed that "Promoters of agricultural biotechnology insist that genetic engineering is just a faster and more precise way to improve crops than traditional plant breeding methods" (Charman 2006, p. 1). Although the assertion that transgenic engineering science is a part of evolutionary forces is relatively widespread, the support for it needs clarification.

One interpretation of this evolution argument attempts to show that science has evolved to produce TOs. This linkage between science and evolution is intended to imply there is a natural progression of science along some form of natural lines. In other words, science producing transgenics is the result of an evolutionary process. Furthermore, the use of the terms similar to "natural progression of science according to the theory of evolution" entails proponents believe the natural is right, or at least, morally good, and science's natural progression is morally good as well. Hence, researchers engineering transgenics are ethically acting in accord with evolution instead of violating it. In addition, since transgenics are the product of natural evolutionary process, they must be morally good. By equating transgenic technology with what is natural, there is a perhaps unintentional attempt to create a positive emotive response in others; thereby making it more likely for them to be comfortable with the development and subsistence of transgenics, themselves.

Although it is understandable why TO proponents desire to make use of this evolution argument, it is also clear why they make a severe category mistake in doing so. The definition of evolution shows it is a process for species, not scientific research and application. Species evolve over time, but by no stretch of the imagination is science an existing species. Hence, science cannot be an evolutionary process justifying any scientific product.

What proponents should actually mean when they claim "that genetic modification of DNA is only an extension of genetic evolution through natural reproductive means" (RAG 2003, p. 1) is that there has been a progression in scientific knowledge enabling humans to cross natural species boundaries to create transgenic organisms. That is, entities that could not exchange genetic material on their own, with the help of human beings, can now do so. Of course, this interpretation of the claim says nothing about the morality of the scientific development.

The mere progression of science is insufficient to imply any form of naturalness and the further implication that the natural is morally good or right. For example, science and technology have moved to the point at which they can help kill individuals in war and terrorism more efficiently than could have been dreamed of a mere

[51] Also see Fox (1999, pp. 4–5).

100 years ago. Neutron bombs can eradicate a large number of people, while doing little damage to the detonation area's infrastructure and private property. As in the case of something being the product of evolution, the mere fact that it is a technological development of science does not mean that it is a morally good thing. What has to be done in order to make this argument work is to show that the particular scientific progression in this particular case is a good thing, but it will have to do this using other moral principles, such as PMC, and not the natural evolution argument.

A superior but still weak argument for transgenic organisms and evolution is not often used, but deserves at least some limited consideration. This line of reasoning claims that all human actions are a consequence or part of the evolutionary process simply because homo sapiens' existence is an evolutionary result. Hence, human actions are in a very major way a consequence of evolutionary mechanisms. For example, William Leonard argues that a dietary change was a driving force behind human evolution. "We now know that humans evolved not to subsist on a single, Paleolithic diet but to be flexible eaters" (Leonard 2003, p. 1). According to Leonard, one reason humans are bipedal is because those bipedal organisms able to reach more food and use less energy from having this characteristic were more inclusively fit than quadrapedals (Leonard 2003, p. 2). Moreover, evolutionary beneficial "[i]nnovations such as cooking, agriculture, and even parts of modern food technology can all be considered tactics for boosting the quality of the human diet" (Leonard 2003, p. 4). There are even some philosophers who have argued that ethical theory can be partly derived from evolution. Bernard G. Campbell and Brant Wenegrat, for example, contend that human propensities, such as human cognition, serve as some of the fundamental underpinnings of ethics. (Campbell 1995, p. 119; Wenegrat 1995, pp. 139–40). If these observations are correct, then it follows that human actions are not contrary to evolutionary process, but are in accordance with them at minimum. Hence, humans creating transgenic organisms are a result of evolutionary mechanisms, at least in part.

Furthermore, humans, as any other organisms, influence their environments and the biosphere, which in turn influences evolutionary changes.

> Modern agriculture is intrinsically destructive of the environment. It is particularly destructive of biological diversity, notably when practiced in a very resource-inefficient way, or when it applies technologies not adapted to environmental features (soils, slopes, climatic regions) of a particular area. (Royal Society of London et al. 2000)

It has already been stated that humans have probably been limiting diversity and changing the environment ever since they evolved from their ancestors. In this argument, humans are a natural force just like any other, only with a greater potential to affect the environment. The conclusion is that TOs are part of evolutionary processes because they are an end of human activity, which is itself an evolutionary result. Hence, transgenic organisms do not violate evolutionary processes because they are created in accord with such mechanisms, and if that which does not violate evolution is good, then TOs are good as well.

However, the conclusion that transgenics are morally good merely because they do not violate evolution and are part of evolutionary processes shows the defect of this version of the evolution argument. The mere fact that an object or action does or does not transgress evolution entails nothing about the morality of the object or action. Those people with greater inclusive fitness, for example, are not morally better *ceteris paribus* than those less fit. Hitler's eugenics program was designed to produce what his regime thought the fittest of human organisms, but the values indoctrinated or held by those involved in the programs were morally reprehensible and repulsive. The mere fact a human being is not a "pure Aryan" does not mean the person is any less of a moral being or morally defective in some way. Being more inclusively fit merely entails the person's genes are more likely to be directly or indirectly passed on to the next generation, not that the person's moral character is better or his actions more likely to be morally right. Hence, evaluating objects or actions according to their "fit" with evolution and its mechanisms is unlikely to convince any reasonable person that transgenics are morally justified on these grounds.

3.7 Conclusion

The fear of transgenic and other new technology and their products could stem from a psychological condition of human beings pithily captured by Douglas Adams. Adams maintained that there is a set of rules governing human beings' intuitive reactions to technology.

1. Anything that is in the world when you're born is normal and ordinary and is just a natural part of the way the world works.
2. Anything that's invented between when you're fifteen and thirty-five is new and exciting and revolutionary and you can probably get a career in it.
3. Anything invented after you're thirty-five is against the natural order of things.

The actual chronological ages Adams uses to delineate the different responses are false, since many of the most vehement opponents of transgenics are in their late teens and twenties. However, if the ages are interpreted as psychological, then the rules hold.

Although the eight definitions of unnatural fail to do their intended work, it is clear that an examination of them is useful in advancing the ethical debate over transgenic organisms. It is clear that the moral value of transgenic organisms and the activity of creating TOs can be determined, at least in part, by the foreseeable consequences of such products. Moreover, the mental states of those creating transgenics have a significant role to play. If researchers and other relevant parties' intentions, motivations, or mental attitudes were primarily evil or the actions were not performed with the proper respect for the persons affected by them, then the creation is morally wrong. For example, some transgenics could be created with the motivation of greed and the intention to destroy another's market share merely for the creators' benefit. Their creation could be unethical not because they are unnatural, but because of other moral considerations. However, if the motivations, intentions, and other relevant factors are primarily good, which seems to be what happened in the Golden Rice case,[52] then the action can be morally right. Therefore, moral principles, such as RPU and QCI, are much more likely candidates to be practical in the debate than is the unnatural/natural argument. The former more realistically help advance arguments because they better tie into the obvious moral considerations, such as the need to keep the bio-system functioning. It is obvious that those areas should be the focus of the debate over the morality of transgenics rather than their supposed unnaturalness.

At this point, no room has been found in the Practical Moral Code for the environmental moral principle incorporated within the Anti-Transgenic Moral Code, and the American Society for Biochemistry and Molecular Biology and Association for Computing Machinery's professional codes of conduct. Actions are neither right nor wrong because they conform to or break, respectively, some rule that obligates us to protect the environment in some way or to act naturally. Perhaps the best way to address the legitimate moral concerns of those who want an environmental/natural element in a universal moral code as well as taking into account the opinions of others is to develop an axiology for PMC which makes it a moral factor in all decision processes. In Chapter 4, the natural/environment will be assigned intrinsic value in a hierarchical system.

[52] Although this seems to be a matter of debate, with some on the anti-transgenic side claiming it is a coldly reasoned publicity ploy to make transgenic organisms more acceptable in the marketplace.

Chapter 4
A Natural Theory of Value

In modern industrialized countries like the United States, the assumption is often that things have only instrumental value unless there is evidence to the contrary. This is in sharp contrast to other cultures that make the opposite assumption; they assume that everything is deserving of respect, and that we ought not to use things as mere means unless there is a good justification for doing so. (Hinman 2000, p. 563)

4.1 Introduction

As was seen in Chapters 1 and 3, many people think naturalness or natural processes have some role to play in ethics.[1] What is more murky is whether it is possible to incorporate these beliefs, or some form of them, into a rational and consistent theory of ethics or a practical moral code people can use to solve their ethical dilemmas about technology.

To find a legitimate function for naturalness to play in technology, biotechnology, and ethics, especially in the transgenic organism debate, we must practically incorporate the ideas the USDA survey participants and professional organizations had in their individual moral codes, as well as those found on both sides of the transgenics debate, into the practical moral code, while keeping it objective and consistent for everyone to use in moral deliberations. What I propose is that naturalness or nature should be thought of as one of many moral factors in ethics. When making decisions, people need to recognize that the intrinsic value of things in nature and nature itself, in addition to all other ends in themselves, in order for them to know what is morally good and to act as they should.

In this chapter, a universal axiology practical for any moral evaluation, regardless of whether the issue at stake is transgenic organisms or new technology, will be

[1] Unless otherwise noted, the environment, nature, the biosphere, and the Earth, will be treated as synonymous terms. Biosystems are subsystems of the biosphere.

D.R. Cooley, *Technology, Transgenics and a Practical Moral Code*, The International Library of Ethics, Law and Technology 4, DOI 10.1007/978-90-481-3021-4_4,

developed. The value theory is based on a moderate environmental position that everything deserves respect, with some objects deserving more than others.[2]

4.2 Naturalness as Sacredness

The sacredness of nature and the natural is one avenue many have taken to try proving special duties toward the environment (Wood 2003, p. 1). An argument suggested by Robert L. Holmes is based on the infinite goodness characteristic of God and his creations. In Leviticus, God is supposed to have told Moses that he should be holy because God, his Lord, is holy (Leviticus 19:2). In other words, God's characteristic can somehow transfer to other existing objects made by the divine entity, in this case, a human being. Perhaps this conveyance is accomplished along the lines of Locke's private property argument in which a person mingles his labor which is also his property in whatever he produces (Locke 1982, p. 18). If holiness transfer serves as a precedent, then other features might as well. If God is omnibenevolent, according to Aquinas, then goodness must be part of his work product because the creator imbues his products with his own essence. Therefore, each object to be found in nature has sacred qualities in them to a greater or lesser extent, and should be treated in the proper way.

Other religions have similar theoretical elements of instilled sacredness in one or more parts of the environment. Some Native Americans and others, for instance, claim that nature deserves reciprocity and respect because it is imbued with spirits (Cajete 2002, p. 150; Hinman 2000, p. 564; Wood 2003, p. 1). Since spirits are sacred objects meriting respect, anything incorporating them would deserve the same kind of treatment. Furthermore, Gregory Cajete argues that early Native Americans believed that nature is a necessary part of human beings (Cajete 2002, pp. 148–9). Native Americans understood that "an intimate relationship between themselves and their environments as the essence of their survival and identity as people" existed (Cajete 2002, p. 149). If people consider themselves to be distinct from nature, they have failed to recognize both their true identity and the values of the environment as a whole and its components. Hence, an adequate ethics would have to acknowledge that the environment deserves as much respect as a person *because* it is part of each person or it has spirits warranting respect.

Chung-ying Cheng argues along a similar ontological line. According to Cheng, humans and human actions are merely different aspects of the Tao, which is the Way of nature's force. In order to act harmoniously with the environment as a whole, people must behave with natural spontaneity, which is effortless action characterized by least effort with maximum effect (Cheng 2006, p. 203). That is, the best results are

---

[2]Given the axiology, Reasonable Person Utilitarianism, and the Quasi Categorical Imperative, it follows people are obligated to use things sparingly (Thiroux 2004, p. 436). Frugality respects each object's value, while at the same time produces an outcome that reasonable people would reasonably believe is likely to be among the best.

produced in the most efficient manner with the action's agent being in perfect synchronicity with herself and her environment as one. Even though the Tao defies fully rational discussion and evaluation, once again, nature is clearly seen as essential to identity. In order to understand nature and oneself, each person must recognize these facts.[3]

The sacredness argument's final version is based on the notion of stewardship, or caring for some valuable thing that we do not own but for which we have responsibility. A central tenet of Hinduism is that all life is sacred, and Buddhism prohibits the taking of animal life and stresses the intimate connection of all things (Reiss and Straughan 2001, pp. 81–3). If either is correct, then the sacred needs to be cared for in the way that each person should respect the sacrosanct, and animals need humans to avoid killing them because they are intimately connected to who we are as humans as we are to them.

Many of those from a Judeo-Christian heritage hold that the land is a gift from God requiring people to be faithful, grateful, humble, neighborly, kind to one another, generous to strangers, honest in trading, and just. The Evangelical Lutheran Church in America, for example, believes its social statements on economic life and caring for "creation" must be guided by the central themes of sufficiency, sustainability, livelihood, effects on all people, justice, and stewardship which it defines as serving "the best interests of creation's integrity in imitation of God's care for us" (ELCA 2004, pp. 3–4). Accepting the land gift from a divine source entails that people cannot use it in any way they want. Although they are given the right to subdue or dominate the earth, they must also carefully steward the land so as not to abuse the munificence of a caring divine entity, who has not given them free and clear property title to the gift, but only lifetime loan of it (Berry 1990, p. 277; Hinman 2000, p. 564; Thompson 1995, p. 72; Harrison 1992, p. 395; Hart 2006, pp. 119–20).[4] Others add that land is an inheritance to be preserved for future generations, who will have only the same lifetime claim to it (Berry 1990, p. 277). In other words, the land is not sacred because it is imbued with spirits or part of what it is to be a person; it is sacred because it is a gift from a divine entity to each person, regardless of when the latter exists. The stewardship view leads to a similar set of duties to the land – humility, understanding, and respect – as mentioned by Cajete (2002, p. 150).[5] Proper land care leads to very good results for the land and the individuals responsible for it. Besides satisfying God's commands and people receiving

[3]Linda MacDonald Glenn mentions a related view – the "Circle of Interdependence"(Glenn 2003). Although this more holistic position does not integrate people and the environment into it as do Native American and Taoist views do, there is still an emphasis on a "non-hierarchical interdependent relationship between man and the Earth and also of its inhabitants" (Ibid.).

[4]In the 14th formulation of the Unnatural Argument against agricultural biotechnology, Comstock uses the term "creation." The use of the term implies that those who find the argument attractive believe that nature is the result of an intelligent design and a designer.

[5]Callicott offers a different view of the relationship of human beings to nature. Instead of dominating nature in any way, human beings are to be "plain members and citizens' of nature" (Callicott 1999, p. 188).

what they need, there is an additional benefit to proper stewardship, viz. a symbiosis of all things in nature working as they were intended to (Harrison 1992, p. 395). This in turn leads to people who are less selfish because they understand that they are not the focal point of all value in the universe, while also grasping that they are important and valuable entities.

To bring reason into sacredness, the best we can do is to accept Kierkegaard's view of a legitimate faith combining reason and non-rationality. Reason can take people only so far in the process of forming permissible religious beliefs, but because evidence is lacking, it cannot justify their belief in one religious proposition over another. That is, reason clears the ground by eliminating beliefs defeatable by evidence, such as literal Creationism. Fossil records, imperfect adaptation, extinction, geology and so on provide sufficient verification of the proposition that the world was not created in six days in the order stated in the two inconsistent stories in Genesis. Reason might also show that the existence of a divine entity with infinite characteristics of goodness, power, and knowledge is impossible.[6] According to Kierkegaard, once reason has done its work, people may believe whatever proposition is left, which will require a leap of faith on their part.[7] Whichever way their faith takes them is reasonable for that person in that situation and at that time. Hence, although reason is part of the process, the non-rational leap makes the entire procedure non-rational as a whole. And therein lays the problem with this approach.

Although sacredness might be a key component of environmental ethics, and by extension, transgenic organisms and other technology, it would be difficult – if not impossible – to use any version of it in a rational argument.[8] The reason why should be obvious. Religious beliefs and sacredness are based primarily on non-rational faith.[9] A person can have legitimate faith in something without any evidence, such as believing Lutheranism is the one true religion. However, rationality requires evidence. Unless one religion or other has sufficient rationalist or empirical evidence for it, which might be impossible to find, then it cannot rationally be believed to be the correct one.[10]

---

[6]The Problem of Evil argument does not prove that a being that knows everything that has occurred or is occurring, is infinitely good, and would like to eliminate all evil but does not have the power to do so does not exist.

[7]There might be a further criterion to pass before it is permissible to believe a religious proposition. Religious beliefs should not only be permitted by reason, they should be morally permissible to believe. For example, any religious proposition advocating discrimination on morally irrelevant grounds should not be believed.

[8]Although it might be useful to convince a large portion of the population to act ethically toward the environment (Callicott 1999, pp. 193–4).

[9]Paul Thompson states that one of the major problems with the sacredness argument is the reluctance of those steeped in the Enlightenment tradition of ethics to accept the sacred as a useful tool and those who accept the sacred having great difficulty ascribing sacredness to what is normally thought as profane (Thompson 1995, p. 9).

[10]This is one reason that we can reject religious arguments for or against transgenic organisms. Some religions, such as Buddhism, Hinduism, and Jainism believe that genetic engineering directly violates prescriptions of non-interference and non-harming or that it disrupts the spiritual process

If we are trying to discuss any moral issue to reach some sort of consensus, then equating the sacred with the natural will stymie our goal. Not everyone will agree with each other person's non-rational view of the sacred. To be consistent, we would not accept another's religious preferences as a definitive argument against us; hence, we cannot use ours against theirs. If we want to persuade anyone to a particular way of thinking, we must use objective, publicly accessible evidence, not faith, to make the appeal. Since there is no evidence as to which religion is the correct one – beyond that which allows us to reject clearly false ones, such as those requiring violations of ethical duties, treating persons unequally, or harming others unnecessarily – then a particular religious view cannot be used rationally to support any case. So even though a particular religion might be right about nature's sacredness, there is no way rationally to know it or employ it in decision procedures. Thus, the nature-as-the-sacred claim's inability to fulfill this requirement requires that it be set aside.

4.3 Beginning an Axiology

Since Reasonable Person Utilitarianism and the Quasi-Categorical Imperative comprise the operational, practical moral decision making principles of PMC, it is necessary to know which things have intrinsic value and which do not.[11] First, for RPU, we must discover what a reasonable person would examine in trying to find which action is likely to maximize utility, which of course entails weighing intrinsic value in some way. A bit of this work was completed in Chapter 2. Moreover, QCI requires proper respect be given to those things with intrinsic value because "It is commonly agreed that something's possession of intrinsic value generates a *prima facie* direct moral duty on the part of moral agents to protect it or at least refrain from damaging it" (Brennan and Lo 2003, p. 2). In order for both principles to be usable, a practical theory of value will be developed.

Axiologies, or theories of value, are vital to ethics because they provide necessary information to understand how any moral theory, principle, or code works. Every theory "needs a foundation, something that instrumental goods aim toward" (Russow 2002, p. 98). That is, in order to have instrumental or extrinsic value, then a theory must have at least one end in itself or thing with intrinsic value. Intrinsically valuable things help "craft" the theory, although they do not determine its entirety (Russow 2002, Ibid.). For example, if pain and pleasure are the only intrinsically disvaluable or valuable things, respectively, then an adequate moral theory must be built upon those foundations. Under hedonic act-utilitarianism, for instance,

---

of reincarnation (Fox 1999, p. 147). However, the mere fact that people believe this does not make it true. In addition, we are obligated to respect people's right to practice their religions, but if their religions negatively affect others who have not freely chosen to be members, then those who would cause the harm must prove why their actions are rationally justified.

[11] Brennan and Lo provide an excellent overview of value in nature (Brennan and Lo 2003).

the only things that count in determining an action's value are all the pleasant and painful consequences of the action. If virtues and vices replace pleasure and pain, the resultant theory still maximizes utility, but weighs virtues over vices in some manner for each alternative open to the agent at that time.

To provide clarity by eliminating different terms being used for the same meaning or different meanings being assigned to the same term, William Frankena's intrinsic and inherent value definitions will be adopted. For Frankena, something is intrinsically valuable if and only if it is good in and of itself or good because of its own intrinsic properties (Frankena 1973, p. 82). In other words, if something is an end in itself, as Kant would define it, then it is intrinsically valuable. Frankena, for example, claims that pleasure experiences and certain other states of mind are intrinsically valuable when they are instantiated by thinking beings. Below, I will argue in favor of including these and other intrinsically valuable entities in the practical axiology.

Although they are sometimes confused, inherently valuable entities are neither intrinsically nor extrinsically valuable. Frankena considers inherently valuable objects to be things good because the experience of contemplating them is good or rewarding in itself (Frankena 1973, p. 82). Works of art and things of natural beauty cause people to have pleasure or other intrinsically valuable consciousnesses, thereby making the former inherently good while the latter is good in and of itself (Frankena 1973, p. 81). The pleasure from appreciating a Renoir is intrinsically valuable, for example, while the Renoir itself is inherently worthy. Moreover, anything able to cause entities to experience instantaneously an intrinsically valuable state of mind from normal contemplation of it is not extrinsically valuable. For inherent worth, the simultaneous intrinsically valuable experience is not a means to an end, which requires the end to occur after the means, not at the same time.

The distinction between intrinsic and inherent value is often overlooked by many of the brightest thinkers in theoretical and applied ethics. Mark Sagoff, for example, mistakenly conflates the two value types together when discussing the failing of interdisciplinary scientists who consider nature as a rational system "to model and administer for the collective good rather than as an object for moral instruction and aesthetic appreciation for every individual" (Sagoff 2008, p. 207). Asserting that the universality of the positive experience entails that nature has intrinsic value because of what it produces. "We do not find the aesthetic object valuable because we enjoy it; rather, we enjoy the object ... because we perceive it to be valuable" (Sagoff 2004, p. 18). Clearly, Sagoff believes that nature is not merely extrinsically valuable, but it is the experiencing of the object that seems to be an end in itself and not the object itself. Nature when properly contemplated is inherently, not intrinsically, valuable. Given the necessary difference between the two things – experiences versus object – Frankena's distinction is useful for clarifying how value works in an adequate value theory.

In order to bring as much objectivity as one reasonably can into an axiology based ultimately upon intuitions and the following Benefit of the Doubt principle explained below, the ranking of intrinsic value will be supported by using the evaluation method Mill described in *Utilitarianism*.[12] In support of his argument for a quantitative and qualitative distinction in hedonism, Mill stated that intellectual pleasures are more likely than physical ones to be preferred by those who have experienced both (Mill 1988b, pp. 8–9). Since those who have known each are probably better educated than those who have familiarity of physical or intellectual pleasure only, the majority opinion of the former will capture reality. Pleasures – and pains, for that matter – can be differentiated and correctly ranked by their quality by the knowledgeable, who must attempt to avoid elitism by also incorporating the ideas of all stakeholders – all those who might be injured or have their autonomy adversely affected by the technology – into the knowledgeable's deliberations. The result of taking Mill's democratic approach is the axiology used by RPU and QCI is rooted in what thoughtful people actually think has value rather than one bearing little on reality.

Although their individual axiologies might be slightly different, the practical value theory's components and rankings are based upon the majority opinion from the USDA survey results, the professional societies' codes of ethics, the environmental ethics' canon, and others who are stakeholders in biotechnology.[13] The survey participants, regardless of which faction to which they belonged, i.e., Neutral, Pro, or Anti-Transgenic, uniformly adopted the value of certain moral considerations and moral principles, although the emphasis they placed on each varied according to group identity. Hence, there is at least some evidence that Hume was correct when he claimed that human sympathy is universal, and is a basis for human morality (Hume 1948a, p. 178). There also seems to be good evidence that a commonality exists between all people on moral ideals and how to apply them. Moreover, the commonalties of responses in the survey and codes of ethics show that different societies have identical sets of virtues, although they rank the virtues in different ways. Each set has the same virtues in it, but arranged in ways appropriate for the particular society (Hume 1948a, p. 205). For example, a robber society will value justice as any other society does, but it is a justice suitable for a society composed of thieves rather than one of law abiding citizens (Ibid.).

Ranking objects according to what thoughtful people actually believe will probably have better ethical results than pure abstract theorists alone can achieve. Since people already think that intrinsic value exists and objects possessing it can be graded, then they are more likely to do the right thing as determined by this axiology than by a value theory inconsistent with their fundamental beliefs. For example, if a

[12]Glenn would reject the hierarchical nature of my axiology in favor of a continuum in which full scale Kantian moral agents are on one end, while pure property is on the other (Glenn 2003). Between the two classes of things are those objects with fewer rights than Kantian moral agents but more rights than pure property.

[13]Rollin claims that assigning intrinsic value to nature is incomprehensible and unhelpful for two reasons. First, human beings have intrinsic value, and by ascribing value to nature, we set ourselves up for instances in which there is conflict. Conflating the problem is the absence of any mechanism to decide which thing should win the conflict (Rollin 2006, p. 153). My axiology and how it works will address both of these concerns.

particular biosystem is ranked as more important than people, it then follows that we can sacrifice people to protect it, if doing so is needed. Perhaps the best biosystem with the greatest diversity is one without people in it, which seems to be what some environmental science has been discovering. However, most people would reject such a foundation belief and its result. Given a choice between preserving the system as it is, which requires the death of moral agents, and changing the system to better support moral agents, most people would select the latter, in part, because it results in the protection of what they believe has greater intrinsic value. As has been stated before, if people's values are incorporated into a decision, then they are more likely to adopt and enforce it than if the values are not. The reason why is obvious. Since the positive values draw them to the values, while negative values impel them away, they automatically are inclined toward any solution based on their beliefs.

4.3.1 Nature as a Moral Factor

The process of developing an axiology including nature's values should start with the work of one of the leaders in biotechnology ethics. Gary Comstock has created a moral principle based upon Aldo Leopold's classic Land Ethic, which is considered to be one of the first ethics recognizing nature's independent value. Comstock states that:

> An action is Morally Right for the land when it (a) promotes the health of the ecosystem, (b) protects or adds to the diversity of the species in the system, or (c) in some other way protects or adds either to the integrity and stability of the whole or toward the system's progress toward its goal, which is a mature equilibrium state for the system. (Comstock 2000a, p. 204)

Although its subtlety is more intuitively appealing than the Unnatural is Unethical argument, what "morally right" means in the principle is not as clear as it might first appear. The ambiguity of the term requires definitional investigation and refinement.

First, merely satisfying a, b, or c of the principle cannot be sufficient to make an action ethical. If it were enough, then performing clearly wrong actions, such as killing innocents by allowing them to starve to death, is morally right even if there is only minimal improvement in the ecosystem's health, diversity, or in any way that "protects or adds either to the integrity and stability of the whole or toward the systems progress toward equilibrium" (Comstock 2000a, p. 204). For example, if human persons are endangering the diversity of species in a system, which humans are now doing, then it would be morally right to terminate all those entities causing environmental degradation. As already has been argued, making matters worse would be when humans are trying to eliminate some contagious disease species for the benefit of humans and other animals. Due to the reduction of diversity, the action would not be morally right to do. These morally repulsive results of the principle show it cannot plausibly be stating a sufficient condition for moral actions.

The term "rightness" might merely be stating a necessary, not sufficient, condition of permissible actions. In order to behave ethically, every person each time she acts must do something to "advance" the biosphere in some way. That is, there is a positive requirement to help nature that entails that people actively have to pursue benefits. Protecting, promoting, and adding each necessitate that all actions produce positive results that are not required by the less burdensome negative limitations on actions which affect the environment, such as not harming or compromising the health, diversity, integrity, stability, or progress of the ecosystem to equilibrium.

The result of defining the principle in this manner is many morally right actions are classified by the principle as unethical, provided they do not accomplish at least one of the required positive environmental consequences. For example, heroically saving the lives of innocent people, finding love, and pursuing the realization of individual potential as a person are unethical if they do not also protect, promote, and add to the environment. The examples of right actions that are misclassified by this interpretation of "rightness" show that the difficulty in making the land ethic principle a necessary condition of normative ethics, *viz.* these particular actions and others similar to them are right or wrong independent of beneficial/harmful results to the environment. In fact, there will be times when it is conceivable that performing any of the above actions might harm nature, yet most people would still legitimately believe them to be ethical. Since it is possible to think of morally permissible actions that do not fulfill the alleged necessary condition, then the condition is contingent at best. Hence, this interpretation cannot legitimately integrate nature's value in a practical ethical theory or code of conduct.

The most plausible reading of "morally right" is that the actions prescribed by the principle require recognition and respect for the intrinsic value of the land or ecosystem. That is, when evaluating the outcome of alternatives open to the agent at a particular time, a reasonable person would incorporate all the consequences in which intrinsic value is involved, including but not limited to increased or decreased human person, animal, and biosystem worth and values of intrinsically worthy experiences, into her utilitarian calculations – as required by Reasonable Person Utilitarianism. Furthermore, the Quasi-Categorical Imperative compels the agent to respect all intrinsically valuable things as ends in themselves for the actual worth that they have. Hence, not only are people to be respected in order for actions to be morally right, but a great number of other things as well, including sentient creatures, living organisms, potential persons, and nature in part, e.g. biosystems, and as a whole, i.e. biosphere.

What I am arguing for here is a very broad, inclusive axiology. Often, we are myopic in our evaluations, generally taking into account only the persons affected by our actions. Sometimes, even if moral agents include potential persons and sentient creatures in their analyses, they ill treat other objects that deserve respect. In the rarest cases, living organisms are also part of the evaluation process, and even more seldom if ever, is the land or nature included. The too narrow focus causes unacceptable results. By some actions' impacts on other intrinsically valuable objects being ignored to the latter's detriment, we imperil our moral character by willfully or negligently doing the wrong thing because we are not approaching the situation with the depth of understanding required by reasonable people. What we need to do is to assume that things have intrinsic value until it is reasonable for us to believe otherwise.

Assuming universal intrinsic value does not lead to the conclusion that all intrinsically valuable things must be treated or thought to be equally worthy; it is still possible to make distinctions between the worth of various objects. That is, although many entities can have intrinsic value, different ones have different values. The difficulty, of course, is gaining general agreement as to what those values are. The solution to ranking employed in this work uses the Benefit of the Doubt principle, common moral sense, and how controversial it is to claim that the entity has intrinsic value. The less contentious objects are more likely to be ends in themselves; hence, they are ranked as having greater worth than those that are more controversial. Human persons are the least divisive, thereby making them the category of things with the highest intrinsic value whose interests should generally be pursued before the interests of other intrinsically worthy objects, although the biosphere which contains all intrinsically valuable things has the highest aggregate value. I will buttress this conclusion with common moral sense, which is based upon the views of environmental ethicists and the general public. For example, the majority of environmental ethicists, philosophers, and the general public assign the highest intrinsic value to persons for a variety of reasons we will examine later; thus it *prima facially* follows, first, that persons are valuable as ends in themselves, and second, that they have the greatest intrinsic value of all entities to be evaluated.

Besides establishing which things have intrinsic value and their rankings, it is important to explain why they have the value that they do and how to evaluate it. As Frankena's intrinsic value definition entails, an object's worth comes from a number of sources, including the telos of flourishing *and* the properties and states of affairs flourishing instantiates. For example, being in a state of pleasure can make a specific entity more valuable at that moment in time, all things considered, than being in a neutral mental state or a state of pain. Being a particular type of thing can also be intrinsically valuable. Sentience, for instance, makes whatever possesses the characteristic an intrinsically valuable entity on those grounds alone.

It is at this point that some potential for confusion should be eliminated if possible. In Chapter 3, teloi were dismissed as being useful in ethical decision making due to serious concerns about their existence and whether we could ever know with any high degree of probability what the functions are. That is correct. However, although evidence for teloi existence was found too tenuous to support the unnatural argument, for the purpose of developing an adequate axiology using the Benefit of the Doubt what was insufficient justification before can now work to our advantage. The difference, of course, is that we are no longer trying to assert that something is wrong or bad based merely upon the fact that the telos was altered or interfered with by genetic engineers. Instead, a telos merely has value to be considered when making reasonable decisions.

To review quickly what was asserted about teloi in the last chapter, the function of a living thing is determined, in part, by its definition and membership in various classes. What the individual essentially is partially creates the object's set of ends.

Sentient creatures, for example, are defined as creatures capable of feeling pleasure and pain; hence, the end for a sentient creature is to feel pleasure and avoid pain. Anything unable to feel either pleasure or pain cannot be sentient and have this end. Other intrinsically valuable traits that confer an end include being a human person, a potential person, and a living thing.

A group's telos is also determined by what is the best for those entities, i.e. what causes the entities to flourish. The telos for each person, for example, is the best that a person *as* a person can achieve.[14] If we adopt Aristotle's more than plausible claim that happiness is the ultimate goal of all human life, to which we add that happiness can be achieved in a variety of ways, then human flourishing is tied to happiness. Moreover, a happy life is always a good, minimal utility or maximal life, which I contend is also an ethical life. Hence, the more ethical a person is, the more the person flourishes and is able to fulfill her telos as a human person.

In order to uncover the telos of a particular living entity, it is important to classify it. Some entities have multiple teloi because they belong to many classes of things with at least one telos each or a class that has multiple teloi. Some may have only one telos because they only belong to one class of thing with one telos. Finally, there are entities that have no telos at all. Of the first type, human persons serve as a good example, As will be shown below, persons have teloi from being a person, a sentient creature, a living organism, and perhaps, being a human being – provided that there is a telos for that. On the other hand, a plant has only teloi of flourishing as a living organism and its species' ends, again if there are such things. Rocks and other non-living matter have no ends other than what a sentient creature or person assigns them, e.g. an elk might classify a boulder as an object on which to rub oneself.

Once the individual entity has been classified fully, then it is possible to evaluate its worth, although this need not be done precisely. The object can be ranked according to how well it fulfills its teloi and instantiates the properties which make the individual intrinsically valuable. Once we identify something as a person, for example, we can then start evaluating his life to see if he is a good moral agent leading an ethical existence. If not, then he is not fulfilling his function as a person. His life will be worth less than that of a person who is satisfying her end as a person, all other things being morally identical. According to what the majority of people who have experienced both are likely to say, the former's existence can be more pleasurable, but it can never be of as high quality as that of someone who flourishes.

4.4 The Benefit of the Doubt Principle

Although the intrinsic value of some entities is not all that controversial, others are more problematic. Even the most cursory examination of the environmental ethics' canon on intrinsic value reveals that many intelligent, reasonable people have widely

---

[14]The telos for an individual person is determined, in part, by the person, herself.

divergent views on not only the definition of intrinsic value, but also which entities have it, why they have it, and to what extent they possess it.

Consider, for instance the debate over whether nature or some parts of it, are ends in and of themselves and how much value the objects have. Influential ethicists, such as Holmes Rolston III, argue that nature or the environment has intrinsic worth independently of the existence of human beings (Rolston 1983, pp. 151–2, 2000, p. 601, 2003, p. 144).[15] J. Baird Callicott agrees that nature has such value, but maintains that it is dependent on human existence (Callicott 1989, p. 173, 1999, p. 15). Others have also given nature some form of intrinsic worth but on other grounds. In addition, a number claim that nature has worth equaling that of human beings, as Sessions and Devall do, or moral considerableness, as Elliott labels it (Sessions and Devall 1990, pp. 311–12; Elliott 2000, p. 572; Snyder 1990, p. 234; Nash 1990, p. 266; Cheng 2006, p. 202). Tom Regan has gone so far as to assert that rocks, rivers, and glaciers might be experiencing subjects of life that we will never be able to prove have that status. However, if they are truly experiencing subjects, then they are individuals, for Regan, having inherent for him, but intrinsic for us, value (Regan 1982, pp. 220–2). Each of these views would require us to take greater care toward non-human entities than we would to objects with mere extrinsic value.

On the other side of the issue, various ethicists, mostly anthropocentrists, have argued forcefully that nature merely has instrumental value to those things with intrinsic worth. Several from the very early history of environmental ethics, such as Thoreau, Marsh, Olmsted, and Muir, reasoned that human beings need nature in various ways, but never claimed that nature has dependent or independent intrinsic value (Thoreau, pp. 36–7; March, p. 41; Olmsted 1990, p. 46; Muir 1990, pp. 94–6). More contemporary philosophers and environmental ethicists agreed with their position (Stearns 1979, p. 611; Vogt 1990, p. 168; Marshall 1990, p. 161; Brower 1990, p. 250; Berry 1990, p. 277; Thompson 1983, p. 85; Thompson 2000, p. 578; Russow 2002, p. 109; Rollin 2000, p. 434 and 1996, pp. 58–60; Livingston 2003, pp. 6–7; Baxter 1974, p. 143; Cahen 2003, p. 114). Joel Feinberg, for example, claims that the fact that nature has no real interest entails that it cannot have rights (Feinberg 1979, p. 586). For Feinberg, the existence of rights indicates the subsistence of intrinsic value and the non-existence of rights indicates the absence of such worth. Lily-Marlene Russow has also argued that there is no convincing argument for nature having intrinsic value (Russow 2002, p. 112). If it did have such worth, according to this line of reasoning, then it should be provable in much the same manner as we establish that human beings are ends in themselves. The puzzle is if nature is intrinsically valuable, then why do so many thoughtful people disagree about it, when they do not generally disagree about the intrinsic value of persons?

Although it might be possible to find a convincing a priori or a posteriori argument in favor of or against an object's intrinsic value, the fact that so many

---

[15] See also Rolston's *Conserving Natural Value,* especially Chapter 6, and *Environmental Ethic,* especially the last part of Chapter 5 and all of Chapter 6, for longer versions of the works cited in this section.

intelligent people reasonably disagree on the subject provides weighty evidence for the claim that a compelling proof is impossible. Even those things most would consider uncontroversial intrinsic worth holders, namely human persons are not immune from attack. Comstock describes a radical form of ecocentrism, which believes that individuals exist for the sake of the whole, and if a species threatens the stability of the system, e.g. nature, then it is morally required to exterminate the threatening species' members (Comstock 2000b, p. 203). In addition, some ecocentrists believe that humans should return to the hunting/gathering social development stage because "to farm at all is to act senselessly, recklessly, and arrogantly" (Comstock 2000b, p. 205). If the number of human beings currently on earth cannot be sustained by hunting/gathering, then it is a moral duty to limit radically human reproduction until this level of sustainability is achieved. If such sustainability is impossible, then since humans are disruptive environmental elements because they cause extinction, pollution, and other destabilizing events, it might well follow that humans have to commit mass murder and suicide in order to protect nature (Marsh 1990, p. 41).

Instead of spending more time and effort on trying to discover a proof capable of convincing all reasonable people of the value or disvalue of an entity instantiating a characteristic, a mental state, or pursuing a telos, in the interests of trying to solve problems facing communities now, it would be best to formulate an argument satisfying the most fundamental principles and moral ideas held by most thoughtful citizens. I propose using the Benefit of the Doubt (BD) principle and a ranking system of intrinsic value capturing the moral intuitions or ideals underlying our ability as persons to have morality in the first place (Gert 1998, Chapter 4; Merriman, http://www.iserv.net/~merriman/natmoral.htm, p. 1).

The Benefit of the Doubt principle[16] assigns intrinsic worth to an object when there is no compelling evidence to prove that the object lacks such value. BD is consistent with the Hinman quote opening this chapter. Instead of believing that objects lack intrinsic value until they are shown to have it, we should adopt the Eastern position of assuming that everything has intrinsic value, until it can be shown not to have it. Put formally,

> **The Benefit of the Doubt principle =df.** In cases in which it is an open question to reasonable people acting as reasonable people as to whether an entity has intrinsic value, assuming the entity has intrinsic value is required provided that the assumption does not unjustifiably harm those entities clearly having intrinsic value.

Basically, the idea is to shift the burden of proof onto those who would deny intrinsic worth, much as we do in most developed countries' criminal system against the prosecution. Those who support expanding the set of what objects deserve

---

[16]The Benefit of the Doubt principle could be considered to be a variation of the Precautionary Principle, but the former does not focus on preventing risks to all those affected by an action. Rather, the risk to be prevented by BD is to the moral agent of the action. If we require that intrinsic value for an object be proven before we give it respect as an intrinsically worthy being, then the moral agent is at risk for doing the wrong thing because he failed to recognize value when he should have.

respect should not have to overcome what might be a biased view held by those in power or by the majority of individuals because they have unfounded and unwarranted preconceptions. It is in fact better overall for objects to be assigned intrinsic worth and be wrong, then to assume they do not have it and be mistaken, as happened in the cases of slavery, the oppression of women, and biological science when people held a Cartesian mechanistic view of animal physiology.

The BD principle is based upon Pascal's Wager, which is a risk analysis of what a rational person should believe.[17] Pascal thought it impossible to prove the existence of a God that is infinitely powerful, good, knowing, and present, but it is still rational to believe in such an entity if he is also the source of moral codes of conduct, judge of actions being in accord or discord with the codes, and rewarder/punisher of good and bad people.

According to Pascal, regardless of his actual existence, belief in this God is rational because it is prudent. If God exists, then the reward is heaven. If God does not, then the person would have lived a better – more ethical – life than he otherwise would have. Belief, then, has a 100% chance of a positive outcome. On the other hand, not believing in God is unlikely to be in the person's rational self-interest. If God exists, then agnostics and atheists receive eternal damnation. If God does not exist, then atheists and agnostics will have saved some time and resources by not praying, going to church, or doing what religious texts require, but they will have lived a less worthy life than if they had believed. According to this line of reasoning, those who do not believe in Pascal's God do not act as ethically toward themselves or others as believers do; hence, their lives are not as good as they would have been through faith and obedience to religious prescriptions on behavior and characters (Pascal 1966, pp. 149–55). Thus, non-belief yields a 100% chance of a negative outcome in this wager. On the grounds of self-interest, it is better to believe.[18]

The Benefit of the Doubt principle and its results are similar to Pascal's Wager. First, in every case of intrinsic value, we can either assume that an object has intrinsic value or it does not. This is a claim about what we can know. On the other hand, the object actually has intrinsic value or it does not. This is an ontological claim about reality; not about knowledge. If we stipulate that something has intrinsic value and it does, then we will incorporate its true value into our moral thinking and judgments as we should. The reasonable person can make accurate calculations of potential utility, while at the same time, every object deserving to be respected is. If we assume intrinsic value, but the object does not have it, then the reasonable person will be mistaken in her utility calculations and she treats things unworthy of respect with respect, but things actually needing to be treated as ends in themselves will be given their proper due.

---

[17]To be more useful, Pascal might have made it a reasonable person standard.

[18]There are insurmountable problems with this argument for belief in God, but there is no time to address them here. Suffice it to say that without any evidence of the last three characteristics Pascal attributes to God, then it is merely non-rational to believe the Wager has any evidentiary weight.

Some might argue that I have being too dismissive of concerns about conflicts between those things assumed contentiously to have intrinsic worth and those things with less controversial classifications of such. Actually, if we assign intrinsic value for things that do not have it, we will most likely act unethically to those things with intrinsic value. That is, if we suppose that trees are ends in themselves, for instance, then we will not have the ability to treat intrinsically valuable people in the manner QCI requires of us for two reasons. First, we would be too tired to do what we ought to do if we spent all of our energies respecting everything we could not adequately prove lacks intrinsic value. The problem is much the same as has been mentioned with the Biblical command to love one's neighbor (agape). Loving one's neighbor in the proper way requires a great deal of work, especially if one's neighborhood is large. If a person devotes himself to agape, he will have little time to care for his or his family's interests. Hence, the agent would have to sacrifice something of great intrinsic value to perform his religious duty. What makes matters worse is if he has to lose something of enormous intrinsic value for something having mere instrumental value, even though there might be many of the latter. If trees do not have intrinsic worth, then the agent would be unable to carry out his duties to himself and others because he was too tired from respecting things that are not ends in themselves.

BD's second possible problem is that respecting everything that potentially has intrinsic worth makes it impossible to respect anything with that value. Janna Thompson claims that if everything has intrinsic value, then the term becomes meaningless. First, if all objects are equal in worth, then there is no reason to select one thing over another (Thompson 2000, p. 577). Furthermore, if all things are deserving of respect because each is equally intrinsically valuable and there is no reason to select one over the other, then respect becomes an empty term. Hence, instead of being able to show the proper level of respect to those things deserving such treatment, "respect" will mean very little, if anything. Something is not an honor if everyone receives an equal amount of it, much the same way it would be meaningless to compete in the Olympics if everyone interested was allowed to compete and win a gold medal regardless of her actual placement in an event.

Making mistakes in utility calculations based on being too inclusive is not of great moral concern because the axiology's hierarchical structure will make most mistakes insignificant on the whole. The intrinsically valuable things' ranking is based upon the probability of an object having intrinsic worth. Persons are in general more valuable than trees; so, we as people will have the same classifications for actions and policies as we did when we assumed trees and plants had instrumental value alone. For example, if it is a choice between killing a tree to save a person and allowing a person to die in order to save a tree, when only one can be done, then it is obvious, *ceteris paribus*, the intrinsic value of the tree needs to be recognized while the person's value trumps its life. The higher the probability, then the more emphasis we should give their interests over those items with a less convincing case for being an end in itself. Therefore, it is unlikely that those things with obvious moral worth will be forsaken in pursuit of the interests of something with less intrinsic value. Of course, we might expend some resources on the wrong things when they could have been better spent on those objects that are ends in themselves, but overall it is permissible for us to presume intrinsic value even when none exists in reality.

Second, it is not at all clear that respecting many different things demands greater effort than that required to respect all persons. If we demand that moral agents treat others in the proper way – which I take to mean they respect the intrinsically valuable individuals, have primarily good intentions, motives, and attitudes – then expanding the circle of respect-worthy objects is easily done. Sentient creatures, living organisms, potential persons, and nature have intrinsic value. The only additional energy a moral agent would have to expend is broadening her mind to recognize the intrinsic value of all relevant things.

Third, assuming that something has value when it does not might be in error, but generally it is good for us to be more, rather than less, inclusive. If we treat more things with the respect intrinsically valuable objects should receive, even when they do not merit it, we can better understand our true importance and place in the universe. If we believe we are not the only things possessing intrinsic value, and are obligated to treat many things beside ourselves with respect, then we refocus from the self alone to the self as part of the overall whole. We have better reason to put aside selfishness when we begin to realize that we are not superior to all other things and that other objects have value beyond that which is determined by how useful they are to us. This is the sole place Regan and other's speciesism hyperbole proves itself useful.[19] By not thinking of ourselves as the only species or set of objects with intrinsic value, we will not overvalue ourselves to the point at which we become arrogant in thinking that everything else either serves us or has no value at all. The new focus forces us to take the interests and value of many other things into account in decision making than we would if we merely thought human beings alone possessed intrinsic value. Hence, we become more empathetic and thoughtful people than we otherwise might have been.

Besides the benefits we individually and as a social community receive by the expanded moral view, other objects profit as well. If people believe too inclusively that certain objects and classes of things have intrinsic value, it becomes more likely that people will take better care of them. On the one hand, mere instrumental goods deserve little respect for being a means to an end and no respect for being an end in themselves. They can be discarded as soon as a better way is found for achieving the same end much the way a dog that has served its master for 20 years can be put down once it is no longer useful, according to Kant (Kant 1999, p. 210). However, if these things are thought to have intrinsic value, then behaviors are more prone to change. Objects that can be owned, such as animals, plants, and land, will be better maintained because they are not considered to be mere instrumental goods. Since people have to respect entities with intrinsic value, they must recognize the value of the objects. Recognizing intrinsic value leads to better treatment of those things with

[19]Specieism is defined along the lines of racism. It is the favoring of one's species over other species on morally irrelevant grounds.

such worth due to the regard with which we automatically treat such objects. Hence, assuming intrinsic value regardless of whether or not it truly exists is guaranteed to lead to positive consequences.

As in Pascal's Wager, there are two other outcome possibilities for the Benefit of the Doubt principle. If we do not assume that something has intrinsic value and it does not, then nothing deserving respect is treated unethically. Moreover, that which does not have intrinsic value is not treated as if it does, and the reasonable person can more accurately calculate utility. This outcome has the positive results of more efficient use of resources – e.g., time will not be spent performing unnecessary utility calculations – and not believing what is false. The person has more true beliefs in her belief set than she would have had she in fact believed that the object had intrinsic value when it did not. There is also the slight danger the person will not give proper attention to something with actual intrinsic value because she is focusing too greatly on something without such value, but as stated earlier that is unlikely to occur given how objects are ranked in this axiology.

On the other hand, if we assume something does not have intrinsic value and it does, then it will be treated unethically. Consider slavery in the United States and elsewhere around the world. When adopted, the US Constitution stated a black man had three-fifths the value of a white man. Hence, in legal disputes, black men's interests were always less important than those of whites. What made matters worse was that no woman was considered to be worth any percentage of a white male. In consequence, the US Founding Fathers and those who followed the prejudices of the time treated blacks and women as they should not. If white men had merely used the Benefit of the Doubt principle for women and blacks, then such moral outrages would much less likely have occurred. If they had assumed that blacks and women have equal intrinsic value to white men, until they had compelling evidence to the contrary, then they would not have denied the former groups the basic freedoms guaranteed to white men by society and government.

It should now be clear why it is better to think incorrectly that an object possesses intrinsic value than not to assume such worth when the object is an end in itself. No one would disagree that things with intrinsic value ought to be respected. If we use BD, then we can fulfill our duties to treat things as ends in themselves when they actually are ends in themselves. We avoid the unethical result of devaluing intrinsically valuable entities when there is no need to do so. Although it is generally suspect to make an appeal to general intuitions, if confronted with the alternatives of treating plants, organisms, and nature as intrinsically valuable things or treating them as only instrumentally worthy, many would select the former, especially if they are attracted to the Precautionary Principle. The latter principle prescribes that people know the risks of an action or policy before it is implemented, rather than implementing it, and then discovering severe problems that would otherwise have been avoided. The Benefit of the Doubt principle also minimizes the risk of not recognizing intrinsic value and acting unethically. It is safer to assume intrinsic worth and be mistaken, then not to assume it and be wrong. In both cases, playing it safe is preferable to risky behavior.

Even though there is great benefit to be garnered from BD's adoption, there are still criteria for its proper application. In order for the BD principle to be plausibly used, then it should render classifications consistent with itself and the already adopted normative principles from PMC.[20] The principle must avoid presuming that an object has intrinsic value when it would cause us to fail in our obligations to satisfy RPU and QCI. Assuming intrinsic value when insufficient evidence exists for rational choice cannot cause us to treat anyone as a mere means or make a reasonable person miscalculate which alternative has an outcome likely to be the best or tied for best. For example, assigning intrinsic worth to artificial objects might cause a reasonable person to believe that saving a great number of them is more important than saving a single person's life. A person might unethically rescue two Vermeers from a burning museum at the cost of a guard's life, for example, when the former have only inherent worth. Great care in reigning in biases and unfounded intuitions, therefore, is called for when using BD.

4.5 Intrinsically Valuable Things and Their Rankings

For all its positives, the Benefit of the Doubt principle will generate a value theory that will make the task of evaluating actions and moral duties more difficult than if a simpler axiology is adopted. If hedonism is true, for example, then the only intrinsic values in the world are experiences of pleasure and pain. All a reasonable person would need to do is to surmise what the possible pleasurable and painful consequences of the various alternative actions will be, and then evaluate them using a cost-benefit analysis. A more complex value theory requires greater work both for the evaluator and anyone asserting the axiology's appropriateness. The complex value theorist has first to defend the inclusion of each class of thing on the list, and then explain how an object's value is calculated, including but not limited to commensurability. For example, W.D. Ross's axiology is, from least to most worthy, desert appropriate pleasure and pain, right opinion and knowledge, and virtuous disposition and action (Ross 1988, pp. 134–41). Ross went on to say that any object from one level is incommensurable, i.e., their values cannot be compared, to an object from any other level (Ross 1988, p. 150). No matter how much desert appropriate pleasure is produced by an action, for example, it can never equal the smallest amount of knowledge or virtue. Other philosophers have also ranked the value of categories based on intrinsic worth and other considerations (Warren 1983, p. 110; Elliott 2000, p. 569; Rolston 1983, p. 588; Thiroux 2004, p. 436; Frey 2000, pp. 440–2). The axiology developed here will follow their lead.

Most, if not all, axiologies depend significantly on their author's intuitions, which can be problematic. Generally what an author already thinks is intrinsically valuable is axiomatic for his ethics. Of course, as has been pointed out previously, reasoning based on intuitions is notorious for being biased in many ways and situations. Janna

[20]Practical Moral Code.

Thompson, for instance, argues against the intuitive claim that there is a universal need for nature which even Kant assumed existed, on the grounds not everyone has a need to see nature in the wilderness (Kant 1999, p. 210; Thompson 2000, p. 579). Paul Taylor disputes what many take as a self-evident truth. According to Taylor, assuming the superiority of human beings is mistaken, in part, because the mere fact that most humans believe homo sapiens is superior is no indication of anything other than human bias to their own kind (Taylor 1986, p. 22). In a number of works, if intuitions are mentioned at all, they are rejected on the grounds of being too prejudiced or unsupported by science (Elliott 2000, p. 574; Comstock 2000b, p. 217). Intuitions, then, are usually considered to provide little, if any, legitimate evidence for a position.

However, some intuitions play a necessary role in ethics, in much the same way axioms do in mathematics. If there are no starting points for a value system, then it is impossible to talk about the value system at all. As Kurt Gödel showed for complex mathematical systems, a system's axioms cannot be defended within the system but are assumed so that there are practical tools to operate within the system. For example, to be intelligible, something as simple as "1 + 1 = 2," must have a stipulated definition of which "1," "+," and "=" are as well as which system is in use. In a binary system, it is impossible for "1 + 1 = 2" because one unit added to one unit actually equals 10. For "1 + 1 = 2" to be true, then the system must be greater than binary. Therefore, in order to get the enterprises of math or ethics off the ground, there must be some vital, un-provable axioms stipulated.

I propose using very general, universal intuitions from environmental ethicists, USDA survey respondents, and other biotechnology stakeholders as axioms. The intuitions are perhaps innate motivations or subjective yet universal valuations (Merriman 2003, p. 1; Callicott 1999, p. 16, respectively). Regardless, they are very basic ideas, such as pain is bad and pleasure is good. Other intuitions included in this axiology are a little more complex, such as deserved pain is better than undeserved pain or the environment has intrinsic worth.

The Benefit of the Doubt principle is very useful in developing the axiology. According to BD, in the face of inadequate evidence, every thing that at least one reasonable environmental ethicist, philosopher, or survey participant has claimed possesses intrinsic value will be considered to have such worth, until the time in which it is compellingly proven not to possess it. However, in the interests of time, I will focus on the least controversial assertions, and temper the more contentious of the claims by creating a ranking appealing to many thoughtful people.[21] Those items having the least controversy will be ranked higher than those that are more controversial. Again the idea is to develop a moral code that will lead to adequate, defensible decisions that all reasonable people can appreciate and understand, but with which they need not agree.

[21] Mill's method of ranking the quality of various pleasures and pains can be expanded for this work. The result is a more democratic process.

4.5.1 Persons

Besides the value of the biosphere as an organic whole, the most intrinsically worthy and least controversially valuable entities, all things considered, are those instantiating the properties of being a person (Guha 2000, pp. 125–6; Rollin 2006, pp. 440–2; Elliott 2000, p. 569; Warren 1983, p. 110; Thompson 2000, p. 577; Feinberg 1979, p. 583; Stearns 1979, p. 611; Frey 2000, pp. 440–2; Baxter 1974, p. 143). That is, an entity instantiating the property of being a person is intrinsically valuable because of the property instantiation. Since dirt, for example, does not have the properties associated with being a person, then it is not intrinsically valuable on these grounds, but might be valuable in some other way, e.g., as a means to an end.

Since being a person is intrinsically valuable, then it is vital to define what such a being is. The fast, simple definition is a person is a moral agent, which means:

> **Moral agent** =df. Any being that possesses those capacities by virtue of which it can act morally or immorally, can have duties and responsibilities, and can be held accountable for what it does. (Taylor 1986, p. 14)

What follows from this definition is that the entity must be able to make judgments, have the ability to engage in moral deliberations and then make choices based upon the deliberations, have the resolve and will-power to implement the choice, and finally, have the capacity to hold herself accountable to others for failing to carry out her choices (Taylor 1986, p. 14). In addition, moral agents can be good or bad depending on which virtues and vices they have. Of course, having these high order faculties entails that there are lower level abilities including but not limited to possessing and using an abstract language, reasoning theoretically, remembering and using her history, recognizing her identity, and having a moral code to apply in deliberations. Furthermore, persons are beings capable of self-fulfillment and happiness. They are creative, active community members, capable of shaping their own destiny in a social context, and having a wide range of skills and knowledge relevant for the purpose of flourishing.

4.5.2 Persons Versus Human Beings

Given the characteristics of what it is to be a person, there is no reason to believe that moral agency and personhood is limited to the set of humans. It is perfectly plausible to believe that there could be other entities in the universe possessing these abilities and having the same intrinsic worth as other person qua person. If it is true that primates or other creatures on Earth are capable of performing the relevant functions, then they too would be persons and of equal moral value to human persons. This is one of the concerns raised about creating human/non-human chimeras for research purposes. If the chimera develops or has the potential to develop the capacities of human persons, then it would be as unethical to use it for clinical trials as it would for human persons and children. Since we value persons too highly to allow them to be treated in the same manner as animals, then researchers must be very careful to discover what the human/non-human chimera's moral status actually is before they can proceed with their work. The main point here is, however, that not all persons need be human beings.

It is also clear that not all human beings are persons, which is something that often gets lost in the debate over value and the environment. Using Conceptualistic Pragmatism from Chapter 3, a human being is any entity having the proper set of chromosomes or DNA recognized as belonging to the species homo sapiens and those characteristics convention attributes to the species. Given this definition, every entity from a conceptus to a deceased homo sapiens is a human being because it has the proper set of traits to be a species member. However, even though concepti and dead humans are human beings, they are not persons because they do not possess the abilities to act as moral agents. For example, neither has the capacity to perform the proper theoretical reasoning required to make actions morally right or wrong.

Although some have claimed that being a human being is intrinsically valuable, the Benefit of the Doubt principle cannot be used correctly to support the assertion. There is compelling reason to reject the claim that species membership is an intrinsically valuable characteristic. *Being a species member* is a morally neutral property many confuse with morally valuable properties, such as being a person, being sentient, or any of the functions that members of a species have. For example, some argue that non-human animals can be sacrificed to preserve human beings because membership in the latter species is more valuable than membership in the former. The problem such thinking encounters is that if there were intrinsic value to possessing species membership, then membership in any species would have an equivalent intrinsic value; therefore each member of a species would have equal worth to any other member of any other species *qua* species membership. Distinctions made between individuals and species would therefore be unjustifiable on the grounds of treating likes as unlike. A sponge being a sponge, for example, would have identical intrinsic worth to a human person being a homo sapiens on these grounds alone. In order to eliminate these inconsistencies and morally repugnant conclusions, it is important to find the actual causes of intrinsic value and what allows us to make ethical distinctions between those entities that are ends in themselves in some way. The axiology developed here attempts to fulfill these goals.

What allows us to classify individual human beings as more or less valuable are the qualities they possess other than being a human, such as deserving punishment, having certain types of consciousness, being alive, being capable of achieving the telos of humans or person, or being a person. For example, a living human baby is more valuable than a dead human adult, not because they are different species, but rather because one is living and the other is not. Hence, in one way, I am endorsing Regan and Taylor's claim that all human beings *qua* human beings have equal intrinsic value, while rejecting what appears to be their position that there are no other moral factors justifying classifying one homo sapiens as being more or less valuable than another (Regan 1989, p. 222; Taylor 1986, p. 77). The axiology from the Benefit of the Doubt principle as employed here and elsewhere is very complex and might appear to draw too fine a line in places, but it seems better to reflect the actual, legitimate distinctions people draw – or should draw on practical grounds – when evaluating various entities' intrinsic value.

4.5.3 *The Intrinsic Value of Persons qua Personhood and Individual Persons*

The intrinsic worth of persons as members of a class and as individuals is the result of a number of features moral agents have. First, any entity that is a person is intrinsically valuable for that reason alone. Increasing each moral agent's intrinsic worth is the possession of other intrinsically valuable characteristics. As will be discussed below, being sentient, being alive, and having the telos/function to be a person and to have a good life as a moral agent all make a person more intrinsically worthy than merely being a person. Moreover, individual moral agents can be more or less valuable based upon certain mental and physical properties they possess at particular times. A person *qua* person is intrinsically valuable because he has the characteristics of moral agency, while an individual person is more or less valuable as an individual person based upon the actual experiences he has now or in the past.

William Frankena claims that a number of properties are intrinsically valuable if they are instantiated in moral agents. The members of Frankena's list are not in themselves ends; rather, "what is intrinsically good is the contemplation or experiencing of them. In themselves, they are inherent rather than intrinsic goods" (Frankena 1973, p. 89). Frankena has recognized that the basic fact that property instantiation, not the properties themselves, have intrinsic worth. Pleasure, for example, can only be real provided that someone is experiencing it. In other words, there must be a state of affairs in which an entity is pleased about something in order for an intrinsically valuable state of affairs to exist.

According to Frankena, there are many different states of affairs types having intrinsic value. His list of intrinsically valuable things is long but well worth consideration:

a. Life, consciousness, and activity.
b. Health and strength.
c. Pleasures and satisfactions of all or certain kinds.
d. Happiness, beatitude, contentment, etc.
e. Truth.
f. Knowledge and true opinion of various kinds, understanding, wisdom.
g. Beauty, harmony, proportion in objects contemplated.
h. Aesthetic experience.
i. Morally good dispositions or virtues.
j. Mutual affection, love, friendship, cooperation.
k. Just distribution of goods and evils.
l. Harmony and proportion in one's own life.
m. Power and experience of achievement.
n. Self-expression.

o. Freedom.
p. Peace, security.
q. Adventure and novelty.
r. Good reputation, honor, esteem, etc. (Frankena 1973, pp. 87–8)

As can be seen from the list, Frankena does not limit himself to particular instances of mental states, such as being pleased about something at a particular time. Instead, the state of being alive is intrinsically valuable. Furthermore, one's life as a whole can be intrinsically valuable, including but not limited to having virtues or dispositions to act ethically in certain ways given certain stimuli.

Moreover, Frankena realizes that a person's intrinsic value does not reside solely within her physical and mental states and properties. Rather, some intrinsic value arises when the moral agent is in her environmental context. For example, a person feeling peace and security is intrinsically valuable for Frankena. However, it is impossible for a person to feel either without taking into account her surroundings. In order for her to feel peaceful and secure, she must think that the entities in the area, including other persons, are of no real threat to her. If she is afraid of bears, for instance, and thinks one is in proximity to her, then she can feel neither beneficial state. Hence the agent's assessment of her situation and its effects upon her are integral to her overall intrinsic worth.

When evaluating a particular person's intrinsic value, it is necessary to look at two more ways of her being valuable through her being a person. Even though each person qua person has identical worth from possessing the property of being a person and having the telos of flourishing as a person that all persons have, individual mental states of an agent can alter the intrinsic worth of the person as a whole, as well as the overall value of her life as a moral agent. I will begin with the former.

In everyday life, we make distinctions between the values of particular experiences. Some experiences are clearly better than others, such as having deserved pleasure when one is recognized for doing a heroic action versus having unmerited pleasure when one is wrongly thought to have performed the same action. Since there are better or worse experiences, it reasonably follows that the individual's overall intrinsic worth can alter according to the experience the person has. Therefore, the value of the individual experiencing one or more of Frankena's intrinsically valuable things at a particular time is greater than if she were not experiencing it at that time.[22] Again, since it is impossible to have the experience without the entity experiencing it, the experiencer must be more intrinsically valuable at the time she is having an intrinsically worthy experience than she would be in different circumstances.

One way of understanding this point more clearly is to consider a metaphor of a beautiful frame. Some frames have been recognized by the art world and others to be aesthetically valuable for their own sakes. The artwork the frame holds can lessen

[22]Since many of the state of affairs Frankena lists can be experienced by higher order animals or all sentient creatures, then their experiencing of them at particular times will make them more intrinsically valuable at that time.

or enhance the beauty of the frame, as the frame can alter the aesthetic quality of the artwork. If a poster of a 1970s TV show was to be placed in the frame, the frame's aesthetic worth as a whole would diminish or be lost entirely. However, if the right painting is selected for the frame, then not only is the aesthetic value of both frame and painting preserved, they are enhanced by becoming a pleasing organic whole, much as the experience and mind of the individual are a totality. Hence, a person, like the frame as a whole, can be more worthy because of the experience she is having at a particular time than someone who is not having a more negative experience.

Certain states of affairs can also make a person *qua* the immediate experiencer intrinsically disvaluable at particular times. That is, certain individuals, such as Hitler, may have negative intrinsic value because he instantiated too many intrinsically disvaluable experiences or contemplations. If Frankena's list of intrinsically valuable things is inverted, then it becomes a list of intrinsically disvaluable things.

a. Death, unconsciousness, and inactivity.
b. Sickliness and weakness.
c. Pains and dissatisfactions of all or certain kinds.
d. Unhappiness, discontentment, etc.
e. Falsity.
f. False opinion of various kinds, lack of understanding.
g. Ugliness, disharmony, disproportion in objects contemplated.
h. Unaesthetic experience.
i. Morally bad dispositions or vices.
j. Mutual dislike, hate, enmity, antagonism.
k. Unjust distribution of goods and evils.
l. Disharmony and lack of proportion in one's own life.
m. Weakness and experience of failure.
n. Lack of self-expression.
o. Loss of freedom.
p. Strife and insecurity.
q. Sameness.
r. Bad reputation, dishonor, lack of esteem, etc.

If a person instantiates one of the properties, then he is less valuable than someone who does not instantiate it, all other things being morally identical. For example, a person who is being vicious is less valuable at that particular time and way than someone who is feeling nothing at all. Since those who instantiate the positive property are more intrinsically valuable than those who do not, it follows that the former are even more valuable than those who instantiate the disvaluable property. If placed on a scale the result is roughly:

(-)	(0)	(+)
___*___	___*___	___*___
Person with disvaluable property	*Person with neutral property*	*Person with valuable property*

The ranking of value follows common perceptions. A person who is instantiating an inherently valuable property is more valuable, *ceteris paribus*, than the same person experiencing nothing or an inherently disvaluable property. According to RPU, overall, we should be trying to maximize intrinsically valuable experiences, while minimizing intrinsically disvaluable ones.[23]

Besides the particular, immediate, discrete experiences a person has, the telos of persons *qua* persons is intrinsically valuable to possess and to fulfill. As has been stated before, the telos or function of a person as a moral agent is to have a good, flourishing life. According to Frankena, a good life is a:

> "mixed life,"...consisting of activities and experiences of the kinds listed earlier, that is of activities and experiences that are enjoyable or both excellent in some degree and enjoyable. [That has] form, not just in the sense of pattern, but in the sense of being inspired by a certain attitude, posture, or "life-style." (Frankena 1973, p. 92)

In other words, happiness is the end of all human life, which might be achieved in a variety of ways. Moreover, a happy life is always a good or ethical life. Hence, the more ethical an individual person is, the more the person flourishes and is able to fulfill her telos as a person, and the more intrinsically valuable as an individual she is.

Moral agents' overall lives can be made intrinsically better or worse in other ways. First, the number and pattern of evil/good experiences and right/wrong actions can alter an agent's intrinsic value. As has already been stated, a life of excellence, which requires certain patterns and life-styles, is more intrinsically valuable than one that is not (Frankena 1973, pp. 93–4). A moral agent who has harmony and proportion in her life, for example, is more valuable than one who lives in disharmony and to extremes, all things considered. In general, those who instantiate more properties in certain patterns and ways will be more valuable than those who do not.

Furthermore, according to Frankena, and this seems to be true in general, the ordering of the intrinsically valuable experiences in each life will affect the overall intrinsic worth of the life. If every valuable state of affairs is grouped at the beginning of the agent's life, while the end is disvaluable, then even if the worthy discrete experiences outweigh the disvaluable discrete experiences, it does not follow that the life was good on the whole. The life has to be evaluated on the grounds of how well the person fulfilled the telos of being a good moral agent, which is determined by whether or not the life was overall happy. Perhaps the person has decided to commit some terrorist action that injures many people in different ways. Being a good moral agent is not only an end of life, it has to be maintained until life ends.

---

[23]What will also change the value of an individual is if an evil person has positive experiences such as pleasure when he does not deserve it. An evil person having a positive experience might make it much lower in intrinsic value than if a good person had the same experience. Furthermore, if a good person has negative experiences when she deserves good ones, then the value of her having the experience is much lower than it would have been if she had deserved it. An evil person receiving a disvaluable experience might be less intrinsically disvaluable than a good person having the same experience (Ross 1988, pp. 136–7).

Given the fact that particular experiences and lives vary, we can justifiably talk about some people being more valuable than others, not as persons qua person which entails equality, but on the grounds of the life they lead as individual persons and their particular experiences' value. Moral saints are more worthy than those who only fulfill the basic minimums required to make them minimally decent moral agents. Immoral devils are less worthy than either the saints or decent agents. However, saints, morally decent agents, and devils have equal intrinsic value as persons qua persons because each is equally a person. The differences between them are the values arising only from their particular life and its experiences.

The axiology is now capable of giving some guidance to applications of RPU and QCI. The ranking of experiences and lives within the category of personhood allows us to make choices on ethical grounds. If people as individuals are equal in intrinsic value, then they must be respected as such. Furthermore, each person as a person, but not necessarily as a particular person, deserves equal respect to any other person as a person. However, we can favor one individual over another based on the intrinsically valuable instantiated properties other than being a person. Saints should be given more respect than ordinary people or unethical devils. Moreover, we ought to behave in such a way that a reasonable person would reasonably believe our action will maximize utility. Therefore, if we are forced to choose between saving two people, one of whom has an excellent life and one that does not, then in order to respect the person with the excellent life and likely produce the best outcome, we should rescue her instead of the person with a less worthy life.

4.5.4 Ranking in Comparison to Other Intrinsically Valuable Things

Although it is still relatively early to examine this issue, it is possible to begin sketching out the intrinsic value of people in comparison to nature. As will be shown below, nature as a whole has intrinsic value, but a difficulty arises in comparing and positioning nature on the hierarchical list of worth according to its ecosystems and the other intrinsically valuable things. My proposal is to order biospheres according to how well they serve the function of providing the other intrinsically valuable entities in nature with the goods each requires to have intrinsically valuable experiences. Furthermore, the ranking will depend on how capably each ecosystem "recognizes" the intrinsic value of every type of entity and its categories. A biosphere having the goods needed for all the intrinsically valuable entities to achieve their ends and instantiate the most intrinsically valuable properties is more valuable than those that do not. Basically, the ordering will be based upon the general principle that the more positive ends obtained and the more positive properties instantiated in the biosphere, then the more instrumentally and intrinsically valuable the biosphere is.

Moreover, biospheres can be sorted according to how well they allow things with greater intrinsic value to achieve their goals and have intrinsically valuable experiences over those things with less value. A biosphere which prevents persons from reaching their goals but is very useful for animals and other intrinsically worthy entities is less valuable than one allowing all intrinsically valuable entities to achieve their teloi. If only one class of entities' teloi can be achieved, then the biosphere satisfying a class of entities with more intrinsic value generally has greater worth than one providing only the goods necessary to achieve the functions of a class of entities with lower intrinsic value. For example, suppose a biosphere subsists in which sentient creatures other than persons cannot fulfill their telos as sentient creatures because resources are being diverted to fulfill the function of persons as persons. The moral agents need to eat the sentient creatures to remain alive. Next imagine there is an alternate biosphere in which the telos of persons qua persons cannot be achieved, but the goal of all sentient creatures can be. The intrinsic worth of the two biospheres is different. The first has greater value than the second because it serves the needs of entities with higher intrinsic value at the same time some of its parts are entities with higher intrinsic value because their higher level teloi have been fulfilled than exist in the second world.[24]

There are limitations on what moral agents may take to fulfill their teloi, have valuable experiences, and so on. The only resources each person requires from the environment are those that permit her to lead a good life. If something is good for an entity, then the entity can flourish (Cahen 2003, p. 137). A good life is not one of excessive expenditure, luxury, or pleasure, but one in which the person achieves the ultimate goal of being a good moral agent and being happy. From Chapter 2 it is clear that good lives require roughly:

1. Enjoyable, aesthetic satisfaction of physical needs,
2. Caring relationships,
3. Satisfying work,
4. Exercise and appropriate care for both the mind and the body,
5. Effective methods of entertainment and relaxation,
6. Security of person, possessions, finances, and psyche, and
7. Education facilitating practical critical reasoning.

Satisfying these seven conditions provides all persons with the opportunity to have individual intrinsically valuable states and overall organization. In addition, working toward these states and organization can give the person many intrinsically valuable experiences. Basically, these goods provide each moral agent the minimal share of commodities required to allow him to take advantage of his opportunities (Baxter 2002, p. 143). If he does not use them to achieve a good life for himself, then he has squandered his opportunity. If he flourishes, then he has fulfilled his telos – he must also maintain the good life if he can – and his duty as a moral agent. In addition, each person has an obligation to see to it that the existence of all human

[24]Of course, if the moral agents are all evil, then the second world will have greater intrinsic value than the former due to the fact that the former are filled with people who are experiencing things that they do not deserve.

persons continues at the high level of intrinsic value necessary for each to have a good life (Stearns 1979, p. 611).

Excess is not morally permissible because it is not necessary for true flourishing of any individual, including moral agents, for two reasons. First, Reasonable Person Utilitarianism states that an action is permissible if a reasonable person would reasonably believe that the action is likely to maximize utility. Using more than one requires to the detriment of other intrinsically valuable things is unlikely to maximize utility. Having a good life is the pinnacle of being a person. Therefore, overindulgence does not enhance life's value, although it can reduce it because entities using resources to excess when they should know better are instantiating disvaluable properties and experiences. In addition, a better distribution of resources will allow more intrinsically valuable entities to fulfill their telos or instantiate intrinsically valuable properties than would occur as a result of some being given excess while others go without the goods they need. Second, the Quasi-Categorical Imperative is unsatisfied. By using or giving excess, persons are being overindulged or treated in a manner not befitting their true worth. They are being elevated beyond their actual merit which does not respect who they are. Furthermore, those entities not receiving what they need, due to too much being given to others, are devalued. Hence, although persons as persons and individual persons are in the vast majority of cases more intrinsically valuable than any other thing, they still have duties to other intrinsically valuable things, and there is still a limit upon what can be sacrificed to satisfy the desires of human persons.

4.5.5 A Different View of the Intrinsic Value of Persons

Some environmental ethicists have argued that human beings or persons have been assigned an exaggerated amount of worth, when there is no evidence to support such a position. Paul Taylor states that there is good rational reason to reject human superiority claims (Taylor 1986, p. 134).[25] The ability to reason, which is used by many to justify placing human persons at value's pinnacle, for Taylor does not confer greater inherent worth on those entities instantiating the property, where inherent worth is what makes an object worthy of respect (Taylor 1986, p. 73).[26] Rather, the ground of all living things' value is the "capacities [and] the fact that those capacities are . . . interrelated functionally so that the organism as a whole can be said to have a good of its own, which it is seeking to realize" (Taylor 1986, p. 148). Furthermore, even if possessing reason conferred greater worth, there is no reason to conclude that anything having it is more intrinsically valuable overall than something without

[25] Taylor's argument begins with him defeating three classical views of human superiority: the Greek's essentialist view of humans; Christian monotheism, and Cartesian Dualism and the claim that animals are automata (Taylor 1986, pp. 134–6). It might have been more helpful in the long run for Taylor to have examined stronger positions than these rather than addressing widely discredited views.

[26] It is unclear whether Taylor's "inherent worth" is equivalent to "intrinsic value."

it. The mere fact that an entity has a greater number of capacities does not entail it has more intrinsic worth than those with fewer abilities.

Although initially plausible, Taylor's claims can be interpreted in at least two different ways. The first is consistent with the axiology being developed here and one of the Practical Moral Principles. If Taylor is merely stating that any entity with inherent worth is worthy of respect and inherent worth is equivalent to intrinsic value, then QCI and Taylor would be in agreement. According to QCI, anything with intrinsic value *because* it has intrinsic value requires the proper esteem form anyone affecting it.

However, if Taylor is making the much stronger claim that being a person is not a more intrinsically valuable characteristic than that conferred by being alive, he is in serious error. The conclusion that human persons are not superior to other life forms does not follow from the three premises he states – as he himself acknowledges – nor can it ever be proven, given the evidentiary standard he requires for those who argue for human person superiority. Taylor claims that:

P1. Humans are members of Earth's community of life.
P2. All living things are related to one another in order of interdependence.
P3. Each organism is a teleological center of life. (Taylor 1986, p. 154)

From these three assertions, Taylor states that it is reasonable but not necessary to conclude that humans are not superior to other forms of life (Taylor 1986, p. 154). However, reasonableness/rationality requires evidence, and there is nothing in the premises inconsistent with the contrary proposition that humans are superior to other forms of life. If the weight of evidence for both propositions is perfectly equal, as seems to be the case here, then it is neither reasonable nor rational to assume either conclusion without further evidence or a principle designed to resolve these matters – such as the Benefit of the Doubt principle.

If Taylor is claiming that human persons have equal intrinsic value to sentient creatures, then he is either trivially right or clearly wrong. First, human persons qua sentient creatures are equal in intrinsic value to any other sentient creature qua sentience because the property is equivalently instantiated in all cases. Intrinsic value does not come in degrees for possession of this property any more than it does for the possession of the property of being alive. Since human persons are equally sentient with all other sentient creatures, then on these grounds alone, QCI demands we respect each equally as sentient.

On the other hand, if Taylor is arguing that humans as moral agents are not more intrinsically valuable than other living entities as whole entities, and should be treated as such in general, then he is grievously mistaken. Taylor's claim – the possession of moral rights by human persons does not entail that there is inequality between moral agents who can possess such rights and other living things – cannot be proven (Taylor 1986, p. 261). Moreover, the Benefit of the Doubt principle, which assumes intrinsic value for anything alleged to have it until the claim has been disproved, and the view of the vast majority of environmental ethicists, philosophers, and survey respondents rejects his position. Moral agents are commonly believed to be more intrinsically valuable than any other living organism.

If most people are asked to choose between saving a hedgehog and a person, they would save the person no matter how nifty the hedgehog is. If they did not, then they would be condemned widely as monsters. Although poll arguments are often mistaken about morality, if we want to create an ethics most citizens are willing to use to solve their and others ethical problems, then it should reflect their views as much as is reasonably possible.

The evidentiary standard Taylor insists upon for those holding that moral agents are intrinsically superior to other living entities creates a problem for him. As has already been stated, ethical systems need axioms – starting points – which by definition cannot be established within the system itself. They must be stipulated because they cannot be proven. If Taylor is asking for any kind of proof beyond the Benefit of the Doubt principle and public sentiment, then he is requiring too high a standard. No argument will be able to show with absolute certainty or beyond all reasonable doubt that anything has intrinsic value. Therefore, discussions of ethics would be mere speculations incapable of providing any rational argument, conclusions, or processes to help people do what is right, be better people or create better societies.

Using the Principle of Charity, I will assume Taylor is merely stating what the axiology under development asserts, and not something rejected by both BD and a large majority of reasonable people.

Every intrinsically valuable entity is a moral subject, including moral agents. Although Taylor's definition of a moral subject – any being with regard to whom others have duties and responsibilities – is too narrow for use here, I want to draw on the framework of it as a principle to distinguish those things with intrinsic value from those without (Taylor 1986, p. 15). In order to do that, the moral subject's definition will be changed to any being with regard to which others have at least *prima facie* obligations and responsibilities. Anything with intrinsic value will always be a moral subject because its value must be recognized in the proper way. However, just because something is a moral subject does not entail that any moral agent will ever have an actual duty or responsibility to it, such as in the case of trees that will never be directly or indirectly affected by moral agents. Taylor seems to recognize this fact when he discusses the moral obligations of moral agents toward entities with inherent worth (Taylor 1986, p. 75). By defining the following intrinsically worthy entities as moral subjects, not only will the differences between moral agents and other entities be preserved, but it is becomes easier for people to respect moral subjects as greater than mere objects or things, thereby helping them to satisfy both QCI and RPU.

4.5.6 Potential Persons

A potential person is any entity having a real chance of becoming a person provided it is given the proper environment in which to achieve that end.[27] In order

[27] Many have taken a great deal of time to discuss the problem of future generations. In order to avoid the controversy, I will stipulate that we have every reason to believe that there will be future

to simplify the axiology by limiting the number of potential people and beginning "real chance's" definition, the potential people that I am referring to are those currently in production. A potential person is any entity, between the conceptus to the moment the entity becomes an actual person, undergoing the process of becoming a person.[28]

What it means to have a "real chance" is difficult to state; however, it is possible to approximate a working definition for the term if it is defined practically. If we are to build a practical ethics, then we must use information from the world as it really is, rather than spending a great deal of time considering examples which are metaphysically possible, but causally impossible or highly improbable, such as anencephalic infants or the human kitten examples used in the abortion debate. When something has a real chance of becoming a person, then becoming a person for that entity is causally possible or can happen given the laws of nature, and the probability of the entity achieving this end is something a reasonable person would think sufficiently high to classify the entity as a potential person.[29] For example, it is causally possible for a kangaroo to have a kangaroo baby, but causally impossible for it to have a human baby or a potential person without some essential supernatural intervention. Given our technology, humans are the only ones able to produce more humans, which can become persons.

The intrinsic value of each potential person arises from four sources: being a potential person, having the telos of a potential person, being alive, and being sentient, when it finally occurs in the development process. The intrinsic value of possessing the characteristics of being sentient and alive will be discussed in the sections on sentient creatures and living organisms. What will be addressed in this section is the value of the former two characteristics setting potential persons apart from actual persons, animals, living organisms as living organisms and all other entities.

Mary Ann Warren recognizes the proper value of potential persons. Although Warren develops her argument in terms of rights, her work captures both why potential persons have intrinsic value and why the value is less than that of persons/moral agents, while being greater than that of animals. First, Warren states that potential persons are learning practical knowledge which will allow them to become fully autonomous individuals (Warren 1983, p. 121). That is, they are developing the skills and abilities which will eventually permit them to be moral agents; this entails

---

generations of moral agents that require resources in order to achieve their moral agent potential. Our duty to them will be to make the world have the capability of producing these resources by advancements in science, reduction of environmental degradation, and all other necessary measures.

[28]Actual persons might at times become potential people, and then resume their former status. Someone is a non-permanent coma is a potential person because he has lost the abilities to be a moral agent. He cannot act; hence, on those grounds alone, he is not an actual person.

[29]It is possible to get sidetracked into a discussion of what the probability must be to satisfy this condition. However, any probability stipulated will be open to the charge of arbitrariness. Since one of the ultimate goals of ethics is to establish a theory that most people would correctly use to solve moral dilemmas, and most of these people are reasonable, the vagueness of acceptable probability levels is not vital to eliminate here.

that they have intrinsically valuable abilities other entities do not have. Non-human animals can and do learn, for example, but are incapable of learning to be a fully autonomous agent.

Second, Warren claims that other moral agents strongly feel that potential persons have higher worth than animals (Warren 1983, p. 121). I take this to mean that persons must be able to recognize intrinsic value in certain entities otherwise the former would not be able to perform their moral duties. Since moral agents understand potential persons as intrinsically valuable, then it follows that potential persons have intrinsic value. Moreover, Warren's claim is consistent with this axiology's approach to incorporate what people actually think is intrinsically valuable, unless there is sufficient evidentiary or moral reason not to do so.

Third, according to Warren, if we do not extend rights (intrinsic value) to potential persons, then too few of them would ever become responsible, morally autonomous adults, which communities need in order for the individuals to flourish at their capacity and to improve society overall with useful members living good lives. In effect, they would not receive the moral entitlements needed to provide them with the goods they require to fully develop their potential personhood (Warren 1983, p. 121). Although the latter seems to make intrinsic value assignment a mere means to the end of creating persons, it is more than solely a practical consideration. Given the greater intrinsic worth in being an actual moral agent, there is still value in being a potential person. In order to increase intrinsic value in the world and show proper respect to the potential person whose telos is to become a person, then if the actions satisfy both RPU and QCI, potential persons should be given whatever goods are necessary to make them full moral agents. In addition, each potential person has intrinsic value which it shares equally with all other potential persons merely because they have this property.

4.5.7 Ranking in Comparison to Other Intrinsically Valuable Things

Although potential persons have intrinsic value, they are less valuable than actual persons for the obvious reason that the former are mere potentialities, while the latter have achieved full moral agent status. The only argument for this claim is based on what most people select when confronted with the moral dilemma of saving a potential person or a person when only one can be rescued. Suppose a pregnant woman living an ethical life cannot safely carry her fetus to term. If she tries to do so, then she will die, but her fetus will be able to survive because of the additional development time within her womb. If the woman aborts now, then she will survive, but the fetus will die.

If we are reasonable persons, then the choice is obvious and tragic.[30] The woman is a moral agent and the fetus is not. The former is a fully integrated member of the

[30] At one of my stem-cell ethics presentations, I met an individual who asserted that a person's life could be taken in order to save a frozen embryo no larger in size than the head of a

community of persons, with all that property entails. To lose her is to lose an entity that is a moral agent, who instantiates all of the qualities of a person, is able to experience everything on Frankena's list of intrinsically valuable experiences, has those experiences, and who has an ordered life. She also not only has the telos of a person, but she is fulfilling it in the proper way. The fetus, on the other hand, has the potential to develop these characteristics, but cannot have done so as of yet, by definition of potential personhood. The fact one exists while the other does not entails the former has intrinsic value the other does not beyond being a potentiality.

On the other hand, potential persons are intrinsically superior to sentient creatures because of the fact the former can become a person. The potential to be a person is more valuable than merely being able to experience pleasure and pain; becoming a person allows the entity to be the most intrinsically valuable sort of thing. Of course, sentient creatures can flourish and have good lives within the scope of their property of being sentient. However, this life at its best cannot match the value of the life of a person at its best. Potential people's worth is increased by their potential. Moral agency or personhood eventually allows the entity to have a good life and experience more intrinsically valuable things than mere sentience can allow. Hence, the potential to be a moral agent confers greater intrinsic value than being sentient.

The intuitive attraction of this claim can be seen in the use of sentient creatures in research intended to help the unborn. It is key component of medical ethics that no medical research is performed on humans before it is successfully tried out on animals to find toxicity levels and other factors that might harm humans. For example, researchers know that surgery to repair defective heart valves and other ailments can be performed successfully on the unborn to better effect than after the fetus has developed as fully as it will in the womb, been delivered, and then is operated upon. This medical procedure could only have been ethically discovered, tested, perfected, and approved through research on animals. Furthermore, society would never allow research to be performed on potential persons without it first being conducted on sentient animals. That is why Phase I testing can only be performed after animal trials have shown a product sufficiently safe for use on humans (World Medical Association 2000, B11).

Of course there are those such as Tom Regan and Peter Singer who might claim that testing on animals before humans is a form of specieism (Regan 1989, p. 221), but it has already been shown that the specieism' charge in these situations is illegitimate. Although it is true that no individual species member as a species member has any greater intrinsic value than any other individual species member as a species member, the justification for using sentient animals is based in part on the fact they do not possess moral agency or potential moral agency. Animal trials can be vindicated by comparing the intrinsic value of other things, such as potential personhood, to sentience, and then drawing conclusions on which takes priority in the situation.

---

pin. This seems a common view in some anti-abortion groups, but not a generally held belief overall.

Therefore, if we claim that it is permissible for animals to be used for testing unjustified on potential persons, we are not necessarily committing specieism. If potential personhood is more valuable than sentience, then sentient creatures can be sacrificed for potential persons.

However, there are limitations to the sacrifice, which are tied into the relative values of the entities involved and the procedures being studied. The only permissible trials are those required so that potential persons can be enabled to achieve their telos of becoming persons and having a good life. Procedures that merely augment – for example, increasing the attractiveness of the fetus – are generally inadequate to justify the intrinsic value loss from the death of sentient animals, their pain and suffering experience states, and their inability to have flourishing lives. This is one of the reasons it is standard to judge the morality of any animal testing on the grounds of its necessity, minimization of pain and suffering, efficient use of animals, and so on. Hence, requiring animal testing in certain cases does not entail that likes are not being treated as unlikes, rather less intrinsically valuable entities are being respectfully used for the benefit of more intrinsically valuable things.

Since it is the potential for having a good life as a person and being a person which makes potential persons have their intrinsic value qua potential personhood, what potential moral agents need from the environment is whatever is necessary to become a person and to have a good life as a potential person and person (Feinberg 1979, pp. 596–7). A new infant does not have the traits necessary to be a moral agent, but he has the capacity to acquire them, and "his inherited potentialities are moving quickly toward actualization even as we watch him" (Ibid.). As long as doing so does not take something an actual person needs for her good life, potential persons require from the environment whatever allows them to achieve their potential, and they should be given these resources. These goods will satisfy the potential person's physical and mental needs, make her life worth living, and help create an individual that can enter into contracts, make promises, be trusted, and be blamed for what would be called the "moral failures" in a person (Feinberg 1979, p. 583). In other words, a moral agent. Hence, a biosphere capable of providing these goods, as well as those required by moral agents, is better, all things considered, than one that can supply one class' needs without fulfilling the others.

4.5.8 Sentient Creatures

Although more controversial than those for the existence of persons or potential persons' intrinsic value, many environmental ethicists and philosophers claim that animals or sentient creatures have intrinsic worth.[31] Callicott, for example, maintains there is a "widely shared intuition that non-human species possess intrinsic value" (Callicott 1989, p. 131). If Callicott is correct, then not only will sentient creatures have intrinsic worth, but living organisms as well. Joel Feinberg takes a

[31] A large number of philosophers, such as Baxter, deny it.

different approach by arguing that animals can have rights because they have or can have interests (Feinberg 1979, p. 587). Such rights might be the right to life, to be left alone, or to not be given unnecessary pain. Lilly-Marlene Russow limits the animal category to sentient creatures, but agrees with Robert Elliott that people have moral duties toward the former because they have interests and aesthetic values (Russow 2002, p. 109; Elliott 1983, p. 572). The question now is what type of interests or characteristics gives sentient creatures their intrinsic value.[32]

As in the cases of potential persons and moral agents, the intrinsic value of a sentient creature qua sentient creature comes from several different sources. First, of course, sentient creatures are intrinsically valuable because they have the ability to feel pleasure or pain. Further value can be found in these entities in whether or not they are actually feeling pleasure or pain at a particular time. Intrinsic value also comes from having a telos of flourishing based upon the presence of pleasure and absence of pain, which is different from the telos of a person qua person for obvious reasons. Furthermore, we can rank the intrinsic value of animals by whether or not they have more or less of the inherently valuable experiences, if they are having a particular feeling at a particular time, and on the fulfillment of their teloi.

All sentient creatures have equal intrinsic value on the grounds of sentience because possession of the quality is conceptual: the creature either has it or not, without reference to gradations of possession. On these grounds, human persons are equal in intrinsic value to any other sentient creature because both posses the sentience quality equally. However, human persons qua persons and sentient creatures qua creatures can have widely divergent values due to differences in other intrinsically valuable properties, experiences, and lives they lead.[33]

4.5.9 Ranking in Comparison to Other Intrinsically Valuable Things

Sentient creatures which are not persons are less valuable than persons because of the fact "they cannot reciprocate the hope and expectation of respect" between moral agents (Warren 1983, p. 119). In other words, although sentient creatures can experience a number of the intrinsically valuable states that persons can, the former cannot feel all of them for the obvious reason that they are not persons. They do not respect each other in more than a metaphorical sense, while persons actually can and do have these mental states. Since sentient creatures have fewer abilities, can experience fewer of the intrinsically valuable states, and merely have a telos of a

---

[32]There are two reasons that I use the term "sentient creatures' rather than "animals." First, not all animals, such as sponges, have sentience. They would have equivalent intrinsic value with any other living thing, but have less value than something that can feel pleasure or pain. The second reason is that something that feels pleasure and pain might not be an animal, although one has not been discovered.

[33]It is clear that not all sentient creatures are persons and not all persons are sentient. Some moral agents are insentient because of rare nervous disorders or other medical abnormalities.

sentient creature, then they are less valuable than persons who can and do experience more of the intrinsically valuable states and have a more valuable telos.

What should also be noted is a significant difference between persons and sentient creatures in regard to pleasure and pain experiences. More precisely, the desert of a person that results from the actions of the person is a factor in the intrinsic value of the person experiencing pleasure or pain, while desert is irrelevant for sentient creatures qua sentient creatures. For moral agents, just desert alters the intrinsic value of the experience of pleasure and pain in the following ways:

1. Positive desert enhances the intrinsic value of pleasure.
2. Negative desert mitigates the intrinsic value of pleasure.
3. Neutral desert neither enhances nor mitigates the intrinsic value of pleasure.
4. Positive desert aggravates the intrinsic disvalue of pain.
5. Negative desert mitigates the intrinsic disvalue of pain.
6. Neutral desert neither enhances nor mitigates the intrinsic disvalue of pain. (Feldman 1986, pp. 12–18)[34]

For sentient creatures the focus is merely on the experience of pleasure or pain, which any sentient creature by definition can have. Since sentience is different from moral agency and moral agency is a necessity for just desert of this kind, the notion of desert for sentient creatures as sentient creatures is meaningless. Desert is relevant only when the individual is capable of deserving something which requires the individual has moral agency, which he has used to do something making him worthy of punishment or reward. Hence, talking about desert for sentient creatures makes as little sense as talking about desert for rocks and trees.

A moral agent's desert can make him even more valuable than he already is or reduce his value to below that of other animals. First, assuming that people and animals feel pleasure in the same way, the value of relevantly similar experiences will be identical for each. For example, the act of sexual congress as a sexual action will generate virtually the same amount of pleasure in any sentient creature qua sentient creature. Hence, human persons and all other sentient creatures have equivalent intrinsic value when experiencing this pleasure. What alters the value of the experience is the desert of the human person. If the person does not deserve the pleasure, then the intrinsic value of him receiving it is less than an animal receiving the pleasure. Neutral desert for a person entails his pleasure's intrinsic value is equivalent to any other sentient creature's. Finally, if the person merits positive desert, then the value of his pleasure is greater than that of an animal having the same experience (Ross 1988, p. 138). Since the former deserves this pleasure for some ethical action he performed, then his receiving what he warrants must be better than an animal merely having pleasure from sexual congress.

[34] W.D. Ross works out a similar but less complex version of these principles (Ross 1988, pp. 136–8).

Pain and desert work much the same way as pleasure. That is, desert can enhance or mitigate the intrinsic disvalue of a particular sentient creature being in a state of pain. Since desert is an empty concept for animals, the value of their being in pain at any time is equivalent to a person who has neutral desert being in pain. If a person merits pain, then, of course, his receiving what he deserves is less disvaluable than if he had neutral desert. Finally, if the person warrants pleasure, but receives pain instead, then the pain's disvalue is much greater than it would have been had a person with neutral desert received it.

Since the present axiology also claims that the organization of the experience episodes a person has in her lifetime have intrinsic value, it should be obvious that some people can be so evil or have done so many bad or wrong actions that their lives are worth less than those of animals. Suppose we are required to choose between the life of a moral monster, such as Hitler, and a cat's. Given a cat's pain and pleasure is equivalent to that of a person deserving neutral desert, and the moral monster deserves pain as punishment, then it follows that the cat's life as a sentient creature is more worthy than that of the moral monster. If we were required to select between which to rescue from a burning building, PMC would tell us to save the cat because it has greater intrinsic value, will likely produce more utility, and will respect worth appropriately. More will be said about the ranking of evil people in relation to other intrinsically valuable entities in the section on Nature as a whole.

Each sentient creature needs two things from the environment. First, since they are sentient, these types of creatures need pleasurable experiences when possible and enough painful experiences to teach them appropriate lessons so that they can flourish. The pain they experience is not good in and of itself, but rather it is an instrumental good to the fulfillment of the sentient creatures' telos. Based on this fact, the amount of pain any particular animal suffers should only be enough so that they learn the lessons they need to in order to flourish. More would merely harm them for no good reason to themselves. Finally, in those situations in which pleasure is impossible, then sentient creatures require neutral rather than painful experiences, if neutral experiences are achievable.

Furthermore, in order to have the best life, every sentient creature needs the minimization of pain and the maximization of pleasure as far as is practical. When pursuing a sentient creature's good life, it is vital to remember that pain minimization is more important than pleasure maximization. The imbalance results due to its being possible to have an acceptable life without pleasure, but impossible to achieve that life with constant or excessive pain. Sentient creatures can survive and reproduce – the teloi of living organisms – even if they do not receive pleasure, but fulfilling those two tasks is impossible if they are always suffering. As anyone who has suffered a migraine can attest, pain reduces or eliminates a sentient being's ability to do much of anything except try to escape the negative state of affairs. Therefore, the best type of environment for sentient creatures is that which allows them to fulfill their functions and experience as much pleasure as possible, while at the same time serving the needs of the more intrinsically valuable entities.

4.5.10 Living Organisms/Entities

According to a number of philosophers and environmental ethicists, the mere fact something is alive entails that it has intrinsic value (Regan 1982, p. 167; Sessions and Devall 1990, p. 312; Elliott 2000, pp. 572–3; Rolston 2000, pp. 588, 592–5; Callicott 1989, p. 161).[35] A living thing, therefore, is more valuable than if it merely existed as a rock or any other non-living thing.[36] From the difference in intrinsic value it follows that if there is a choice between saving from destruction something alive and a non-living thing, then choosing the non-living thing qua non-living thing would cause a loss of intrinsic worth that would not have occurred if the living entity was saved. Furthermore, if being alive makes an entity more valuable than merely existing, then all living organisms as living organisms will have equal value. Human persons and plants will be equally valuable in regard to being alive, but can have different values on other grounds.

There is an intuitively appealing example in support of living organisms as living organisms' intrinsic value. In "Why We Have No Obligations to Animals," Kant writes admiringly of Leibniz's behavior toward a tiny worm the latter had been studying. According to Kant, Leibniz "carefully replaced it with its leaf on a tree so that it should not come to harm through any act of his" (Kant 1999, p. 210). Although Kant believes that moral agents have only direct duties to other moral agents, it is clear that he thought there is a natural feeling for living organisms. Leibniz would have been sorry for undue harm to befall the entity, which implies he believed that a living worm is better than a dead one. Kant's example shows it is at least plausible for reasonable people to think that dead organisms are less valuable than living organisms because the latter have the property of life.

Other philosophers have been more explicit in their claim that living organisms have worth in and of themselves. According to Tom Regan, all individuals have equal intrinsic value provided that they fulfill certain conditions. First, the individual must be an "experiencing subject of a life, a conscious creature having an individual welfare that has importance to [it] whatever [its] usefulness to others" (Regan 1989, p. 220). Although it might seem clear that Regan is only referring to sentient creatures being experiencing individuals, he also claims that rocks, rivers, trees, and glaciers might qualify as well, although we might never be able to prove they have that status (Regan 1989, p. 222). If entities appearing to most as non-sentient are actually experiencing individuals, then it follows that living organisms, such as plants, sponges, and other entities unable to feel pleasure and pain have intrinsic value for Regan as well. Moreover, each individual has equal worth from

[35]Rollin claims that non-sentient things neither have intrinsic or inherent value nor are they the direct objects of moral concern. His reasoning seems to be based on the possible bad result that if nature had such value, then some people would state that nature is perfect as it is, and then try to keep it in stasis (Rollin 2006, p. 151). However, it is hard to see how intrinsic value equates to perfection in any way; so Rollin's concern appears unwarranted.

[36]Teitel claims that all genetic carriers have worth (Teitel 2002, p. 32). It is difficult to know if dead genetic carriers have worth or the same worth as living ones.

being such an individual to any other entity with the identical characteristic (Regan 1989, p. 220). Using the Benefit of the Doubt principle and common conceptions of the value of living things, we can reasonably conclude that living organisms are intrinsically valuable *qua* being a living organism.

The intrinsic worth these entities have comes from instantiating the property of being alive and having the telos of flourishing as a living entity. Aristotle's vegetative soul provides a useful definition of what flourishing is for this class of things. According to Aristotle the vegetative soul has the function of survival and reproduction (Aristotle 1941a 413a:30–34). Hence, living entities, such as human persons, animals, and plants, have intrinsic value for at least two reasons: being alive and having the vegetative telos.[37]

The intrinsic value of a living entity qua living entity can vary according to how well the entity fulfills its teloi. The better an organism is at surviving and reproducing in its environment, then the more intrinsically valuable it is. The less suited the organism, the less valuable it is intrinsically as a living organism. Those entities, for instance, genetically mutated so that they cannot procreate or their chance of dying is greater than what is normal for their species are less valuable than members having nominal survival and reproduction chances.

4.5.11 Ranking in Comparison to Other Intrinsically Valuable Things

Of course, at this level, all living things in nature as biosphere are going to have equal intrinsic value. Each living organism from virus to human person shares equally in being alive, although their individual intrinsic value might be vastly different based upon other characteristics the entity instantiates, such as being sentient. The result is if we are comparing the value of a human person qua living organism to that of a plant qua living organism, then we would have to conclude the two's intrinsic value is identical for this characteristic.

4.5.12 The World's Ecosystem (Nature) or the Biosphere as a Whole

Many environmental ethicists give some sort of moral status or value to nature as a whole or its parts, which can be useful in formulating an ethical theory or principle making nature part of moral decision making (Elliott 2000, p. 572; Snyder 1990,

---

[37]Regan attempts to show that non-sentient objects, such as cars, can have inherent goodness. An entity has an inherent good if it has a teleological end, or a good X is something that fulfills the definition of being an X. Some cars would be better than others, even if no one is interested in them, due to their ability to fulfill the definition of being a car (Regan 1982, pp. 176–80). Of course, the problem with such an argument is that it equivocates on "good." A good car has nothing to do with morality or being morally good, rather a good car is one that fulfills the definition of being a useful car. If Regan's function definition is correct, then he would have to accept that Lee Harvey Oswald was a good presidential assassin because he was successful.

p. 234; Nash 1990, p. 266; O'Neill 2003, p. 132; Sessions and Devall 1990, pp. 311–12). Of these ethicists, some are ecocentrists or holists, who believe that the ecosystem and all its parts have intrinsic worth which must be recognized and respected (Callicott 1999, pp. 273–5; Hinman 2000, p. 562; Thiroux, p. 427). Other holists hold that Gaia is a living organism,[38] species and populations, not individuals, are where intrinsic value resides, or they take a neo-vitalist stance that life is something beyond mere physics and chemistry (Volk 2003). Holmes Rolston III, for example, contends that the ecosystem as a whole has intrinsic value even if no human being existed (Rolston 1983, pp. 151–2, 2000, p. 602, 2003, p. 144, 1994, pp. 176–7).

Other environmental ethicists are biocentrists, who merely claim that all living things, which might or might not include living communities, are owed respect (Hinman 2000, p. 561). Of course, it is easier to prove biocentrism true than ecocentrism because the latter has much greater scope and often counter-intuitive assertions. However, what is important for the practical axiology is to be useful for generating consensus or defensible positions any thoughtful person can accept as reasonable on difficult ethical issues, e.g., new technology. In pursuit of this goal, it becomes mandatory to incorporate as many of these ethicists' beliefs as sensible, while remaining consistent with what other thoughtful people have said, including those who argue that nature is merely good as a means to the ends in themselves.

In the environmental ethics canon, nature can have two types of value: intrinsic and instrumental (Russow 2002, p. 95). In order to understand what is a good system and what is a bad one, we need to incorporate both worth types into the definitions of a good or bad biosphere. A good biosphere is one allowing each individual member – in order of its intrinsic value rank – at least the bare minimum it requires to achieve its basic needs and the opportunity to satisfy its objective and subjective potentials, with human person having a minimal utility level life.[39] A better biosphere will have the same characteristics in addition to supporting everyone having a maximal level of life and flourishing. Moreover, the number of intrinsically valuable experiences and lives a biosphere can support are also factors in its value calculation. The reason is obvious. The more individuals having more intrinsically valuable experiences or instantiating more inherently valuable properties, such as those in Frankena's list, then the more intrinsically valuable the biosphere must be, *ceteris paribus*.

Biospheres' intrinsic worth can be further evaluated based on their stability, integrity, diversity, and beauty (Nash 1990, pp. 263–6; Rolston 2000, p. 601; Russow 2002, p. 109; Sagoff 2004, p. 8 and 2008, p. 207). Suppose a particular global environment permits its members to achieve their basic needs and offers

[38]Tyler Volk argues that Gaia has a physiology much like a breathing organism.

[39]Sandler states that virtue is the key to evaluating the environmental impact of new agricultural biotechnology. If the technology negatively affects the environment's ability to produce what is needed for the development and maintenance of virtue or if the technology violates any of the relevant virtues for human/natural environment interactions, then it is worse than technology that does not have these bad results (Sandler 2005, pp. 220–1). Although the ability to sustain virtues is part of the overall assessment of an environment's value, as stated before, good lives require more than virtue.

opportunities to satisfy each entity's telos, but only fleetingly. Assume a different system has the same characteristics, but is stable. In this case, the latter entity would be more intrinsically valuable than the former because of its long term worth as a whole. Furthermore, nature's aesthetic inherent value, the intrinsic values of the individual experiences of that beauty, and the value of those entities having a better life because of the beauty experiences will also affect the biosphere's value as a whole. The more beautiful a system is, then the more it will cause people to have intrinsically worthy experiences immediately accompanying contemplation of the aesthetically pleasing. Hence, although beauty is not intrinsically valuable on its own, it does play an important role in nature's intrinsic value. The values contributed by a system's integrity and diversity will work the same way. In general, the greater they are, the more valuable the biosphere.

Although a stable nature as a whole is intrinsically important, it is clear that maintaining stability does not take precedence over the needs of something that is a person, potential person, sentient, or living. If it is a choice between the biosphere remaining static and one of the latter four having a need met – admittedly an artificial example – then the latter should be served over the former because a new, stable environment with all necessities satisfied is superior to a static environment in which intrinsically worthy individuals are injured. That is, if the stability, integrity, and diversity of nature have to be harmed in order to satisfy the flourishing requirements of an intrinsically valuable entity, such as a person, potential person, or sentient creature, then the system must be altered accordingly.

There are numerous instances of being forced to choose between stability and other significant environmental elements and other intrinsically valuable objects. For example, John Helgeland of North Dakota State University recounted when he and another hunter were asked by a poor farmer to help control the prairie dog population on his farm. Since they had no natural predators in the area, the prairie dogs had become overabundant. In consequence, they had denuded the farmer's fields of vegetation, making it very difficult for him to support his family. The two hunters killed so many prairie dogs, the farmer's family had to remove them in wheelbarrows. The following year, as a result of the slaughter of the prairie dog colony, the farm had recovered enough to support the family.

The example's point is twofold. First, it shows that the intrinsic value of biosystems and spheres can be compared according to their intrinsic values.[40] Although the system with an overabundant number of prairie dogs had many more entities with intrinsic value, the system itself was less valuable than one with fewer of the animals. The latter has stability and integrity the former does not. Due to the uncontrolled population, the prairie dog community was destroying what would have otherwise been a renewable resource for them and other intrinsically valuable entities.

[40] If Janna Thompson is referring to instrumental value when she claims that a degraded system is not bad if it will result in an even greater state, then she is correct. However, there is no reason to believe that the same holds true of intrinsic value. We can and do compare the degraded state to the greater state to determine which is more intrinsically valuable, which of course is true by definition.

The example's second point is that an ecosystem which can support people in such a way they can have what is necessary to have a good life is more valuable than a system that merely allows more entities to stay alive. If we were merely concerned with the intrinsic value of the parts of a system, then we should encourage as many intrinsically valuable things to exist there as possible, as long as their lives meet the subsistence level of life.[41] If the value of the parts of the system are equivalent to the value of the system as a whole, then the more intrinsically worthy entities we add, *ceteris paribus*, the more valuable the system – at least until that moment when it is destroyed from overpopulation.

However, there is a significant problem with assuming a cardinal calculus for intrinsic value. We can and should make distinctions about the worth of biosystems, a biosphere, or nature as a whole. The world in which intrinsically significant things have more valuable lives is better than that in which physical needs are barely met but there are a lot more intrinsically valuable parts. Although the more valuable lives that there are does generally increase the value of the world as a whole, it does not entail that the world's value is equivalent to the sum of its parts. Nature is an organic whole with emergent properties, it can affect its parts, and the organic whole's parts are systematically related and internally related (Callicott 1999, pp. 273–5). That is, the organic whole with all of its components and their relations to each other creates something with its own intrinsic value which, much like the value of human person's parts to the human as a whole, is not necessarily equal to nor proportional to the sum of its parts.

When trying to calculate a part of nature or nature as a whole's intrinsic worth, the error many philosophers and environmental ethicists make is separating people from the biospheres in which they live or the biosphere the people live in from the people, themselves (Glaeser 1995, p. 143). First, people are not independent of nature for they cannot live without it (Vogt 1990, p. 168). In addition, potential people, sentient creatures, and living things cannot be removed from their environment, nor is nature with these entities that exist in it distinct from the entities. As Warren states:

> Because of the holistic nature of biotic systems, it is impossible to determine the value of an organism simply by considering its individual moral rights [intrinsic value]: we must also consider its relationships to other parts of the system. (Warren 1983, pp. 125–6)

Another way of characterizing this ontological concept is to think of the interrelations of beings and nature as an "internal relation of [a being] to his surroundings based on the integrative interdependence and a harmony between [each being] and the world" (Cheng 2006, p. 201). When we evaluate the actual intrinsic value of humans, animals, and other entities, it must always be in their setting as integral parts of nature.

Second, nature as an organic whole's overall value cannot be measured without being aware of its parts and their relations to each other. As G.E. Moore states, "It is plain that the existence of any such [intrinsically valuable] part is a necessary condition for the existence of that good which is constituted by the whole" (Moore

[41] See Chapter 2.

1988, p. 29). In other words, the whole's value would not exist nor would nature have the identity it does without the intrinsically valuable components. Besides all living entities needing the environment, nature as we know it *would not be* nature as we know it without humans, animals, and the other component entities of which it is, in part, comprised. Therefore, asking if nature has intrinsic value independent of either human persons or its other intrinsically valuable components is necessarily fruitless.

Rolston mistakenly claims otherwise. He states that "the possibility of valuations is carried to us by evolutionary and ecological natural history and such nature is already valuable before humans arrive to evaluate what is taking place" (Rolston 2000, p. 602).[42] The problem with this assertion is that if human persons do not exist, then it is impossible to know if intrinsic value subsists in the biosphere. The only way we can examine such an entity is by entering it in some way to evaluate it; hence, there is always some form of human person observation and assessment in even those cases in which the example is a world without human evaluators.[43] Moreover, as physics have shown us, when we observe the universe, we change it, and it changes us. Our mere existence as moral agents qua moral agents creates a new entity (Callicott 1989, pp. 161–9). Hence, it is impossible to value a world unless we do it from the viewpoint of a human person valuer, and thereby alter it by the observation and evaluation (Callicott 1999, p. 224).[44]

While it is possible to look at the parts of nature as a whole to determine why each has intrinsic worth, e.g. sentient creatures and sentience, and rank the intrinsic value of individuals and groups based upon their possession of the qualities affecting intrinsic value, e.g., consciousness and individual desert, it does not follow that nature as a whole has a different intrinsic value on its own. Nor does it follow that the intrinsic worth of the whole equals or is proportional to the sum of the parts (Moore 1988, p. 28).[45] Nature as a whole's intrinsic value could be more, less, or equal to the total of its parts depending on what the components are and how they are related to each other.

Given this approaches' non-linearity, we must find other ways of trying to capture nature's organic wholeness. What metaphor can help us understand nature's value?

---

[42]Marti Kheel adopts a somewhat similar argument against Rolston. Her care ethics for human and non-human animals rejects Rolston's attempt to transcend "personal ties of affection...[and] our self-centered 'animal' natures, in favor of a more expansive awareness...[and] renouncing 'sentimental' ties to other-than-human animals" (Kheel 2008, p. 227).

[43]This problem is why G.E. Moore's isolation test for intrinsic value is impractical. We can never have the object under consideration totally isolated because of the fact it is under consideration (Moore 1988, Section 112).

[44]Many people assume that human beings are ethical subjects and only human subjects can establish value (Rolston 2000, p. 588). Although the assumption is true in the sense that people are intrinsically valuable and establish value in their own lives and for other things, there is no reason to assume the mere fact we can examine the parts of a whole independently and explain a bit about a value system tells us anything about the value of the whole.

[45]Warren states that it is more intuitively plausible to say that non-sentient things do not have intrinsic value (Warren 1983, p. 128).

Although it is clear that nature is not a person because it fails to instantiate the necessary characteristics of personhood, it can be compared usefully to a properly functioning human body (Moore 1988, Section 21). In order for any of us to be the person we are, we must have an integrated, working body. Blood must flow so that oxygen can be delivered to cells. The heart has to pump the blood, the stomach and intestines must digest food for energy, and so on for all of the other bodily entities, their relations to others and their functions. We can fully understand the workings of each part of the body or even the working of the body as a whole without ever once understanding what it means to be a person or the value of such an entity. The working body and person are different in a significant way from the parts of which it is comprised.[46] The former are organic wholes, while the latter are essential components to a working body.

If organic wholes have intrinsic value which might neither equal nor be proportional to the sum of its parts, then using the Benefit of the Doubt principle, we must assume that nature as a whole has intrinsic value until it can be shown that it does not.[47] What is in favor of the assumption is that the parts of nature as a whole do indeed have intrinsic value of their own, and we cannot understand them fully without putting the parts in their proper context, which is as part of nature. That is, we cannot know the true value of a human person, potential person, sentient creature, or living organism without placing the entity into it environmental context. For example, a plant is a desired plant in one context and a noxious weed in another. Nor can we understand the value of nature without its individual components being placed in their contexts.

Given that nature has intrinsic value because of the existence of evaluators and its being a specific organic whole, it is now possible to state a plausible nature principle of sorts which is consistent with the views expressed by the USDA survey participants, the professional codes of conduct, and a variety of philosophers and environmental ethicists.

> **Nature Principle:** Nature or the biosphere as a whole has intrinsic and instrumental values that are moral factors, which must be taken into consideration in any adequate decision procedure on technology or other actions that affect the biosphere.

Incorporating nature's intrinsic value into ethical calculations is to assume that nature's value imposes an additional variable to be factored into ethical reasoning, moral theories, and principles. Human persons have a duty to make nature capable of supporting as many intrinsically valuable entities as possible, with all that entails, including but not limited to, maximizing the number of intrinsically

---

[46]The reverse can also be true. The value of an entity as a whole need not be the same as its parts. Janna Thompson argues that each person has intrinsic value, but it would be a distribution fallacy to believe that each part of the person also has intrinsic value (Thompson 1983, pp. 580–1). In other words, objects having no intrinsic value can be put together in such a way the whole does have intrinsic value on its own.

[47]Lehman claims that a proper understanding of the parts of a whole is best done through understanding the whole itself (Lehman 1995, pp. 118–22).

valuable experiences, instantiations of inherently valuable properties, and good, flourishing lives. At the same time, moral agents must maintain the biosphere's stability, diversity, and integrity in so far as it is required to maximize the above and respect all things with intrinsic value.

Reasonable Person Utilitarianism and the Quasi Categorical Imperative are the controls in using the Nature Principle in moral evaluations. When applying the principle, we first see how an action will affect those entities at higher levels of intrinsic value. If the action serves or is neutral in regard to their interests, then it is morally right – perhaps even obligatory – to do what is in the interest of the environment as a whole, as long as doing so helps or does not hinder goal achievement for those higher level entities. If it would be best to preserve the integrity, stability, and diversity of nature for the sake of all intrinsically valuable entities, then RPU and QCI declare such actions are morally right. Moreover, nature can be altered only so long as a reasonable person would reasonably perceive such as action as likely to produce the best results and everything affected by the action is treated with the respect it warrants. Hence, nature cannot be harmed for trivial reasons, such as providing goods for human persons in excess of what they need for a good life.

Nature's instrumental value is dependent upon how well it helps to achieve the ends of persons, potential persons, sentient creatures, living things, and itself. The value of each level in the axiological list is different from the other levels' value, with the level of persons being the most valuable of nature's components and having the highest priority for the pursuit of its interests, while the biosphere is the most intrinsically valuable thing in general but with the lowest priority for pursuit of its interests. If there is a conflict between levels for resources or how the environment should be evaluated for stability and integrity, then the highest priority level involved in the conflict triumphs over the lower, all things considered. Some exceptions are evil people deserving severe punishment having their interest of achieving a good life for persons *qua* persons trumped by the interests of a potential person, sentient creature, living organism, or nature itself. Furthermore, if a biosphere had some sort of mechanism which human persons could use to enable all other living things to survive at the cost of moral monsters dying, then it would be a far better biosphere than one without it, according to RPU and QCI. The moral monsters have far less moral worth than good people, potential persons, and so on; therefore, a world in which the formers termination causes more of the latter to exist and flourish must be better than one in which moral monsters are alive.

4.6 Is Nature or the Biosphere Alive?

Not many people claim that nature is a living entity on moral par with human persons. Gary Snyder, however, maintains that nature is as alive as any person and has intelligence in its biomass, which he defines as stored information (Snyder 1990, p. 234). Furthermore, Snyder contends that human persons and nonhumans, for example, mountains, should be represented equally in a democracy because they have interests to pursue (Ibid.). In other words, mere existence creates equivalent moral worth in all things that must be taken into account when making decisions affecting any existing thing. This would be an example of taking the Benefit of the Doubt principle to its logical extreme.

The equality position can be rejected using an empirical question and William Baxter's argument against flora and fauna being moral agents. First, it is not at all clear that nature is alive in the way each person with the organizing principle of mental life is. That is, nature is not a moral agent in the fashion that people have been described previously. We cannot hold nature accountable for wrong actions because it is not something capable of acting, although actions and physical events do occur within it.

Second, in order to be represented in a democracy, an entity must be capable of voting. Capable in this sense does not mean suffrage for flora and fauna in which the society creates a law giving such entities a legal right to vote, rather it is the mental ability to make a choice and the ability to signal to others what the decision is Baxter (2002, p. 144). The former is as useless as giving all men the right to be pregnant in the same mode that women can be. Although it treats everyone equally, it is a distracting waste of effort that could be better spent on supporting more plausible candidates. Since non-persons are incapable of at least one necessary criterion for citizenship, then it does not make sense to claim they deserve equal representation.

Others do not go so far as to assert that all existing things have moral worth, but they do question the supremacy of human person intrinsic worth or the priority of pursuing their interests over that of others. The deep ecologists, George Sessions and Bill Devall, make a similar claim when they write, "We can make no firm ontological divide in the field of existence: that there is no bifurcation in reality between human and non-human realms ... to the extent that we perceive boundaries, we fall short of deep ecological consciousness" (Sessions and Devall 1990, p. 311). Furthermore, they hold that people and living objects are equal and have an equal right to flourish (Ibid., p. 312). The fact that human persons as moral agents are capable of much more than other living organisms, e.g., rationality, use of language, and being ethical, in no way confers greater intrinsic value on them or any priority to fulfilling their needs.

Janna Thompson maintains that those who equate human persons to animals, biosystems, or nature are hypocritical. According to Thompson, the actions of those who profess a love of nature often show that they value their own happiness and welfare much more than nature's intrinsic value. If they truly believed the environment is equal to any person, then they should not value their goods over the goods of non-human entities. Since human beings leave such enormous "footprints" on the environment and other creatures do not, then on the grounds of equality, humans should cease acting in the ways that they do. Deep ecologists, because of their fundamental beliefs, ought to act as role models of right living. However, since deep ecologists do not sacrifice their commodities for non-human goods, then there is no legitimate reason to doubt that they believe they are more important than the others. Hence, even those who make equality claims seem to actually believe that there are rankings of intrinsic value, with human persons being less intrinsically valuable but more important than nature (Thompson 1983, p. 91).

4.7 Arguments Against Nature's Intrinsic Value

Other philosophers and environmental ethicists have argued that nature is merely instrumentally valuable. That is, the environment does have value as a means to an end, but it is not intrinsically valuable. I will consider some of the leading proponents of this position below.

Comstock maintains that there are two problems with assigning intrinsic value to biosystems. First, the goals, or teloi, of biosystems have to be externally directed because they do not have a unifying mental consciousness in any way similar to that of moral agents. Comstock writes that the intrinsic value of an object depends on whether the object's function is internally directed. Those entities with externally directed endpoints can only be instrumentally valuable things (Comstock 2000b, p. 208). Comstock is not alone in his thinking. William Baxter and Stephen Clark argue along the same lines (Baxter 2002, p. 144; Clark 1994, p. 114).

Comstock's second argument is that biospheres have spatial identity problems, including but not limited to the vague boundaries of the system, which cause them not to be ends in themselves (Comstock 2000b, p. 210). That is, "the problem of identity is that the boundaries and 'natural' equilibrium states of the ecosystem are so difficult to identify ... ecosystems have several 'identities'" (Comstock 2000b, p. 212). Generally, the borders are arbitrarily set by moral agents, who might or might not understand the true nature of them, rather than there being any actual or real limits all can clearly identify. For example, a conservation group might consider a specific wilderness area, such as a rain forest, a discrete biosystem, while others do not. However, as Comstock points out, there are outside objects entering the system at all times and internal objects leaving it. Whenever the wind blows, rain falls, or water migrates, elements migrate into or out of the system. In effect, in the real world, there is no self-contained entity which is a biosystem.

Although not a perfect solution, the first and second problems can be resolved in moral decision procedures if we limit discussion of nature to the biosphere as a whole, as I have been doing. Comstock is correct that it is impossible to have discrete identities for biosystems as part of the overall whole; hence, it is impossible for them to be intrinsically valuable qua biosystem. However, if the biosphere as a whole is examined, then there are fewer problems. First, it is a discrete enough entity, even though objects, such as dust and light, enter and leave it. But objects, such as food, enter and leave moral agents; so, the fact that the Earth is not a closed system should not unduly concern us. Second, there cannot be an external direction in the biosphere other than those imposed by natural laws and objects entering it. No outside moral agent seems to be altering it to fit her purposes.

Moreover, the biosphere can be internally directed, not in the same way that moral agents are but in a manner recognizable to business ethicists. The biosphere could be considered analogous to a business entity, such as a corporation. We consider businesses to be moral entities, even though they are not organized the way persons are. Businesses, social organizations, and nations are merely the people who comprise them, the rules organizing and governing them, and their assets and deficits (Feinberg 1979, p. 591). The environment will be less of a moral entity than a business or government because it is made up of so much more than an individual business, no matter how large and international the latter is, and the former has less organization and control on the part of the moral agents it contains than does the discrete corporation. However, nature still has parts that are moral agents, and, of course, people have moral duties to the moral entity of which they are members, much as they do in business entities.

Since we often talk about employees' duties to achieve the internally and externally created goals of efficient competition, it makes practical sense to discuss people's duties to the environment in much the same way. It is possible for individual persons acting in their "official capacities" to have obligations to the corporation and all agents affected by their official actions. Hence, it is possible for individual people to have moral duties in their official capacities as moral agents within the society or nature (Feinberg 1979, p. 591). In business, it is a fundamental employee obligation to maintain the corporation's health or to act in its best interests. Similarly, some of each moral agent's duties are to help the environment sustain the intrinsically valuable components and the whole in the ranked order of their value and importance, while still respecting the integrity, beauty, diversity, and stability of nature as a whole.[48]

A second potential attack on the assignment of intrinsic value of nature comes from Lilly-Marlene Russow's reminder that everyone who makes an assertion shoulders the burden of proof for it. If a person claims a proposition is true, then it is his duty to prove its truth rather than someone else's obligation to show the statement is false. Therefore, those who assert that nature has intrinsic value are obligated to justify it, rather than someone else having to disprove the claim.[49] The Benefit of the Doubt principle, on these grounds, should not assign intrinsic value to an individual or class until it has been shown to have that value. Given the reasonable person standard and the requirement to take evidence seriously, this criticism has the potential to pose serious problems for PMC.

Since Russow's argument is illustrative of both the difficulty in proving the existence of intrinsic value and an unachievable proof standard, I will take the time to formalize it here. Russow states:

P1. If nature has intrinsic value, then harming nature directly by decreasing or destroying its intrinsic value must be possible.
P2. The examples surveyed to date rest on harms to individual sentient creatures rather than direct harms to nature.
P3. Absent a new counter-example, the so-called harms to nature are more fundamentally understood as harms to individual sentient creatures or diminution of instrumental value as judged by those creatures.

---

[48]Inanimate objects, because of their complexity, aesthetics and mere existence, can have intrinsic value according to the Benefit of the Doubt argument, but it is not really a useful notion (Elliott 2000, p. 573). A rock has as much intrinsic value as the dust into which it is ground for paving. The only possible differences would be if there is intrinsic value from aesthetic value or potential aesthetic value. The best way of thinking of value for these types of things is to say that they are intrinsically valuable for the ecosystem

[49]Stephen Clark has the same position (Clark 2006, p. 184).

P4. *There exists no convincing argument at present that nature can be harmed.*
C. There exists no convincing argument at present that nature has intrinsic value. (Russow 2002, pp. 111–12)

Russow's gold standard of establishing a claim does not require the absolute certainty of Descartes' Foundationalism. Rather her criterion seems to be located at the level of non-deductive inference. A good argument of this type's conclusion probably and plausibly follows from its premises, but does not necessarily do so. An inductive argument, with adequate sample size and representation, is an example of a good non-deductive inference.

There are at least three problems with Russow's argument. First, no intrinsic worth claim can meet her proof standard. Extremely intelligent individuals have been working on the problem for a number of centuries, yet have not managed to convince a consensus of people, much less a unanimity. Plato, around 400 BC, discussed virtues as valuable. John Stuart Mill claimed that qualitative pleasures and pains are the only things with intrinsic value, while years before him, Jeremy Bentham maintained that quantitative pleasures and pains are the right answer. In the 20th century, W.D. Ross argued forcefully that deserved and undeserved pleasure and pain, knowledge, and virtue are incommensurable intrinsic values. Many other objects, such as nature, have been proposed by a number of reasonable persons thinking reasonably (Callicott 1999, p. 224). Since universal agreement, or anything even approaching it, among these thoughtful people is lacking, there is good reason to believe that arguments in support of any one or more things having intrinsic value are never convincing in the way required by Russow.

In fact, it would be reasonable to conclude that from all of the effort expended to date there must be something about intrinsic value making it impossible to establish conclusively that anything has this type of worth (Callicott 1989, pp. 158–9). As was stated previously, my position is that intrinsic value's existence is an unprovable axiom because it has to be assumed in order to have a system of ethics. Reasonable people can thus create different reasonable systems from their reasonable intuitions and adoption of moral theories, although it is likely that they will have commonalities as was seen in Chapter 1.[50]

Second, Russow incorrectly assumes that nature is not an organic whole with identity over time. That is, because she focuses solely on individual sentient creatures and ignores other types of intrinsically valuable things and their relations to each other and other entities within the biosphere, her position seems to be that at any one point in time, nature as a whole's value is equal to the sum of its parts at that moment. However, to make such a claim requires that evidence already available has the power to convince reasonable persons. The fact that Callicott and Rolston

[50] I do not maintain that each system is equally legitimate. The systems can be evaluated using what thoughtful people think are intrinsically valuable-general intuitions-and which intuitions are likely to yield a system that respects people and intrinsic value (QCI) and is likely to be best overall (RPU).

disagree with her position shows the assertion's support is not convincing or compelling on the grounds of non-deductive inference.

Furthermore, if nature is an organic whole, then its value might neither equal nor be proportional to the sum of its parts. If nature has identity over time, which seems a reasonable assumption given the fact that we speak of it being different as it changes during the seasons, through human intervention, and so on, then the intrinsic value of nature can change as well. Moreover, each biosphere is intrinsically valuable *because* it has as one of its components entities that evaluate (Callicott 1999, p. 260; Cahen 2003, p. 133). Their existence makes the organic whole intrinsically valuable. If we think of the biosphere as analogous to a human person, then it becomes apparent how the former has intrinsic value even though parts of it might not.

Third, we assign nature intrinsic value to the biosphere or nature because of the Benefit of the Doubt Principle. Since we have no adequate argument establishing that nature is not intrinsically valuable, and it is a matter of debate as to nature's actual value, it is better to error on the side of caution by assuming the existence of intrinsic worth. In fact, there is reason to believe ecosystems might have value because they are communities of life, whose members have entwined destinies (Rolston 2000, pp. 598–602, 1988, 1994). Or, perhaps it is true that "as morally sensitive rational people we should recognize some state of affairs independent of our needs and interests [such as diversity, naturalness, integrity, stability, and harmony] as valuable or disvaluable" (Thompson 2000, p. 579).[51] Possessing one or more of these qualities is what helps gives a biosphere or nature its particular value. A biosphere with more of the properties is more valuable than one with fewer, all things considered.[52] Although there is no compelling argument to accept these characteristics as having a role to play in an adequate axiology, for the Benefit of the Doubt, we should assume they do until it is shown they do not.

In addition to those who are opposed to the position that nature has intrinsic value, many others have argued against assigning such value to one or more of type of entities upon which I have used the Benefit of the Doubt principle. One of the most notable is Bernard Rollin, who maintains that sentient creatures have intrinsic value because "those animals who are conscious also have fundamental interest arising out of their biologically given natures (or teloi), the infringement upon which matters greatly to them, and the fulfillment of which is central to their lives" (Rollin 2000, p. 432). Non-sentient entities, on the other hand, are lacking at least one of the intrinsic-value-conferring-characteristics sentient creatures possess. Moreover, according to Rollin, if we attempted to designate intrinsic value to such things as non-sentient living entities or the biosphere, then we would be unable to use traditional morality. Traditional morality offers no way to show non-sentient

[51] It should be noted that JL Thompson is merely characterizing a position on intrinsic value she does not hold.

[52] This answers Janna Thompson's question about why environments can have different intrinsic values.

natural objects have intrinsic value; if we would assign such worth, then we would have the Herculean task of having to reconfigure morality to account for the new parts (Rollin 2000, p. 433).

Rollin argues that there is no need to develop a new moral code including a more inclusive, complex axiology because traditional morality already has a way to incorporate non-sentient natural objects in decision procedures. According to him, "it would be better to nurture public appreciation of subtle instrumental value, especially aesthetic value," instead of trying to show non-sentient natural objects have intrinsic worth (Rollin 2000, p. 434). Rollin believes that assigning instrumental value to non-sentient beings is sufficient to make the destruction of a painting, plant, biosphere, etc. morally wrong. The aesthetically pleasing, instrumental value of the object is enough to render its alteration or destruction impermissible on the grounds of how it serves those entities with intrinsic value, such as people or sentient creatures. At times, for Rollin, it can be the case that the instrumental value can overcome intrinsic value, perhaps when a sentient creature(s) must be destroyed in order to protect instrumentally valuable objects.

Unfortunately, Rollin's argument that traditional morality and a boost in people's awareness of non-sentient objects' value can do all the work needed is defective. Although his use of Ockham's razor for the simpler system would generally be plausible, in this case it seems to be mistaken for a number of reasons. First, life and ethics might at times be more complex than one would like or Ockham's razor allows. For example, if science had followed Rollin's approach, then physicists would never have thought about quantum physics, relativity, and chaos theory because what we already had could get us through most normal life activities. In fact, Newton's physics would have remained supreme, even though it is inadequate in explaining quantum levels and the universe as a whole. The universe's actual complexity would not have been discovered to even the extent human persons have as of now accomplished for the simple reason it was not required to achieve most of our goals. But the fact that a system is simpler and generally useful does not mean that it captures reality or things' essences, if they exist, which is something we must remember while striving for good, flourishing lives and making the best possible ethical decisions.

The second reason to reject Rollin's assertion that traditional ethics will have to be reconfigured for the new intrinsically valuable things is the fact that he inconsistently acknowledges conflicts between things with intrinsic value and things with only instrumental value. As has been shown above, there are times when he claims that instrumentally valuable entities can trump things with intrinsic value. More specifically, there are situations in which it is permissible to sacrifice the intrinsically valuable thing to preserve the entity with instrumental worth.

But there is a tension in Rollin's position which has to be examined. Why should we assume that intrinsic value and instrumental value are commensurable? It makes sense to compare likes to like, but there is insufficient evidence that the two value types are comparable to each other in any meaningful way. It would be along the lines of trying to criticize an apple *qua* apple because it does not have the standard characteristics of an orange. Hence, it is unclear that intrinsically valuable things can be sacrificed for the preservation or creation of instrumentally valuable entities. Rollin needs to explain and defend his answer to how something with only instrumental value *ever* overcomes the interests of something with intrinsic value.

The axiology and principles I propose are actually simpler to use than comparing what might be incomparable values. In cases of conflict between entities, it is easier to evaluate alternatives if all the things affected by the action – namely people, potential people, sentient creatures, living organisms, and the biosphere – have intrinsic value, rather than merely extrinsic. We could measure objects on their intrinsic value alone, e.g., a person as a person, a sentient creature, a living organism, et al. versus an individual dog who is sentient and living. The intrinsic worth qua some property or telos would be identical for all entities of that type of thing, while the intrinsic value of overall whole of one individual might be different from an almost identical individual based on other properties or teloi each possesses. Hence, one person could be more valuable than another person, while a living organism can be more valuable than an extremely evil person.

In addition, the traditional moral framework is capable of incorporating the axiology found through the Benefit of the Doubt principle, with only minor modifications. Rollin claims that we make choices between objects with instrumental or intrinsic value in the current moral framework. Therefore, traditional ethics already is able to handle the comparison of different types of value. This feature allows us to choose between preserving the instrumental value of a forest, for example, and a sentient entity, such as deer, when there is an overabundance of the latter posing a threat to the forest's existence. In this case, it might be permissible to thin or entirely eliminate the deer herd to protect the instrumentally valuable entity. Given traditional morality's ability to make these distinctions, as Rollin claims it does, we can achieve the same result by expanding the set of intrinsically valuable things, while not introducing another value type into the decision procedure. The inclusive axiology uses only the clearly commensurables, instead of trying to compare incomparables. For the example above, the worth of the deer herd and those entities that are in relationships to them with their experiences and properties have to be weighed against that of the trees, those living entities that rely upon them, and the biosystem. In this case, *prima facially*, it would seem that utility would most likely be maximized by the deer herd's elimination, while all things with intrinsic value would be respected for the value they have. Hence, if we treat all things as having the intrinsic values assigned to them by the practical axiology, then the system is already in place to handle comparisons.

Finally, as has been stated previously, when we assign instrumental value alone to entities which might have intrinsic value, then there is always the risk of acting unethically toward them. If non-sentient things really have intrinsic value, then by not recognizing their true worth, moral agents do not respect them in the way they should be respected. Although it might not result in severely evil consequences such as maldevelopment that is the "violation of the integrity of organic, interconnected, and interdependent systems that sets in motion a process of exploitation, inequality, injustice, and violence" or patriarchal oppression of women and nature, each failure to recognize intrinsic value makes it more likely that people will act unethically

(Shiva 2002, p. 180). Without intrinsic value, people do not appreciate the worth of the thing in itself, only what instrumental worth the thing has. Without the recognition of intrinsic value, too many people become arrogant toward nature because it is merely of instrumental value, i.e., its goodness depends on how useful it is to people and other intrinsically valuable entities, which leads them to overstate their own importance in the system to the detriment of all (Guha, p. 125).[53] However, if we use the Benefit of the Doubt principle, we would be less likely to fail to recognize intrinsic value in those places it actually does exist, but which society has so far failed to appreciate.

4.8 Conclusion

The task of identifying and ranking intrinsically valuable things has been involved and somewhat convoluted. The difficulty arises, in part, due to the number of things asserted as having worth in and of themselves and the complexity of ranking not only the properties that are the source of intrinsic value, but those individual objects instantiating one or more of them. In the interests of clarity, a brief list of results is helpful for uses of RPU and QCI in the transgenic organism, agricultural ethics, bioethics, and new technology debates.

4.8.1 Intrinsic Value/Priority Rankings in General

1. Persons/Moral Agents
    a. Properties conferring intrinsic value
        i. Being a moral agent.
        ii. Telos of a moral agent-telos of having a good life and being a good person.
        iii. Being sentient.
        iv. Telos of sentient creatures.
        v. Being alive.
        vi. Telos of living organisms.
        vii. Instantiating any mental state from Frankena's list.
    b. A member's ranking within the class is determined by how well the member fulfills the telos, how good the agent is, how well the member instantiates the various properties, and the experiences the member has.

[53] If intrinsic value is assigned for real properties, then attempts to base value on metaphysical qualities can be avoided (Sagoff 1979, p. 617; Rollin 2000, p. 433). It is more compelling to say that a tree is intrinsically valuable because it is a living organism, than to say that it has value because it is metaphorically "majestic." No one would disagree that the tree is alive, but there would not be agreement on whether or not the tree is majestic.

2. Potential Persons/Moral Agents
   a. Properties conferring intrinsic value
      i. Being a potential moral agent.
      ii. Telos of a potential moral agent.
      iii. Being sentient – if applicable.
      iv. Telos of sentient creatures.
      v. Being alive.
      vi. Telos of living organisms.
      vii. Instantiating any mental state from Frankena's list.
   b. A member's ranking within the class is determined by how well the member fulfills the teloi, how well the member instantiates the various properties, and the experiences the member has.
3. Sentient Creatures
   a. Properties conferring intrinsic value
      i. Being sentient.
      ii. Telos of sentient creatures.
      iii. Being alive.
      iv. Telos of living organisms.
      v. Instantiating pleasure.
   b. A member's ranking within the class is determined by how well the member fulfills the teloi, how well the member instantiates the various properties, and the experiences the member has.
4. Living Organisms
   a. Properties conferring intrinsic value
      i. Being alive.
      ii. Telos of living organisms.
   b. A member's ranking within the class is determined by how well the member fulfills the telos and how well the member instantiates the various properties.
5. Biosphere as a Whole
   a. Properties conferring intrinsic value
      i. Being an organic whole that includes entities with intrinsic value in interdependent relationships.
      ii. Being stable.
      iii. Being integrated.
      iv. Being diverse.
      v. Being able to satisfy the teloi of the intrinsically valuable organisms of which it is in part comprised.

b. A biosphere's ranking within the class is determined by how well its intrinsically valuable parts are able to fulfill their teloi. The more teloi of the more intrinsically valuable parts satisfiable in the long term, then the more valuable the biosphere. In this light, the more stability, diversity, and integrity a biosphere has to play this vital role, then the more intrinsically valuable it is.

In order to determine if a transgenic organism or any other technological innovation should be developed or marketed, the axiology developed here and the Practical Moral Code consisting of RPU and QCI are useful. If the technology satisfies both principles as formulated with the inclusive axiology, then it is morally permissible to develop or market it. In the case of Round-Up Ready wheat in North Dakota, Monsanto's short term goal of lifting its temporary hold on the wheat was unethical. First, given the market's resistance to its introduction, no reasonable person would reasonably believe that it is better for the company to continue to overcome adverse opinion than to solidify it against them. Second, maintaining the moratorium in good faith allows the company to treat all intrinsically valuable entities as ends in themselves. In order to respect moral agents, it is often necessary to respect their decisions, as long as such decisions are ethical.[54] In the Round-Up Ready case, since insufficient buyers have shown interest in purchasing the technology, others are vehemently opposed to it, the technology was not necessary at the time for wheat production, and the possible risks associated with cross pollination and mixing would have been potentially disastrous for the state's agricultural environment, then refraining from marketing Round-Up wheat respects all of those affected by the action, while pursuing marketing would not. Hence, the temporary hold is morally justified, and Monsanto should pursue other strategies, as it actually decided to do.

As will be noted here, each technological development any company is considering creating or implementing will have to be evaluated on its own grounds. Transgenic bacteria, for example, developed as biological weapons will generally fail the Practical Moral Code, while technology designed to improve the lives of intrinsically valuable entities as ranked in the axiology, will generally be morally permissible. If pharmaceutical-agricultural biotechnology makes people lives better, without harming potential persons, sentient creatures, and living organisms, and maintains the stability, diversity, and integrity of the biosphere, then it is morally permissible to create and market it. The ethics of some products will not be as clear as these two examples, but if they satisfy the conditions of RPU and QCI, then they are morally permissible.

In Chapter 5, I will use PMC and the axiology to address some of the most important moral questions about transgenic organisms. What is important to remember is that whatever answers the evidence and PMC give us have to be recognized as reasonable and justified, no matter if they are contrary to our desires.

---

[54]Obviously, if the decision is evil, such as harming the innocent for wicked reasons, then the decision does not need to be respected.

Chapter 5
Applying PMC to a Few Transgenic Technology Issues

5.1 Introduction

Every one of our decisions and acts must be ethical. Is it difficult to live up to this absolute rule? Not so much so for the smaller actions one performs with regularity as when one interacts with one's family and fellow workers. However, for the major decisions in which values conflict in important ways and stakeholders are significantly affected, the ethical going gets much harder. Here is where people find themselves going astray from the right and good, often for what they subjectively perceive as justified reasons but objectively are not. Many times the cause is not that the person is evil or seeking to do wrong for its own sake, but rather the reason is that the person lacks adequate information or a practical decision procedure that can help her make a moral choice, enact it, and be able to justify it to other reasonable people.

New technology is one of the principle areas in which people struggle to find ethical solutions to moral dilemmas. Transgenic organisms are one case in point. Although some estimate that 60–70% of the food on grocery store shelves contains at least one TO ingredient, transgenics are not without their problems. TOs have been vilified as an example of human beings' overwhelming arrogance toward both nature, the poor in the developing world, and agricultural producers. At the same time, others have heralded TOs as the salvation of the developing world, the very same agricultural producers critics cite, and markets looking for more efficient ways to produce foods and goods. At the very beginning of the work on transgenics, ethics lagged far behind the technology's pace, but now that time has passed and information became more readily available, it is clearer what actions can be justified.

In this chapter, the general permissibility for developing and marketing TOs will be examined, as well as some factors considered that can make it morally wrong

---

Some of this chapter's significant arguments can be found in Cooley (2008, 2007, 2004a, b, 2002b). Since they first appeared, more nuanced reasoning and information has come to light that has affected my arguments, and in one case, changed the more wary position on transgenics in Cooley (2002b) to a more positive one.

D.R. Cooley, *Technology, Transgenics and a Practical Moral Code*, The International Library of Ethics, Law and Technology 4, DOI 10.1007/978-90-481-3021-4_5,

to create or market a particular transgenic.[1] I will first reintroduce PMC, which is a practical code that anyone can use to decide, explain, and justify her general and particular positions on TOs, regardless of whether they are in favor or against. The second section is an overview of many of the standard arguments for and against TOs, and the third section draws the conclusion that the permissibility of developing and marketing any TO rests on the action's consequences and whether or not all stakeholders affected by the action are respected. The final section briefly addresses market arguments and legal questions.

In general, I contend that Monsanto and other biotech companies have acted ethically by pursuing TO technology, but that there are moral pitfalls which need to be avoided. However, it should always be kept in mind that, given PMC's nature, it is possible at times to have the same information as another reasonable person and draw a different conclusion. At the beginning of transgenic technology, when evidence was lacking, many different actions were morally right because knowledge of risks, benefits, and consequences was speculative in a large number of cases. As relevant information accumulated, some actions could no longer be justified, while others gained greater support. Thus, if we want to be reasonable people, we will evaluate a claim or argument in light of the actual situation.

5.2 The Practical Moral Code Again

As we have seen, two moral theories are often thought to be fundamental to ethical decision making, Kantianism and utilitarianism. First, when making difficult choices or performing actions affecting many people, we should treat all people influenced by our actions as ends in themselves with certain intrinsic value, and if we do not, then the action is automatically unethical (Kant 1956, pp. 62–3). In order to respect a person in the proper way, one must respect her for being an autonomous agent and her autonomous choices as well, as long as the latter do not illicitly harm another, where illicit harm is unnecessary injury caused another without her acquiescence. The formal version of the principle is the *Quasi-Categorical Imperative (QCI)*,

> An act is morally right only if in doing the act, the agent does not treat any person or intrinsically valuable thing as a mere means.

Although treating a person as an end in herself sounds as if the agent's physical behavior is the only focus of Kantian morality, the agent's mental states are much more important. When acting, the agent's primary intentions, motives, and attitudes toward the act must primarily be good. In addition, the agent is required to have the proper respect feeling for any individual's affected by her actions that any intrinsically valuable entity should receive, as well as respecting each individual intrinsic

[1]For more in-depth examinations of many of these biotechnological issues, see Rollin (1996), Sherlock and Morrey (2002), and Thompson (2007). For an older but still useful introduction, see Fincham and Ravetz (1991).

worth. Human persons, for example, deserve better treatment than living plants, but all intrinsically worthy entities need proper treatment. This means that the agent will adequately evaluate the goals, desires, self-interests, and flourishing of all those intrinsically valuable entities affected by the act to determine which she can assist in, if that is necessary, or avoid interfering in, if that is possible. When conflict situations arise, then the goals that are obligatory or permissible, in that order, to obtain will be the ones the agent will attempt to bring about, while rejecting the impermissible ends.

Even though QCI is a necessary principle for ethical actions, on the grounds of practicality, it cannot be sufficient. Given the contentiousness of issues such as TOs and their impact on many different stakeholders, there must be some form of utilitarianism that requires all agents, groups and governmental agencies to do the best they can at the same time they respect all those affected by their actions as ends in themselves. Luckily, in the transgenic debate, both opponents and proponents already use a form of utilitarianism consistent with QCI. *Reasonable Person Utilitarianism (RPU)* states that:

> An action is morally right only if it is reasonable for a reasonable person to believe that the action would probably have at least as much utility as any alternative to the action, where utility is defined as the value found by subtracting the amount of evil produced by an act from the amount of good produced by the act.

In general, a reasonable person is someone who recognizes that morality is "a creative, cooperative enterprise whose end is to better the world by trying to realize in ourselves and others nurturing goods such as caring, considerateness, compassion, sympathy, and love" (Holmes 2003, pp. 228–9). In the pursuit of the ethical, he desires and works toward achievable, good goals and the means to reach them in order to make himself and other intrinsically valuable entities better, provided that doing so does not surrender something of greater moral value (Kant 1956, pp. 89–91). In addition, the reasonable person correctly and consistently analyzes the value of the data available to him for the particular situation in the time available and incorporates all relevant facts about external world society's rules, practices, and customs; rules and responsibilities associated with specific roles the agent is playing at the time; claims others have on the agent and the agent has on others; value of consequences; and all relevant things into their decision process (Holmes 2003, pp. 227–8). If new relevant information becomes available, the reasonable person re-analyzes the situation to see if his position should be altered accordingly (Rescher 1983, pp. 120–1).

Although unmentioned in Chapter 2, it should be clear that Reasonable Person Utilitarianism is actually two moral principles: one for individuals and one for governments. RPU as stated earlier is for individuals and any community citizen that qualifies as a moral entity, such as businesses in which a group of people are committed to act in concert for one or more ends adopted and pursued by the organization. For these objects, trying to maximize utility by performing an action that at least one reasonable person reasonably believes would maximize utility is sufficient for them to perform this aspect of their moral duty.

Governments and other entities representing large groups, on the other hand, because of what they are and can do, must pass a more rigorous standard than individuals. Governments are required to act in the manner that a simple *plurality* of reasonable people examining the situation and coming to a reasonable conclusion would say is likely to produce the best result. The reasons for this conclusion are extremely strong. First, governments have enormous power to wield in making and enforcing decisions that will affect each citizen's life and flourishing. How to regulate technology, for example, impacts not only those creating it, but those who would purchase, repair, use, or be affected by it in any way. If the impact is negative, then governments need to perform to a higher level than the affected individuals who do not have the power or resources to challenge them significantly. Second, good governments should strive to promote or maintain the flourishing of their constituents as a whole. Although they need not be Platonic ideals, leaders should place the many's good above the good of the few, when both needs cannot. be satisfied, and the government treats everyone with the respect she deserves. Third, governments need to justify their decisions and actions. If they have chosen a course that is agreeable to the plurality, then they will have an easier time defending themselves and having their citizens accept governmental action.[2] An appeal to a democratic approach to decision making will go a long way in assuaging any negative feeling that might be generated, in addition to making everyone feel more incorporated in the community. Practically speaking then, the stricter standard is likelier to produce better consequences as a whole than operating according to whatever any reasonable person would believe in the appropriate way. Although the latter gives greater freedom of choice for individuals, it conflicts with the central purpose of good governments to represent the society as a whole.

Regulators in various countries might want to utilize a practical restraint principle when analyzing a product's acceptability using PMC.[3] Practical restraint principles are different from precautionary principles because the former focus on more values than human or environmental safety and do not justify preventative measures on the mere basis that harm might occur in some possible world. Autumn Fiester's Presumption of Restraint incorporates five criteria that biotech animal projects have to satisfy in order to be morally permissible. Project supporters must demonstrate that:

1. a pressing reason to take the dramatic step of genetically altering life,
2. [a] careful consideration of the potential consequences of the project. . .,

---

[2]In the absence of a plurality, RPU for individuals and governments are identical.

[3]Linda MacDonald Glenn suggests a global or international approach to deciding what deserves legal status and respect (Glenn 2003). Although I have argued for universal truths about value, my concern with Glenn's view is that it will be taken by some to an extreme that she is unlikely to approve. To achieve universality, some might not allow individual cultures to have different values from other cultures. If there are morally permissible differences in these general areas, then it would be better and more respectful to allow diversity to exist rather than to extinguish it.

3. a recognition that unbridled animal biotechnology could create a world we no longer recognize or want to live in. . .,
4. a clear regard for the basic tenet of animal research, i.e., that the benefit must far outweigh the cost, and
5. a strong resistance to debasement and adulteration of sentient life (Fiester 2008, p. 43).[4]

Given the axiology developed in the last chapter, Fiester's Presumption of Restraint can be broadened to include any form of biotechnology, e.g., "animal" could be eliminated and "and intrinsically valuable things" inserted after "sentient life" in the fifth condition.

The practical difference between products and how they are evaluated with this principle will be determined by the values involved in the particular circumstances. For research involving animals, being possessors of intrinsically valuable features, such as sentience and life, and having intrinsically valuable experiences are relevant, while plant research would more generally focus on fewer values. Although the expanded Presumption of Restraint does not replace PMC and the axiology, it can be a useful tool for regulatory agencies to more efficiently utilize PMC because, like Gert's 10 rules, it helps them focus on which values are relevant and which are not.

It is possible to begin sketching out what sort of moral factors should be taken into account when using PMC's normative principles and axiology to decide what should be done about research and new technology. The idea of sustainability offers practical guidance if we judge technology based on its ability to "sustain the capacity to produce [the goods] which are required for satisfying conditions essential for preserving the lives and well-being of human populations for an indefinitely long time" (Lehman 1995, p. 156). In other words, after broadening Lehman's scope of what has intrinsic worth to reflect the axiology developed in Chapter 4, decisions should be made on how they affect the flourishing of what is valuable for its own sake now and in the distant future. To help decide which technology seems likely to accomplish this goal, moral agents must seek out and incorporate the best evidence addressing, among other things, the technology's necessity, environmental, social, and economic impact and viability, oversight by relevant agencies, compliance with regulations, and the potential public demand for and acceptance of it (Fox 1999, pp. 165–6).[5] Will each person

---

[4]The demonstration required to satisfy governments and regulators should be limited to what will satisfy PMC. Excessive documentation and other standards illicitly limit autonomy which would not respect people as ends in themselves, and would fail to be likely to maximize utility. The same conclusion can be reached for lax regulations that unnecessarily endanger intrinsically valuable things or create problems that could have been avoided easily.

[5]Fox uses actual demand and acceptance, but flexibility is required here. If people do not know enough about a technology, then regardless of how the public would respond, the lack of demand would kill most new developments before they ever happen. In addition, new technology might not now be acceptable, but would be in time. How to evaluate these more tenuous pieces of information

conceive the same method for achieving this goal? In most cases, of course not, but as long as she uses PMC she will find at least one right, reasonable, defensible answer.

5.3 Transgenics and RPU

In order to determine if developing and marketing TOs is generally likely to maximize utility according to a reasonable person, both the potential positives and negatives of the plausible alternatives have to be examined. That is, at least one reasonable people must decide if the overall outcome of marketing is likely to be better than doing something else for individual action, while a plurality is required for governmental action. In what follows, I will alternate between the potential negatives and positives consequences that are relevant to a reasonable person's use of RPU and QCI. At the end of this section, I will consider several alternative actions that could be better than what has been done to create TOs.

5.3.1 Negatives/Risks

There are a number of negative possible outcomes to consider in the utilitarian calculus.[6] First, there is a very good chance that biodiversity can be reduced by the corporations producing transgenics.[7] For example, since the seed market is essentially controlled by an oligopoly, diversity is usually the last thing served by transgenic organisms (Rifkin 2007, p. 2). Given creating and marketing new TOs' expense – including sometimes having to overcome fierce resistance to their introduction in certain markets – profits are maximized by developing fewer species. Agricultural producers, as a result, have fewer choices, and monoculture becomes a greater possibility with the environmental fragility that using one crop variety or food species always brings with it. Moreover, transgenics' research and development costs are prohibitive for start-ups companies now that so few corporations control the market; hence, there is little positive reason to believe alternatives to the major companies or

would once again be thrown to the standard of what the reasonable person thinks using RPU and QCI.

[6]There are a variety of people who have claimed various problems with transgenics (Davies 2004, pp. 76–81; Ho 2006, p. 291; Rollin 1990, pp. 300–2; Westra 1993, p. 222). The lists seem to be uniform, which might entail that they are complete.

[7]Mark Sagoff notes that an increase in biodiversity is not always a good thing. More wildlife in certain areas increases impurities and pathogenic microorganisms from their fecal matter (Sagoff 2004, pp. 131–2).

their offerings are likely to happen soon enough to benefit consumers, marketplaces, or environmental diversity (Tabor 1989, p. 333).[8]

Second, a series of studies have raised concerns about TOs' environmental effects. Pesticide resistant crops crossbreeding with weeds,[9] harm to the environment through plant and animal species' destruction, pollution, and dangers posed to human beings and members of other species are the chief issues.[10] Consider the potential problems of one TO in wide use. *Bacillus thurigensis* (Bt) crops have been engineered to produce Bt pesticide internally instead of requiring it to be sprayed on them during their growing season. Among environmental worries is that Bt toxin remains active in the soil for at least eight months after harvest which might make it more likely that people, plants, and animals will ingest or be affected by it (Saxena, et al. 1999, p. 480). In addition, since the toxin is always present, the longer exposure to Bt crops causes pest resistance to increase at a faster rate than previously expected. The result is that new genetic alterations will have to be made in the crops to keep ahead of the resistant pests. Finally, some predatory beneficial insects become sick after eating Bt corn or bugs with the toxin in their systems (Walliman 2000, p. 41). Since, naturally occurring pesticides – the insects – are being harmed, there could be even greater problems for the environment later on with an increase in non-beneficial insect populations and harm to the creatures that eat the beneficial insects as part of their diets. Perhaps of greatest concern is that farmers, especially organic producers, might be forced to adopt transgenic technology because they have lost the pest protection which made their way of producing goods possible.

An additional negative outcome is that TO biotech corporations have three practices that can harm farmers more than help them. First, seed companies tie their transgenic products to their pesticides. In order to receive the pesticide resistant technology's benefits, producers must also buy the chemicals required to use the transgenic technology efficaciously. The tying arrangement decreases opportunities for farmers to mix and match various products for the lowest seed and pesticide prices. Second, in order to protect their genetic patents, in many cases, corporations require farmers to destroy any leftover seed they have at the end of the planting season. Saving seed is forbidden and can cause any holding producer caught without permission to have it to be sued in the court chosen by the seed company, generally one in the city in which the company is headquartered.[11] Of course, since the

[8]Although one of Curtin's primary concerns is private ownership of the major food crops' genomes, he also expresses reservations about two few companies owning the information (Curtin 2005, p. 149).

[9]For the sake of discussion, I will label as weeds all plants which have no value to those planting the crops and which will reduce the amount harvested. See Reiss and Straughan for a much more detailed definition of weed (2001, p. 149).

[10]In 1996, a Danish research team under Denmark's Environmental Science and Technology Department observed the transfer of a gene from a genetically engineered crop to a weedy relative (Rifkin 2007, p. 5).

[11]Monsanto and other companies could have introduced Terminator technology which would make all the seed sterile, but public outcry over the effects to developing world farmers caused the com-

company is a community member that pays taxes, employs other community members, and contributes in other ways to the quality of area life, having a case heard in the locality gives the company a home advantage, as well as making it more expensive for a producer to defend herself away from her farm. Third, although developing world producers would benefit greatly from transgenic crops genetically designed for their idiosyncratic growing areas, there has been more money spent on the development of a frost resistant strawberry than on improving crops developing countries rely upon, such as maize, cassava, and beans (Deane-Drummond 1995, p. 309). The reason for the lack of interest on the part of biotechnology companies is the relative unprofitability of these crops (Altieri 2002, p. 67). Although there is no right for a company to profit, there is also no obligation for a company to develop technology that will harm its ability to survive in a competitive world. Therefore, without some incentive for companies to research orphan crops, this relative neglect looks likely to continue. Fourth, due to the high costs of transgenic technology and TOs' profitability, small sustainable farmers in the developing world face destruction by corporate farms (Fox 1999, pp. 25–6).[12] Basically, this is an argument based on economies of scale and who has the capital to purchase biotechnology and production equipment. For the scale component, the larger the farm is, up to a certain point, then the more inexpensive it is to produce each additional unit of goods. This allows larger producers to make more net profit than smaller producers; thereby weakening the latter's ability to compete in the market. If the additional capital for large producers is factored into the equation, then it is clear why smaller producers can be squeezed or bought out. Of greatest concern to those worried about the fate of developing world farmers is the attempt to create TOs to replace more expensive conventional plants and animals. For example, Madagascar and the Ivory Coast's economies are based upon cocoa and vanilla. If transgenic variations of these crops are developed for growth in the wealthiest nations, then these poorer countries will lose their primary exports. Of course there are other risks associated with transgenic organisms and markets, especially those in the developing world. For instance, S.B. Banerjee has written extensively on the modern "colonization" of developing countries by companies based in the developed world. Instead of making developing world farmers more efficient, reducing their costs, and increasing their meaningful alternatives,

panies to agree not to use Terminator. Ironically, many of the problems of cross-pollination would have been avoided with Terminator technology, but if Monsanto changes its mind, it will be pilloried by environmental groups.

[12]Some people have argued against biotechnology on the grounds that it would harm family farms (Zimdahl 2006, p. 166). According to some family farms should be preserved because they are the backbone of a valuable rural community; if we lose them we will lose an important political-economic entity of traditional values, a cherished symbol of moral values, and a solution to long term natural resource problems (Ibid., pp. 166–7). Lehman rejects the argument on the grounds of no general consensus on what a family farm is, family farms do not conserve resources any better than corporations do, and small farmers tend to adopt minimum tillage options at a slower rate than other groups do (Lehman 1995, pp. 130–7). If we wanted to establish whether or not corporate farming is better than family farming, we would still have difficulty in finding universal standards by which to measure results (Lehman 1995, p. 138).

multinational corporations tend to reduce each as they pursue their short term best interests (Banerjee and Linstead 2001; Banerjee 2002, 2003). Hence, the "promise of transgenic organisms is becoming … a means for making profit in a way that allows further domination of the poorer Southern nations by the richer Northern ones" (Ibid.). If true, this is both a negative for the RPU calculation and violates QCI; unfair economic domination of the poor by the powerful does not respect the formers' value or their decisions.

The best RPU arguments against genetic engineering show that some alternatives will produce at least as much utility as genetic engineering without causing the risks associated with new species or products. If, for example, the argument focuses on food production for an expanding population, we can easily establish that genetically modified foods' introduction is likely to make the situation much worse than it would otherwise have been. Food security, especially for the developing world, is one of the major problems confronting global policy makers (IAASTD 2008, p. 17). There are already 1 billion people in the world who have inadequate intake of food, vitamins, and minerals (Mackey and Santerre 2005, p. 47). If the projections that there will be 10 billion people in the world by the middle of the 21st Century and that demand will increase by 300% in Africa, 69% in Asia, and 80% in Latin America are accurate, then there is a very large potential for even more people to have inadequate nutrition with the production methods we currently employ (Mackey and Santerre 2005, pp. 46–7).

Suppose that, for the benefit of argument, twice as much food can be produced by genetically engineered plants and animals than can be generated using current technology. Second, assume that, as Thomas Malthus and Garrett Hardin[13] recognize, with an expanding population, the world will be faced with a point in time at which there is insufficient sustenance to feed everyone. It follows from these two assumptions that if we do not use genetically engineered food, the starvation time will come more quickly and have much fewer people starving to death as a result. This is a horrific vision, but it pales in comparison to the consequences of allowing genetically engineered food to be introduced into the market. By producing twice as much food, twice as many people can be fed at the bare subsistence level; hence, when more births occur and everyone begins to starve, there will be twice as many people dying from hunger than there would otherwise have been, had the food resources been depleted earlier. On these grounds alone, some reasonable people would ban the introduction of genetically altered food, until that time when it is reasonable to believe that the additional people can be provided for adequately.[14]

Perhaps the most important negative consequences are those associated with costs from complying with some governments' restrictive regulations and overcoming consumer rejection. Given actual circumstances in several large markets, some might argue that it is improbable that TOs will maximize utility. For instance, the

[13] "On Not Feeding the Starving".

[14] Deane Curtin would state that letting people starve to death now to take care of the population problem is a coercive approach to the issue (Curtin 2005, p. 75).

National Grain and Feed Association, which closely monitors market conditions, is concerned about selling its members' products in the European market. It and others are justified in this belief for a number of reasons.

First, the European Union regulations on transgenics and products containing them will likely keep the market relatively difficult for transgenic producers to enter. The European Union regulatory framework for transgenic organisms is Regulation (EC) 258/97. Among the set of rules are strict safety testing and authorization provisions, as well as requirements that all products containing transgenic ingredients be traceable from "farm to fork," and those exceeding 0.9% authorized TOs or 0.5% unauthorized TOs be labeled as such (EC 2003a, pp. 1–9). Furthermore, animal feed is governed by the same principles as food, which entails greater market costs for the higher standard than existed in the past (Pew Initiative 2003b, p. 8).[15]

The new regulations clearly incorporate the Precautionary Principle and the asserted right of consumers to know what is in their food as two of its primary guiding maxims (EC 2003a, pp. 5, 9; 2003b, p. 1).[16] In part, based upon Article 1 of the Cartagena Protocol on Biosafety, the EU claims the right to implement a new framework that "establishes a regulatory regime for [transgenics] after a careful assessment of risks, appropriate control and monitoring measures, and proper information to consumers" (EC 2003a, pp. 1, 9).[17] As a result, the EU's framework is stricter than those employed in the United States and many other countries.

The United States called the EU's rules costly, unworkable, unenforceable, unnecessary, and illicitly discriminatory (Pew Initiative 2003b, p. 12).[18] The United Kingdom's Food Standards Agency predicted that the former system's compliance costs of £93 billion to business would increase to £720 billion when the current EU framework is fully implemented (EFSA,http://www/efsa.europa.eu/en/science/gmo/gmo_opinions.html, p. 1). Although predictions that the compliance expenses of environmental and market regulation tend to be higher than actual ones, observance will cost transgenic producers more than they would have had to spend in relatively open and predictable markets, such as that of the United States (Bailey et al. 2002, p. 246). Given that it is relatively unclear what the compliance expenditures will be, in part due to their overestimation and regulatory unpredictability,[19] the market fact is if compliance makes transgenic organisms' price

[15]If it can be shown that EU regulations are unjustified for products consumed by humans, then it will follow that they are unjustified generally for products consumed by animals.

[16]The Cartagena Protocol on Biosafety seems to be the international standard adopted by the European Union.

[17]The meaning of ambiguous terms, such as careful assessment of risk, is a matter of interpretation. Each of the individual parties involved in the issue will use the definition that best suits its conceptual framework.

[18]In 2004, Germany adopted rules that oblige producers to keep transgenic and non-transgenic mixing to zero percent, which makes it virtually impossible to comply unless there are no transgenics in the supply line anywhere (House of Commons Select Committee on Environment, Food, and Rural Affairs 2004).

[19]The French food agency, AFSSA, wants more testing on BT-11 TO maize, even though it is not intended for human consumption and has been approved by the European Commission. The

too high for consumers, then transgenic companies will not be able to compete efficiently and survive in the long term.[20,21] Hence, there is at least some reason to believe that transgenic products will maximize neither utility nor agent-utility. Since the additional expenditures of complying with the EU's regulations on both separation and farm to fork traceability, including testing, establishing a dual distribution system, and finding alternative markets and re-routing grain and other crops are significant, any overhead savings in the production of food transgenics is likely to be lost in the deficits from regulatory observance.

Moreover, the EU is not alone in imposing greater restrictions to its markets. At least 37 countries, with another 10–12 expected to join, have mandatory food labeling requirements for products containing more than a specified limit of transgenics. Given the available information, the "trend in many [non-US] has been toward more intensive rather than less intensive regulation of [transgenics] in recent years" (Wisner et al. 2006, p. 16). Other countries are restricting access to, if not outright banning, transgenic technology from their markets, which means the potential for profit from the technology is likely to decrease. For instance, China, one of the world's largest markets, began to require much more stringent TO labeling in 2002 than it had previously. The National People's Congress Deputy and China Seed Association President, Professor Wang Lianzheng, stated that China was worried about the safety of transgenic organisms, which justified China's new import regulations on transgenic agricultural products (Kyne 2002, p. 1). In addition, if China's concern about food safety increases, it is possible it will ban some TOs from entering its market.[22] If these trends continue worldwide, regulatory observation costs will increase as additional markets compel dual handling distribution systems and other requirements already adopted by the EU.

Current and future regulation can be expensive for transgenic producers. The European Union, China, Japan, South Korea, Australia, and New Zealand require

---

AFSSA's decision makes it very difficult to estimate compliance costs, since most companies would assume that satisfying the European Commission would have been sufficient.

[20] Although precision in regulation is desirable, uncertainty often plays a role in making regulations stricter than they might in the long term need to be, but it is often thought that it is better to be cautious, than to act precipitously and cause preventable harms. For instance, the EU establishes environmental policy on acceptable levels of pollution, based upon an analysis that is as political as it is scientific, and which has a high degree of uncertainty because of a lack of data and disagreements in the scientific community as to harm and risk (Bailey et al. 2002, p. 248). In order to make the process fairer and help businesses understand what is expected of them, there needs to be a consensus-based, cooperative relationship between regulators and those regulated. Hence, businesses can better estimate how much regulation compliance will cost them, and then make more informed decisions about continuing to invest and market products.

[21] Some might argue that the new environmental regulations are necessary to force businesses to adopt innovations that will benefit the environment, but that the businesses would not have otherwise have adopted. Following the regulations then becomes part of the cost of doing business in that market, which then can be reflected in the price of the products.

[22] What is odd about Wang's statement is that China is becoming one of the leading nations in developing transgenic organisms in agriculture.

labeling, which many consumers say they desire, of transgenic foods (Byrnes 1999; Huffman et al. 2003, p. 481). In addition, Wendy Craik, executive director of the National Farmer's Federation, Australia's strongest farm lobby, stated that strict labeling laws and TO crop controls "could alter the trade picture" (Byrnes 1999). Labeling would compel manufactures to test their products or be able to trace all ingredients back to their origination to determine if any transgenic organisms were used in them. The amount of testing required to guarantee conformity with market standards might also cause "bottlenecking" at enforcement facilities; thereby delaying products from reaching the market as efficiently as they would with a more open venue (Van den Eede et al. 2002, pp. 757–61). Labeling and the "farm-to-fork" traceability system's high costs are reasons that utilitarianism might be against the introduction of transgenic organisms into the market (Graff 2002, p. 1). If the projected observation expenses for EU regulations alone will be eight times greater than they currently are, then compliance expenditures for other markets' regulations will make the overall cost even higher (FSA 2002, p. 1). Since it is very costly to satisfy such elevated standards, the additional expenditures might tip the balance in conventional or organic crops' favor; thereby, further limiting the market for transgenics and the incentive for farmers to produce them.

Furthermore, the desire to see TO products labeled is not unique to areas of Asia, Australia, and Europe. A survey conducted by the University of Maine found that 85% of American consumers wanted to know whether TOs are in their foodstuffs (Goad 2002, p. 1B; Mack 2002, p. 1). If labeling in the US is required of producers, then it becomes even more costly to market transgenic goods. Furthermore, it might be the case that if American consumers know transgenic organisms are used in their foodstuffs, according to transgenic opponents, more people will take the organic route or demand that companies cease putting transgenics into their products.

All is not necessarily doom and gloom for transgenics' chances in global markets. Regulatory compliance costs can be offset if consumers in restricted markets are more likely to buy products containing transgenic organisms than they are to purchase those with solely non-transgenics. This is what has occurred with organic food that costs more but is sought out by enough people to make it profitable. The 2007 organic market's value in the US alone was projected to be $30.7 billion, after a five-year compound annual growth rate of 21.4% (Kiesel and Villas-Boas 2007, p. 5).

However, there is considerable evidence to believe that not enough consumers will purchase products because they contain transgenics. While Americans are more ready to embrace TO technology and products, Europeans are much more reluctant (Lawson et al. 2003; Pew Initiative 2001–6, 2003; Hallman et al. 2003; Siegrist 2003). Granted there has been some positive attitude movement by Europeans toward transgenic products, their general acceptability in the EU market is still in long term doubt (Lheureux et al. 2003, p. 3; Grossman 2007, p. 32). Isabelle Garzon, an advisor to chief European trade negotiator Pascal Lamy, stated that "[Europeans] don't see the value added in genetically modified foods. They don't want it" (Burns 2002, p. 2). As Robert Wisner asserts, the problem stems from the concern that foreign consumers have over food safety and the adequacy of their and other governments' regulatory processes to protect them from harm (Wisner et al. 2006). What makes EU consumers even less likely to select transgenics is they do not have faith in them (Wisner et al. 2006, p. 11). Since many buyers do not trust their governments to ensure them of the safety of transgenic organisms, the fact that their government's scientific and regulatory agencies have already claimed that certain transgenics do not pose unreasonable risks make consumers more concerned about the products than they otherwise would have been.

Second, the strong resistance to transgenic technology and the high probability of loss in pursuing it is exemplified by the actions of other governments around the world. In Zambia, where two million citizens were facing starvation due to drought and floods, the government refused to import TO grain, even when it had been ground to guarantee it could not be planted (Burns 2002, p. 1).[23] Although the government's actions might be the result of ignorance about transgenic organisms, they are still an indication of the difficulty of transgenic products acceptability in the country's market, much less purchased by Zambian consumers.[24] Another reason for the reluctance to buy transgenic products could be based on the food habits of individuals which are based upon intensely held symbolic, emotional, or cultural meaning (Kunkel 2000, p. 233). European unwillingness to adopt transgenic wheat and other food crops, for example, has been tied to the cultural meaning Europeans give to what they eat. Americans, on the other hand, are often thought not to have such connections, which makes it easier for them to consume products made from transgenics. Given these facts, there is very little reason to believe transgenic technology will be profitable in Zambia or other areas of the world sharing its concerns.

Some countries recognized the signs that some TOs are unlikely to maximize utility and agent-utility, and adjusted their crop policies accordingly. Canada, for instance, has changed its approach to restricted markets. In Canada's International Business Strategy 2001–2002 document, the transgenic threat to Canadian sales to Japan, the European Union, and Brazil are addressed, as well as the costs of labeling TO products in China (Team Canada, http://atn-riae.agr.ca/info/can/e3204.htm, pp. 13, 15–17, 31, 23–4). In the report, it appears that Canada has little hope of introducing TOs into the European market, and is instead focusing on developing organic and conventional organisms to export to restricted markets (Ibid., p. 16).

The Canadian approach is justified on the grounds of utilitarianism and egoism. Canada knows if a product is unlikely to maximize utility, then it is irrational to put

---

[23]Kelemu et al. state that:

> While we should debate and challenge new technologies and their products, bringing the GMO debate into food aid in Africa when millions are faced with life-and-death situations is irresponsible. When people are reduced to eating grass, is it ethical to prevent them from consuming GM foods that are nevertheless being consumed by millions of people around the world? Who really would prefer to die rather than eat GM foods? (Kelemu et al. 2003, p. 398)

[24]It is difficult to tell if the starving people would have wanted the government to take this action. However, the fact it did indicates a very large problem for TOs market access.

capital toward producing it.[25] The fact there probably will be considerable compliance costs and a very strong bias against TO technology for some time to come in restricted markets implies that Canada and its producers would be better off producing conventional or organic crops for which there is demand. Hence, acting in Canada's long term self-interests by focusing on non-transgenic technology will be best for it and its producers. Furthermore, if buyers and sellers are efficient utility and agent-utility maximizers, then they would not attempt to introduce transgenic organisms and products containing them into markets until it were clearer doing so would serve their individual and society's interests.

There are various health concerns raised about transgenics. First, the wellbeing of animals that eat genetically engineered crops, and human beings, who eat the animals in turn, might be harmed in some way. Of greater consequence would be transgenic organisms' health effects on human beings when the former are consumed directly, as in the potential case of transgenic wheat flour used in bread (Pustazi 2001).[26] A number of studies conducted by Pustazi, Richard, and others are purported to show that TOs are unsafe for human and animal consumption. The Pustazi work allegedly demonstrates that rats eating Monsanto transgenic corn had changes in their blood composition and reduced kidney size (Wisner et al. 2006, p. 12). Based upon this evidence, TOs inherent character, and some people's affinity for the Precautionary Principle or mere caution in general, transgenics are often assumed to have deleterious effects on health until they can be proven safe.

Although these trials' scientific merit is questionable, other science based concerns can be legitimate for reasonable people. For example, even when evidence for transgenic safety is provided, it is often not as extensive as many reasonable people would like for their utility calculations (Kuzma 2007, pp. 3–4). These reasonable people would hold the technology to the highest standards, and would like to see transgenic food tested in the same way that pharmaceuticals are (Carman 2004, p. 90). Others require less stringency, but still observe oversights in transgenic organisms' evaluations in the general research community and by governmental regulatory agencies. In 70% of Canadian transgenic crop decisions, for example, there is no trial measurement of toxicity or chronic use of transgenics on animals or human beings (Clark 2006, pp. 1, 4). Furthermore, Clark states that the trials that are conducted tend to be scientifically incomplete or suspect (Ibid., pp. 4–5). The result is that in some cases, inadequate information makes it impossible to answer with

[25] Anna Meldolesi argues in "Political will to lift the GMO moratorium emerging in Europe" that the new governments, which have emerged in Italy, Denmark, and France, are more likely to lift the ban provided that strict traceability and labeling laws are in place. However, she has not taken into account the anti-transgenic organism attitude of the populace that might play a strong part against lifting the moratorium.

[26] For example, since some transgenics are created using viruses to splice the desired DNA into the recipient genetic material, then there is a possibility of horizontal gene transfer from the transgenic to the gut of the organism consuming it.

certainty every question about TOs' health impact.[27] As a consequence, a reasonable person might use greater precaution and require that each transgenic undergo extensive additional testing against that which already exists so that all risks are identified and adequately addressed before the transgenic can enter the market.

In addition, we should never forget that inadequate science often has a greater impact on consumers' belief than does competent science.[28] It all depends on what the consumer wants to believe rather than on what the evidence supports. First, some studies have shown that, when placed in realistic buying situations, consumers in the US and EU are less likely to purchase TO products than conventional ones (Frewer et al. 1996; Wisner et al. 2006, p. 1). In one report on transgenic wheat, human participants felt an incentive to buy conventional products over transgenic wheat and products made from it (Wisner et al. 2006). Many consumers make decisions based on partial information, cognitive conservatism, or even poor evidence made stronger by their fear of being injured, motivated, perhaps, by an almost inherent distrust of new technology. Good products can fail or be put on abeyance merely by consumers' fears and businesses' responses to those fears. The Canadian Wheat Board, for example, stated 87% of Canada's wheat buyers required guarantees that the wheat they were receiving was not transgenic. This and pressure from Canadian officials helped Monsanto decide against Round-Up Ready wheat release in Canada, even though the wheat is GRAS. In addition, if consumers are afraid, then they are more likely to pressure politicians into placing restrictions on TOs and other technology. The EU's past and current restrictive regulatory frameworks are prime examples of what scientifically unfounded consumer fears can do to businesses and others.

The potential loss from pursuing transgenic crop technology is large. First, the consequences of not being competitive in the European market are severe, for it is the largest single market in the world, with a population greater than that of the United States. The EU and US share the biggest two way trade and investment relationship in the world, with 2006 US imports at $308.8 billion and exports at $186.3 billion (USTR 2006, p. 235). European Union leaders, as well as those of Japan, informed US farmers they do not want TOs in their food supplies, which creates a serious problem for US corn and soybean producers, since the European and Japanese corn and soybean markets are two of the largest (Babcock and Beghin 1999, p. 12). Furthermore, the European population is expected to grow to 550 million people in the next five to ten years making it an even bigger market to which TO producers will want ready access (Team Canada,http://atn-riae.agr.ca/info/can/e3204.htm, p. 15). Hence, the inability to sell transgenic products in such circumstances is a distinct competitive disadvantage to growing TO crops, which conventional and organic farmers do not encounter.

---

[27] Jeremy Rifkin also wonders how pharmaceutical plants will affect foraging animals, seed eating birds, soil insects, and those creatures which will eat them (Rifkin 2007, p. 4).

[28] Some of the citations seem to be twisting the results to conclusions attacking Round-Up Ready crops that are unsupported by the researchers as in the case of Benbrook and the studies of glysophate and human placental cells by Richard et al. (Benbrook 2005, p. 3).

Second, multinational corporations, such as Cadbury-Schwepps, Heinz, Gerber, Nestle, Pepsi Cola, and Kellogg's, have publicly stated they will not use transgenics as far as possible in countries in which the consumers reject them (Siegrist 2003, p. 44; Edwards 2002, p. 19). Since these companies sell a great deal of their products overseas, their refusal to buy TOs means they will not be leading a drive to introduce transgenics into foreign markets. Moreover, since the companies will want to keep the two crop types separate for their different markets, it indicates that producers will be required to maintain crop homogeneity or purity or face losing large purchasers.

However, not all the trade news is bad for transgenics. Evaluating future market acceptance and regulation receives some guidance from the 2006 World Trade Organization (WTO) ruling against the EU's moratorium on TO approvals and releases. The WTO found that the EU had violated the WTO's trade Agreement on the Application of Sanitary and Phytosanitary Measures by using its version of the Precautionary Principle despite the fact that the EU's own scientists had stated that transgenics pose no greater hazard than their non-transgenic counterparts. This would seem to be the best decision that could have occurred to foster transgenic technology.

However, there are still several drawbacks in the WTO decision that bolster a utilitarian case against transgenics. Since the Complaining Parties – US, Canada, and Argentina – did not request them, the panel drew no conclusions about TOs' safety. This means that transgenic exporting countries will not have a WTO ruling that can overcome the trade barriers other nations have put in place based on their concerns over the safety issue. More importantly for producers, no ruling was given on the EU's pre-marketing approval requirements, including a product by product assessment with a scientific consideration of various potential risks, and whether the labeling and farm to fork traceability requirements are trade barriers that violate WTO agreements. These open questions allow the EU, if it chooses, to continue making it difficult for TOs to be approved, and thereby increase producers' costs. The system required for farm to fork traceability is likely to be expensive in rich countries that have relatively few growers and poses a more substantial burden in poor countries where roughly 95% of the world's farmers live (Peterson et al. 2000). Thus when making its decisions, Monsanto and other producers should know that even if they have the best products supported by the best scientific evidence for their quality, if consumers do not want them and regulations make it too difficult to overcome or enter the market, then their products will fail.

5.3.2 Positives/Possible Benefits

First and foremost, transgenics do not appear to be any more dangerous than their organic and conventional counterparts. Even though additional, improved information is something transgenic producers should pursue for consumers and themselves to make better autonomous decisions, there are several different factors which help reasonable people perform adequate cost-benefit analysis without having perfect knowledge. The fact is many TOs are GRAS or have undergone a rigorous decision/testing procedure by their owners and the regulatory agencies governing the markets in which the transgenics will be sold if approved. Although imperfections in the system, bad actors, and other factors damage the process' reliability, trust in it is not unwarranted, as long as one is willing to critically examine the information supplied and the decision justifications made.

The US's regulatory framework for conventional products – although US standards are not the same as the Precautionary Principle – already addresses many of the EU's concerns about the safety of transgenics. In the US, there are three agencies regulating transgenics or products containing TOs. First, the US Department of Agriculture regulates meat and poultry products. Its mandate states that the agency has authority over genetically engineered organisms that are in any way plant pests, or is an unclassified organism and/or an organism whose classification is unknown, or any product which contains such an organism" (USDA 1997, 7CFR340.1).

The USDA requires biotechnology owners to obtain USDA permits for most regulated articles, and provide formal notification for every transgenic organism within US borders. According to the agency, only the safest TOs can be introduced without a permit. For example, the regulated article cannot be a weed, the genetic material introduced must be stably integrated, the function of the introduced genetic material needs to be known and its expression in the regulated article cannot result in plant disease, the expression cannot cause infection, toxic, or pharmaceutical results, and the genetic material cannot posses a viral or disease danger to plants, animals or humans (USDA 1997, 7CFR340.4). Overall, the USDA addresses each of the concerns the EU has about dangers to the lives and health of humans, animals, plants, and the environment, although not as stringently as some want.[29]

The second US agency regulating transgenics is the Food and Drug Administration. The FDA governs all food or food additive transgenics, with the exception of meat and poultry products. Any new transgenic organism is considered GRAS if and only if "the objective characteristics of the substance [do not] raise questions of safety sufficient to warrant formal pre-market review and approval by the FDA" (FDA 1992, p. 22985). To be GRAS, each new transgenic organism must be evaluated using a stringent decision tree to determine safety. Of the eleven possible pathways in the tree, seven lead to required consultation with the FDA or outright rejection of the proposed organism. Clearly the stringent criteria for approval demonstrate the FDA appears meaningfully concerned about allergens, toxicity and other possible problems caused by transgenic organisms (FDA 1992, p. 22993).

Finally, the Environmental Protection Agency (EPA) has authority over pesticides and pesticide residues in food (FDA 1992, p. 22985). The EPA's mandate was given to it by two federal acts. First, the Federal Insecticide, Fungicide, and Rodenticide Act requires that pesticides used in the United States to

[29]For a representative example of the USDA's extensive review process, see the USDA/APHIS findings on Roundup Ready Alfalfa.

be properly registered and "prescribes labeling and other regulatory requirements to prevent unreasonable adverse effects on health or the environment" (EPA, http://www.epa.gov/oppfead1/fqpa/backgrnd.htm, p. 1). Reasonable risks, because they exist for all products, are insufficient to trigger labeling or other requirements. Second, the EPA controls maximum permissible pesticide residue in food as authorized in the Federal Food, Drug, and Cosmetic Act. The standard for evaluating permissible risk is a reasonable certainty that no harm will result from the aggregate exposure to the pesticide or pesticide residue, rather than the more restrictive Precautionary Principle which requires there be beyond a shadow of a doubt certainty that no harm will occur (EPA 1996, 110 STAT. 1516).

There have been criticisms raised about the US regulatory system's efficacy. Among the most serious issues mentioned are the insufficient numbers of ecological scientists involved in the vetting process, an agricultural agency should not be the lead agent for ecological assessment, ecological assessment should be performed by an environmental agency, the entire process should be more transparent to the public, when evaluating technology, ethical values, such as the public welfare, relative benefits and risks, and appropriate caution, and socioeconomic criteria should have equal weight with scientific criteria (Lehman 1995, p. 155). In addition, the lack of long term impact testing and scientific bias of researchers and regulators has brought into question the adequacy of the current regulation (Kuzma 2007, p. 4). According to Jennifer Kuzma, many people involved in regulating biotechnological products have strong conflicts of interests that create serious doubt about their actions. For example, some regulators have been or will be employees of the companies owning the technology, and safety studies are often conducted by those researchers who developed the technology. The research results being published also appears to be biased in favor of biotechnology. Those researchers who try to make public papers showing grounds for potential dangers run the risk of having their credibility questioned and their careers harmed (Kuzma 2007, pp. 5–6). If all these assertions are true, then the reliability of transgenic and other biotech's product information and regulation is doubtful at best.

When debating an ethical issue, those involved often lose sight of the bigger picture and how other factors in the world will affect the positions they might take. Transgenic and other biotechnology is a case in point. There appears to be too little resources spent on conducting independent safety testing to find potential risks and overseeing the products only when considering the narrow field of transgenics. However, if we broaden the scope of our inquiry, we begin to see that there are real problems that the regulatory system should address prior to concerning itself about possible difficulties for which there appear relatively little evidence. The FDA, for example, is unable to inspect and police enough food imports of conventional products to ensure that dangerous products are not brought into the US (Schmit 2007). Some 100 citizens of Latin American countries have been poisoned and died as a result of tainted mouthwash imported from China (Reuters 2007). Although it would be good to spend enough resources so that all eventualities of a product can be discovered, to do so in the transgenic case would be similar to bandaging a scrape while allowing a deep wound to the jugular vein to remain untreated. In other words, actual dangers take priority over possible dangers until the latter are shown to be of equal importance. To do otherwise would be to act unethically because focusing on the minor concern cannot respect those who are actually going to be harmed by the danger nor can it be reasonably thought to be likely to maximize utility.

Furthermore, although real difficulties with the quality of information available to people making decisions about transgenics and other technology are matters of great concern, we need not rely merely upon federal sources for our evidence. First, producers of transgenics evaluate their products for their potential impact on several different entities. Included in their analyses are the percentage of expressed protein relative to the host organism's overall protein, the history of the protein and its effects over time, whether there is a similarity to known allergies or toxins, the protein's functions, a toxicology study based on one thousand times the projected human or other intended end user consumption, the breakdown of the product under similar digestive conditions, and the safety of the product to the environment as determined by greenhouse and field trials (Mackey and Santerre 2005, pp. 53–6). Even if we pessimistically assume that producers are not concerned inherently about their technology's negative effects on people, animals, plants, and the environment, if they are pursuing their best interests, then they will seek adequate data so they are not sued for negligence.[30]

Second, many neutral scientific societies and studies around the world have shown that certain claims about transgenic dangers are exaggerated. First, the Royal Society has found that allergenic risks of transgenics are no greater than those for conventional crops, the use of specific viral DNA sequences in transgenic plants is negligible, and consuming transgenic DNA has no undue negative effect on those who eat it (Royal Society of London 2002, pp. 1–4). Second, the United Kingdom's GM Science Review Panel, made up of 25 experts, established that current TO plants are unlikely to either become super weeds or pose an immediate threat to human health (GM Science Review Panel 2003, 2004). Furthermore, the United States National Academy of Sciences in a report requested by the Food and Drug Administration found that there was little health risk from transgenic organisms (Powledge 2003, p. 1). Finally, the Royal Society established that many transgenic organisms are substantially equivalent to conventional crops (Royal Society of London et al. 2000; Royal Society of London 2002, p. 5). What this classification entails is that transgenic organisms are not fundamentally different from conventional organisms as a product, although they have very different origins.

Offering additional support, the World Trade Organization stated in its ruling against the EU that many of its risk concerns were unlikely to happen or were

[30] Claims have been made that TO producers intend that weeds become herbicide resistant so that they can sell their new herbicide and seed system to combat the problem, and that the herbicide drift into neighbors' fields will force them to buy their products else their crops be destroyed (Schmitz 2005, pp. 62–4).

just as probably with non-biotech crops (WTO 2006b, 8.5).[31] For instance, conventional and organic plants and animals have never been subjected to the testing being demanded of TOs, even though the former might have ill effects on people.[32] Although it is assumed that they are safe, without adequate scientific trials, there is literally no reason to believe that they are any less risky than TOs. In fact, non-TOs can be more dangerous than TOs. Consider peanut allergies in children. If a transgenic peanut is created that susceptible children can eat without having a deadly reaction – these are in development – then it is safer than conventional and organic products.[33] Moreover, the EU's own experts had "evaluated the potential risks to human health and/or the environment prior to the granting of Community-wide approval, and had provided a positive opinion" (WTO 2006b, 8.9). The findings are supported by the fact that people have been exposed for a number of years to this technology without ill effect.

Paul B. Thompson states that arguments relying on the proposition that no one has been harmed by transgenic technology are fallacies of ignorance. According to Thompson, there is often no evidence of harmful effects because no one has taken the time to investigate all relevant consequences, some of which might be very subtle and hard to detect, of a transgenic product (Thompson 2007, p. 67). At best, it is misleading to tell others that there has been no harm when there is inadequate evidence for such a claim. Hence, as rational creatures, we should withhold our judgment until justified in taking a position.

Thompson's argument is right in one way and wrong in another. First, where Thompson is incorrect is when he asserts that there is no evidence of injuries. Although some proponents are disingenuous when they assert that there are no harms associated with transgenics or their products' use, their views are not universal nor should this straw man of an argument be taken to represent everyone's view. Using the Principle of Charity, a more defensible claim is that there has been no obvious health problems linked to transgenic organisms. If eating StarLink corn or some other TO had resulted in consumers becoming ill or dying, then there would have been immediate evidence that something was wrong that warranted in-depth investigation. Much as what happened in the *E. coli* contaminated spinach case in the US, the sick individuals' past activities would have been traced back to the source of the exposure, and then appropriate actions would be taken to eliminate the risk. But there has never been such a clear case of a TO having a toxic effect on humans or animals, otherwise it would have been in the media, which justifies a

---

[31] Monsanto's transgenic wheat is GRAS are the organic and conventional varieties; so it has the same gluten allergy problems as they do.

[32] Bailey claims that products of conventional breeding would never be able to pass the requirements transgenics face (Bailey 2002, p. 38).

[33] Conventional farming has its own environmental, health, and social problems including pollution from fertilizer runoff, cropping techniques that allow for excessive soil erosion, pesticide pollution, and problems stemming from monoculture production, and loss of jobs after the introduction of machinery and other technology (Zimdahl 2006, p. 78).

claim that no obvious harm has arisen due to transgenics. This provides evidence in favor of transgenic organisms and is not a fallacy of reason.

Thompson is correct if we understand his claim to mean that there has been inadequate testing to discover all the negative effects of transgenic use because doing so is impossible. No one, by definition, can know unanticipated outcomes, nor is it likely anyone will be able to comprehend every potential danger unless the person had perfect knowledge of the product, its role in every conceivable situation, and all the laws of nature, especially those in biology. In fact, there will always be some question as to whether or not a transgenic product is totally safe because everything entails risks of some sort. However, a standard of absolute certainty is unnecessary for ethics. Thompson writes that "When researchers have diligently looked for evidence of environmental or health impact it is unreasonable to neglect that work in public decision making" (Thompson 2007, p. 67). In other words, diligent, ethical scientists are reliable sources of information for making claims about the safety of transgenics. According to the reasonable person' evidentiary standard, their information would be sufficient for the RPU calculation.

Moreover, reasonable people know that TO businesses are not generally suicidal. The alterations to the original organism cannot be so great that it lacks the desired beneficial characteristics, including safety that makes it valuable to the market. The whole point to creating transgenics is that they have the beneficial and not harmful qualities of two different organisms that could not be obtained through conventional breeding techniques. If corn or wheat, for example, became toxic or harmed health and lives from their chronic use, then not only would the product be worthless, no company would introduce it into a marketplace and risk the costly consequences, especially in an atmosphere in which the smallest lapse on their part will be used against them by opponents in all public realms. The potential losses from legal liability alone would be sufficient reason for a prudent person to ensure that dangerous products are never marketed.[34]

Not only have some claims about possible injuries been exaggerated, transgenic plants and animals can, in fact, be part of healthier diets and provide relatively sustainable medical benefits to wealthy and poor countries. That is, they can offer both producer and consumer benefits that will increase the probability of flourishing for those things with intrinsic value. For example, a banana plant is in development which will produce fruit that will immunize those who eat it from *E. coli* infection. Since the plants will be able to continue to produce the genetically altered fruit, developing countries will always have an inexpensive supply, thereby allowing them to focus resources on other pressing issues that negatively affect the quality of their lives.[35] Other consumer-benefit-TOs are in the development pipeline. If transgenics can feed and medically treat people more efficiently than is the case with conven-

---

[34] Again, there will be bad actors who try to escape the confines of what morality dictates, but they exist for all areas of endeavor.

[35] Scientists who deleted two genes in mice to prevent a blood supply being developed in the body to feed tumors, thereby starving the tumors to death, believe that anti-angiogenesis, as this procedure is called, will be useful in battling cancer in human beings (Wade 1999, pp. 1–3).

tional and organic products, then an attractive consequence for TOs is preserving more lives, making them better, and increasing social goods.

The positive evidence for the claim that transgenics do not pose unreasonable risk makes it more difficult to support assertions that transgenics might be harmful. If the standard by which claims are evaluated is based upon rationality, then those who contend TOs pose serious risks must now provide equal or greater evidence to justify their continued allegations against proponents who have evidence in their favor. In the face of evidence to the contrary, the mere possibility that harmful consequences can occur from transgenics is no longer sufficient to warrant preventing research on or banning them.

Besides lack of more than minimal risk posed by transgenics and their possible health benefits, TOs have the potential to increase environmental diversity. Instead of a species' variations being limited by evolution and other natural forces, new variations and even species can be developed to better fit ecological niches. It is true that *Homo sapiens* have eliminated a disturbing number of species, but transgenic engineering allows researchers to create new ones. Genetic engineering has already saved certain crops from becoming extinct by putting new genetic character traits in its DNA, such as the papaya, which lacked variations that would make it immune to the papaya ringspot virus. Bananas are also being rescued. Due to 10,000 years of conventional propagation methods for bananas, there is too little DNA variation in the species; therefore, all the members can be wiped out by a single disease.

In addition, as scientists become more adept at creating TOs, the time and resources required for new products will be reduced. This will limit development costs and generate greater flexibility in situations in which an unforeseeable genetic problem with conventional and organic species must be overcome quickly else the species is lost.

Market competition reduction as a result of mergers and buyouts is a legitimate matter of social concern, but we must be clear about who is responsible for keeping markets fair. Maintaining or promoting biodiversity and competitive markets is the duty of governments who have the legitimate authority to regulate, not businesses'. First, if insufficient market competition is a problem for transgenics, the same ethical approaches to making markets more competitive by government agencies can be used on transgenic producers, including but not limited to breaking up the companies or encouraging start-ups with tax credits or other incentives to bring capital into the market. Second, if biodiversity is a social concern, then governments can encourage it through greater production of transgenics and conservation of existing species through mandated and voluntary sustainable practices.[36]

The response to a lack of market diversity is also useful for answering the Malthusian/Hardinian overpopulation and starvation objection. The argument stated

[36]Conventional farming has lead to monoculturalism. In 1972, US corn with the genetic trait Texas T cytoplasm was attacked by a virus. Since that type of corn was the most extensively planted, the crop was a disaster (Thompson 2007, p. 33). Therefore, attacks on transgenics based on the alleged reduction of diversity are not actually about the transgenics as much as they are about diversity in general for all products.

that transgenics might lead to a greater number of deaths in the future than if we merely allow a smaller number to die when conventional and organic production is exceeded. However, once again, there is a false assignment of responsibility. Transgenic producers have power in marketplaces and different societies, but not enough to alter the cultures of all the countries the producers affect in the international marketplace. In addition, the responsibility to diversify diets and contain population growth must be borne by a variety of agents and agencies, which does not necessarily include the transgenic producers. Governments and the citizens most closely impacted by over-population should bear the primary responsibility for their long term well being; they are the ones creating the problems and can fix them better than can transgenic producers. International transgenic corporations do have a role to play, but it is and should be limited to only what is expected of any corporation caring for its stakeholders, which in turn, serves the market to move capital where it is best utilized. Giving them the burden of solving population problems is unfair when they are not responsible for creating them, others have greater control over the problems, their solutions are not required, and although they are community citizens, creating competition for themselves contradicts their duty to ensure the corporation's competitive health. Moreover, no company should be engaged in limiting the autonomy of people to choose their own destiny – that is a violation of QCI and respecting the autonomous decisions people make. According to social welfare capitalism, transgenic producers ought to be engaged in creating products that allow people to create sustainable agriculture and marketplaces so they can be in control of their health, nutrition, and lives because that is likely to be what these consumers want and beneficial to the corporations' bottom line of agent-utility and being a good community citizen. Thus, a reasonable person can reasonably believe that certain transgenic products can and are likely to maximize utility; they, as a component of a sustainable world, have the potential for making people's lives better in significant ways than they otherwise would have been.

TOs' third benefit is that agricultural producers and those relying upon them can receive adequate returns on their investments. Plants and animals can be created to fit various environmental niches now incapable of sustaining agriculture. For example, transgenic crops can be modified to be grown in soils with high saline content or too short a growing season. In addition, conventional organisms can be genetically modified so that they increase harvest sizes or become more resilient in their current agricultural environments (Westra 1993, p. 215).[37] Third, farmers can spend less time caring for what used to be more fragile crops and animals. Efficiency is increased due to the fact that more goods are generated using fewer people, and then the cost savings can be passed on to the consumer, as well as enriching the farmer. Fourth, governments often bail out farmers when disasters strike or offer subsidies and trade protection, which cost taxpayers and consumers more money than if they did not. With the benefits of transgenics and lower production costs,

---

[37]These genes have also been used on potatoes for the same reason (Reiss and Straughan 2001, p. 153).

such incentives need not be offered; thereby lowering tax bills and putting more disposable income into the hands of those who earned it. This money could then be spent on other products in the market or paying off debt, thereby, stimulating economic expansion. Finally, there is tremendous potential benefit for the developing world. As stated before, transgenic crops can be designed to best suit difficult growing areas. In many developing countries, there are poor soils, extremes in moisture, heat, and drought and a variety of pests that can be addressed more efficaciously with transgenics than exists with conventional and organic production techniques (Paarlburg 2002, p. 56). Assuming that food productivity increases, then food prices will become lower, thereby allowing capital to move toward non-food areas of the economy (Ibid.). If enough resources are shifted, then there is an increased potential for them to be spent in ways which improve the life quality of citizens, including developing their own scientific and institutions' capacities (Ibid.).

The final positive benefit from transgenics is the potential profit for those who control the patents on the new species or sell genetically engineered products. Two-thirds of corn and three-quarters of soybean seed sold by Novartis Seeds, Inc.,[38] one of the three largest seed companies, are genetically engineered technology (Brasher, http://www.dailynews.yahoo.com/h/ap/20000201/pl/biotech_crops_2.html). In the United States, 61% of corn, 83% of cotton and 89% of soybeans grown in 2006 were TO (NASS 2006, pp. 24–5). Transgenic crops have found worldwide producer acceptance if the 8.5 million farmers, over 90% of them in the developing world, using them is any indication. Of the 222 million acres planted worldwide with TOs over 33% are accountable to developing world farmers (Prakash 2006). With more extensive adoption, corporate profits increase, which in turn helps the company executives to continue making the company competitive.[39]

Weighing all potential costs and benefits for TOs is difficult given the number of them in the global market and their enormous impact. However, one can justifiably get a feeling that at least one reasonable person would reasonably state that the benefits outweigh the costs, thereby, making marketing TOs in general likely to have positive utility. Overall, they are good products designed to serve certain needs and they can be approved in relatively large markets, even if it is after a great deal of bureaucratic paperwork. Other reasonable people might disagree with the analysis, but RPU does not require consensus, merely that at least one reasonable person reasonably believes that the action is likely to maximize utility.

5.3.3 Non-transgenic Alternatives

In order for PMC to classify the marketing of transgenic organisms as ethical, a reasonable person must find it reasonable to believe that modifications of this sort would probably have at least as much utility as any alternative to the action and that

[38]Novartis has combined with AstraZeneca to become Syngenta.

[39]See Adam's Smith's *The Wealth of Nations*.

marketing TO products must not treat anyone as a mere means. If the action fails in either of the two requirements, then it would be impermissible to create and sell them. Unfortunately, some TO products satisfy neither QCI nor RPU.

Those who support genetic engineering have sometimes failed to recognize that their utilitarian arguments favoring it involve a false dilemma – perhaps because they do not adopt a more holistic approach to the issue. Usually, proponents exclusively present the alternatives as allowing the genetic industrial engineering system to continue as it has in the past so that people do not starve, better food is produced, etc., *or* stopping the modification of foodstuffs altogether, which would result in catastrophe for farmers and the poor of the world. However, these are not the only two options from which a reasonable person can select in this situation. There is excellent evidence showing that there are alternative methods to genetic engineering, such as better product distribution,[40] education, the empowerment of women to make decisions for themselves, and resolving social problems (Curtin 2005, p. 96; Peterson et al. 2000), which would achieve the many of the same positive consequences without incurring the negative results transgenics are alleged to have.[41] The International Assessment of Agricultural Knowledge, Science and Technology for Development (IAASTD) claims that public and private investment in developing technology and more efficient management systems for goods already in existence would be a more practical use of resources to achieve the same results as those championing transgenics (IAASTD 2008, p. 37). Another alternative that will benefit the developing world would be the reduction of trade barriers and elimination of protective tariffs (Ibid., p. 13).

If there are obvious alternatives that can achieve better results than those of specific TOs, then RPU would not be satisfied by the latter's marketing. Golden rice is one case in point. Golden rice was genetically engineered by the Rockefeller Foundation to eliminate Vitamin A deficiencies in Asia, which resulted from white rice being too great a dietary staple, while food products containing Vitamin A were not enough (Gura 1999, pp. 98–9). Unfortunately, instead of going through the expense of creating, gaining acceptance for, and distributing golden rice, the same dietary results could have been achieved by fortifying white rice with Vitamin A, or even more frugally, by diversifying the food stuffs grown in the area.[42] The benefits of dietary diversification include

> [requiring] a minimal amount of foreign currency[for implementation], it promotes intakes of a whole range of micronutrients rather than singling out and tackling just one, it is sustainable, it fosters community and individual involvement, and can help stimulate local

---

[40] Miguel Altieri makes this claim among others (Altieri 2002, p. 66).

[41] Problems might also arise for reasons unrelated to the transgenics themselves. Since this biotechnology can create unjustified emotions in quite a few people who are expected to accept it, before a product is marketed, it would be wise to educate the relevant sections of the public. Proper education have the beneficial consequences of avoiding people making decisions primarily on emotions, such as exuberance or fear, rather than reason (Riess and Straughan 2001, p. 228).

[42] Fox insists that replacing conventional agriculture with organic farming would save $60 billion on public health annually, but provides inadequate evidence to support his claim (Fox 1999, p. 41).

food economy. Furthermore, this approach does not "medicalise" food and nutrition; rather it enables individuals, families and communities to maintain their own health and nutrition. (Oneworld 2007, p. 9)

Given its benefits and the lack of culturally insensitive features – the golden color was rejected by the very people it is supposed to help – diversification would have been more culturally acceptable to the local populace. Hence, instead of creating transgenic organisms which are unwanted and unneeded, more efficient market solutions should have been pursued by the foundation.

Although the development and marketing of some TOs fail to satisfy RPU and producers should have searched for more alternatives before developing a particular product, it does not follow that TOs in general are unlikely to maximize utility according to a reasonable person. There are times when there is no viable alternative to transgenics as in the cases of papaya and bananas. Moreover, education, the empowerment of women, dietary diversification, and other alternatives might have been unable to alleviate certain problems, culturally insensitive, or prohibitively expensive. Women's empowerment, although laudable, might be too abrupt a change in some areas of the world until other social attitudes have been sufficiently enlightened. In other situations, alternatives might not be as obvious as those in the golden rice case. In these circumstances, where total information is impossible or impractical to obtain, a reasonable person could reasonably believe that particular transgenics are capable of maximizing utility for the situation. Finally, the transgenic organism might be more likely to maximize utility than non-TO products. Given that the second wave of biotechnology is turning to products that create producer and consumer benefits, the probabilities of utility maximization increase. As long as the negative potential consequences are minimized and the positives maximized along the lines of the hierarchy from Chapter 4, then there is good reason to believe that at least some transgenic organisms are likely to produce the best utility of all alternatives to them. The question now arises as to whether or not marketing transgenics can fulfill QCI.

5.4 Transgenics and QCI

Marketing transgenics must respect all people affected by it as ends in themselves. This requirement has to be fulfilled by transgenics in general and in particular in order for any transgenic organism to be *prima facie* permissible. Whether or not transgenic producers can satisfy these conditions depends a great deal upon how those opposed to TOs and uninformed consumers are treated. Of course, it is unlikely that anyone potentially impacted by transgenics products will be able to have all of his needs and desires satisfied to his liking. However, those creating and marketing transgenics can respect everyone they impact even if the latter are injured at times. I will first consider three autonomy arguments against transgenics. The first two are designed to show that transgenics should not be allowed in certain markets, while the third allows entry with strict oversight and labeling. Each argument fails to show that creating and marketing transgenics is inherently unethical on QCI grounds. Although some companies have failed in their ethical duties, it is not the case that TO producers need do so.

5.4.1 Autonomy, Government, and Free Markets

The European Union and many governments have autonomously chosen to impose a relatively preventive framework of regulations on transgenic organisms and require strict labeling and traceability of TOs in products, based partly on citizens' demand (Lheureux et al. 2003, p. 31). For example, one survey found that 95% of European Union consumers do not want transgenic organisms introduced into their markets without their consent (Brasher, http://www.dailynews.yahoo.com/h/ap/20000201/pl/biotech_crops_2.html, p. 3). If producers of transgenics are allowed free access to the EU market as the former desire, then consumers' decisions to restrict their markets are ignored by "the ugly American approach that by God, if we produce it, you'll eat it" (Legislator, USDA grant interview transcript). Although phrased more tactfully, the same sentiment can be seen in Hugh Grant's Letter to Shareowners. Monsanto's former CEO stated that in uncertain markets such as Brazil and the European Union, the company is now "focusing on the processes we can affect rather than waiting for rulings that are often delayed and disappointing... We will be proactive, rather than reactive to legal or political maneuvers" (Monsanto 2007, p. 1). This statement can be plausibly interpreted to mean that the company is not going to wait for citizens and their elected governments to decide on their own whether to allow TOs into their markets, but to try to obtain rulings favorable to the company before negative regulations have been implemented. Under the proactive approach, there a loss of freedom on the part of citizens to determine what enters their markets. The trouble with this approach is that if QCI is taken seriously, then since the buyers have already freely decided what they want and how they want it, TO proponents should not tell the market what it will take merely because it is favorable to transgenic producers.

Transgenic organisms' advocates cannot justifiably respond to governmental restrictions by appealing to autonomy. Control of a country's market should rest primarily with the country's citizens, who are the ones who will be most affected by it and are responsible for it more than any outside agency. China's Seed Association President Wang Lianzheng stated that, "The US is very attentive to human rights, and this is also a human right, the right to chose [TO or non-TO] foods" (Kyne 2002, p. 1). If China, the EU, and others require transgenic product labeling, restrict access, or do not want TOs in their market, then producers must respect consumers' rights to govern their markets, especially if producers desire to have their decisions respected as well.

Take as a case in point Monsanto's position in the matter. In its Pledge, Monsanto mentions its freedom to operate in the market. In the Code of Business Conduct section, the company states it needs to build trust with the community so that it can accomplish its visions and avoid the "nearly insurmountable barriers to [its] freedom to operate" (Monsanto 2004). That is, Monsanto has the right to select what it wants to produce and market. Carl Casale, Monsanto's vice president of North American Markets stated, "The bottom line is that no one should deprive American farmers of their right to choose any approved product that they feel will help them grow their crops with higher yields and lower costs" (Monsanto 2007, p. 1). That is, Monsanto endorses the position that no one may interfere in their farmer-producers' rights to raise and distribute the crops they freely choose. Thus, to be consistent, they and all TOs' producers must demand the same rights for everyone else that they want respected for themselves. Therefore, if consumers and their governments have freely decided to implement costly frameworks for access to their markets, then Monsanto and other producers are morally obligated to operate within those frameworks. Producers can go to less restrictive markets, if they select not to make the effort to comply.

Furthermore, the fact that there is no such thing as a perfectly competitive free market with completely free access and no outside governmental influence has a negative impact on arguments to open markets to transgenic technology. Governments, including that of the United States, have justifiably regulated trade based on safety concerns, fairness, and national interests. They are supposed to protect their citizens from easily avoidable harm caused by defective products, untested pharmaceuticals, and other goods not meeting the mores found in the society. For instance, the US citizens rightly expect its Environmental Protection Agency, Food and Drug Administration, and Department of Agriculture regulate research and production of transgenic organisms (EPA, http://www.epa.gov/oppfead1/fqpa/backgrnd.htm; FDA 2001, 1992; USDA 1997, 2001). If transgenic products do not meet these agencies' criteria for acceptance, then citizens require them to prevent the products from entering the market.

In addition, even though the European Union unfairly blocked TOs' introduction in its marketplace, governments of restricted markets can point out the US's hypocrisy on the issue. The US does not have a totally free market because its government strictly controls some products entry into its market. For example, US sugar tariffs are so high that foreign sugar producers cannot compete with those in the US, even though foreign sugar would be cheaper without the tariffs than sugar produced domestically (United States International Trade Commission 2007, Chapter 17). If TO technology producers are consistent when they demand restricted markets be opened, then they must also support the opening of US markets in the same way. If proponents want European and other markets to act in the ways they desire them to, then *ceteris paribus,* proponents must give Europeans equal authority to tell American markets what to do. This move would be consistent with the US Trade Representative's office complaint that there is a "lack of meaningful opportunity for non-EU stakeholders to provide input on draft EU regulations and standards" (USTR 2006, p. 236). Since it is unlikely that the US government and TO producers in the domestic market will grant such liberty to market outsiders, then the former cannot consistently maintain other markets are obligated to accept transgenic organisms.

When pro-transgenic producers attempt at times to open restricted markets, they ignore the basic fact that democratic governments are freely given their protective powers by their citizens. By making decisions about market access and requiring their governments to implement them, citizens are exercising their right to political participation. Since Europeans have concerns about the safety of transgenic organisms, then, as part of their power to protect their citizens as their citizens desire, EU governments are justified in restricting TOs, until the time in which their constituents freely decide to open their markets.

Besides being entitled to be left alone, the right to political participation imposes a positive moral duty on transgenic producers and companies. Multinationals have an obligation to help protect citizens from being deprived of their fundamental entitlement to political participation (Donaldson 1989, pp. 88–9). According to Thomas Donaldson, corporations ought not to engage in behavior serving "to undermine a democratic system in which publicly elected officials hold a position of public trust" (Ibid., p. 88). Furthermore, expanding Donaldson's limitation on foreign ownership of property, although foreign businesses may permissibly participate in the free market, they should not have so much power as to violate the "right to national self-determination and national democratic control" (Ibid., p. 89). Of course, this will be a restriction on the company to pursue its bottom line, but no one is entitled to illicitly harm other people merely for one's own profit. By convincing or trying to sway governments to open their countries' constrained markets, even when it is clear that the citizens do not want their governments to do so, transgenic producers undermine the democratic system. Correctly or not, citizens will feel that the multinationals' market power give the latter equal or greater influence than the citizens themselves have in creating national policy. Hence, if citizens freely decide their markets are closed, then on the grounds of libertarianism and respect for autonomous agents, TO producers should respect governmental regulation enforcing that decision.

The argument based on respect for autonomous agents is powerful, but although several transgenics' producers have ignored some governments' legitimate concerns and decisions, it does not follow that creating or marketing transgenics has or will always do so. In fact, there are a variety of arguments that show that closing or restricting markets to TOs is disrespectful to potential buyers and sellers. In addition to the undue harm caused producers based on inadequate scientific evidence, nations have the duty to honor their morally legitimate trade agreements.[43,44] The reason why this obligation exists is obvious. First, markets

[43]Some might argue that capitalism is unethical; hence, what is a necessary condition for capitalism, especially social welfare capitalism, to exist is prima facie unethical as well. However, for the purposes of this work, I will stipulate that capitalism is ethical on the grounds that fair competition does often lead to more efficient systems and production of novel ideas.

[44]Of course, some contracts do not need to be honored, especially if they contract for illegal services. Furthermore, if the contract does not offer consideration to one or more parties, fraud has been committed, or one or more of the parties is not competent to give consent, as well as other reasons, then the contract can be voided.

require utility maximization, justice, and freedom of choice in order to operate ethically (Velasquez 1998, pp. 408–10). If investors and companies could not have some sort of knowledge and guarantee about what will happen in the future and at what time, then it would be irrational for them to risk their capital in such a chaotic system. Furthermore, it would be impossible for businesses and governments to plan for even short terms goals if contracts and agreements did not exist. International codes such as those of the WTO agreements "provide [transnational corporations] with better defined environmental and public interest objectives and standards with which to formulate policies and procedures in pursuit of their corporate purposes" (Allison and Prentice 1990, p. 714). In return, countries obtain the benefits from corporations' activities and increased trade to allow for the best type of competition according to market forces within the country, while transnationals fulfill their duties by adhering to the national standards and policies protecting the welfare of the country's populace and environment (Ibid.). Hence, it is vital for capitalism and social utility that contracts and agreements be internally and externally enforced.

Of the various WTO agreements, two are relevant to the transgenic organism debate, markets, and the EU's regulatory framework. First, the WTO's Agreement on the Application of Sanitary and Phytosanitary Measures SPM is designed, in part, so that member states can protect the health of residents, animals, and plants, which in turn guards the member state's health itself. SPM asserts, in part, that:

> 2.1. Members have the right to take sanitary and phytosanitary measures necessary for the protection of human, animal, or plant life or health, provided that such measures are not inconsistent with the provisions of this Agreement.
>
> 2.2. Members shall ensure that any sanitary or phytosanitary measure is applied only to the extent necessary to protect human, animal or plant life or health, is based on scientific principles and is not maintained without sufficient scientific evidence, except as provided for in paragraph 7 of Article 5.
>
> 2.3. Members shall ensure that their sanitary and phytosanitary measures do not arbitrarily or unjustifiably discriminate between Members where identical or similar conditions prevail, including between their own territory and that of other Members. Sanitary and phytosanitary measures shall not be applied in a manner which would constitute a disguised restriction on international trade. (WTO 1994a, Article 2)

In other words, in order to limit trade, there must be objective scientific evidence justifying the restriction. Speculations about possible risks are insufficient grounds to support trade decisions, especially if there is scientific data to the contrary. Moreover, the evidence must support a claim that the restriction is necessary to protect the life or health of humans, animals, or plants.

However, the mandate to defend living beings cannot result in the imposition of illicit trade restrictions.[45] The only instance in which scientific evidence need not be used immediately to justify limiting trade is laid out in Paragraph 7 of Article 5,

[45]Unlike Article XX of the General Agreement on Tariffs and Trade 1994 (GATT), in which member states must show it is necessary to violate GATT rules in order to protect health or the environment, SPM rules do not allow for exceptions.

> In cases where relevant scientific evidence is insufficient, a Member may provisionally adopt sanitary and phytosanitary measures on the basis of available pertinent information, including that from the relevant international organizations as well as from sanitary and phytosanitary measures applied by other Members. In such circumstances, Members shall seek to obtain the additional information necessary for a more objective assessment of risk and review the sanitary and phytosanitary measure accordingly within a reasonable period of time. (WTO 1994a, 5.7)

However, the exception to the science rule does not entail, in the face of opposing scientific evidence, that restrictions can be applied based on theoretical risk. Rather, limitations on trade can be justified if and only if there is inadequate evidence to evaluate potential risks to health and life. Furthermore, the member state desiring to restrict trade has the burden of justifying the constraint in a timely manner, which entails that open ended restrictions are impermissible.[46]

The second relevant WTO policy is the Agreement on Technical Barriers to Trade (TBT). However, the TBT also cannot be used by signatories to justify overly restrictive frameworks designed to keep transgenics out of their markets. The TBT allows member states "to adopt technical regulations and conformity assessment procedures if these have a legitimate objective, such as protecting health or the environment" (Wolff 2001, pp. 2–3). The parts relevant to the regulation of TOs are,

> 2.1. Members shall ensure that in respect to technical regulations, products imported from the territory of any Member shall be accorded treatment no less favourable than that accorded to like products of national origin and to like products originating in any other country.
>
> 2.2. Members shall ensure that technical regulations are not prepared, adopted or applied with a view to or with the effect of creating unnecessary obstacles to international trade. For this purpose, technical regulations shall not be more trade-restrictive than necessary to fulfill a legitimate objective; taking account of the risks non-fulfillment would create. Such legitimate objectives are, *inter alia*, national security requirements; the prevention of deceptive practices; protection of human health or safety, animal or plant life or health, or the environment. In assessing such risks, relevant elements of consideration are *inter alia*, available scientific and technical information, related processing technology or intended end-uses of products. (WTO 1994b, 2.1–2)

One of two legitimate objectives that might permit labeling and the traceability ensuring accurate labeling is consumers' right to know what is in the products they are buying. The additional information allows buyers to make informed decisions about one of the most important areas of their lives, *viz*., what they eat. The second objective is the protection of the health, safety, or lives of humans, animals, plants, or the environment. Labeling, traceability, and maintenance of low levels of transgenic presence in products might be warranted to achieve the latter goal. These considerations will be taken up in the section on labeling below, but first, it is important to show why the EU and others frameworks designed to restrict transgenic market access cannot be justified on WTO grounds.

---

[46] In their cases the pro-transgenic side will use SPM 2.1-3 to argue that the EU is imposing illicit restrictions on trade, while the EU will use 5.7 to justify their actions.

First, the EU does not require that new crops created from artificial techniques, such as radiation mutagenesis, undergo the same authorization standard as transgenics. In radiation mutagenesis, seeds are exposed to radiation to produce mutations. The beneficial mutations are kept, while the bad transformations are destroyed. To date, there are over 1,500 crop varieties resulting from this process (Harlander 2002, p. 161S). Given the fact that radiation and genetic mutations can be dangerous to human, animal, and plant life and health, on consistency of WTO application grounds, the EU should be requiring the same procedure for these products as it does for transgenics.

Second, the EU's approval process hypocritically makes transgenic organisms go through a much stricter authorization procedure than other products that should raise greater concerns. Even though there is more than sufficient evidence to prove that they pose far greater risk to citizens' health and the environment than do transgenics, the EU imports clearly hazardous products into its markets without the same restrictions in place for their approval. For example, in 2005, the EU continued to import US tobacco leaf for tobacco products (USDA 2005, Table 8). It has been more than conclusively shown that cigarette smoking is dangerous to human health, yet the EU still allows its citizens widespread access to the product. At the same time, there is no such study showing that transgenics pose the same health risk.

Tobacco is not the only imported product with potential risks. The EU allows for the trade of pesticides and crops sprayed over time with pesticides.[47] Pesticides are widely recognized as posing known risks to the health of humans, animal, plants, and the environment. Since the SPM and TBT require member states to treat like products the same, then if EU does not require the same strict regulatory framework for hazardous or conventional products, it is illegitimate to require it for TOs.

From these features, it *prima facie* follows that on autonomy grounds, governments and other external forces should not regulate the entry of TOs into the market any more than is necessary to fulfill their limited protection duties.[48] A government's obligation is to provide minimal protection for each citizen's life, liberty, and private property; not to guarantee that each person is safe from all possible threats (Locke 1982; Nozick 1974). Overregulation not only unethically limits buyers' choice and treats them as if they are children unable to make their own decisions, it also prevents sellers from entering the market freely. Since at least one developed country with high review standards has approved transgenic products' introduction into a developed world marketplace,[49] the European Union and other countries should show full faith and allow transgenic products into their markets. Of course, this line of reasoning assumes that US regulation is adequate. But if

[47] See Commission Regulation (EC) No. 1204/2003 of 4 July 2003.

[48] In the United States, there are twenty-nine federal agencies responsible for major consumer protection activities, including but not limited to food, environment, commerce, and energy (Brennan and Kubasek 1998, pp. 486–91).

[49] A caveat is appropriate at this point. The Mellman Group has found that Americans are uncertain about transgenic food safety and the plurality believes that more federal regulation of this biotechnology should exits (Pew 2001–6, 2006, pp. 4–6).

it is reasonable to believe it is, and evidence to the contrary has been included to build the strongest case, then it follows that additional regulation and market access impediments violate buyers and sellers' right to access and their autonomy.

5.4.2 Autonomy, Diversity, and Free Markets

There is a different autonomy argument designed to prevent market access to transgenic products that does not focus on governmental regulations of markets. According to opponents, marketing transgenics has the potential for unethically limiting the autonomy of many individuals in their major life choices. No one would disagree that some of the most important autonomous decisions people make is about their lifestyle, such as being a vegetarian or meat eater. If the market becomes entirely dependent upon genetically engineered plants and animals, then the autonomy of consumers who do not want TOs in their diets will most likely be violated.[50] They will be coerced into eating transgenics or doing without. Moreover, many people have already made it clear that they do not want genetically engineered products to eat or in their markets. In the United Kingdom, for example, 70% of people thought using genetically engineering organisms in food was morally wrong (Reiss and Straughan 2001, p. 47). In the 2006 Eurobarometer poll, 58% of respondents were concerned about the risks/problems posed by TOs in their food or drink (Eurobarometer 2006, p. 15). In addition, when asked directly about TOs, the survey found that 25, 37, and 24% were very, fairly, or not very worried, respectively, while only 10% were not worried at all. If anxiety about a product indicates the likelihood of resistance or unwillingness to buy the product, then TOs are not wanted in EU markets (Ibid., p. 24). Making matters worse for those who reject transgenics is that genetically engineered products are "nearly ubiquitous" in any product containing corn or soybeans. Thus, consumers' autonomous decisions not to eat TOs or products with them have not prevented TO producers from bringing them to the marketplace.[51] Since individual autonomy is necessary for personhood, any interference with it requires a compelling ethical justification, e.g., prevention of undue harm to innocents, otherwise it disrespects the person.

John Stuart Mill's utilitarian argument for diversity and freedom of choice helps support the claim that it is wrong to limit a person's freedom to choose; thereby limiting their autonomy. For Mill, people must not only have the total freedom of forming true opinions, they also need liberty, though limited by unjustified harm to others, to act upon those beliefs. If liberty is curtailed too much, the results are dire. People who do not think and act for themselves are merely "automatons in

[50] Of course, if the market still has a healthy organic sector, then autonomy can still be respected according to QCI.

[51] Laura Westra uses a rights based argument against genetically engineered food that has some similarity to the autonomy argument which I develop. I think that mine is the stronger of the two for it avoids the problems of establishing the existence of moral rights.

human form", whose minds are "bowed to the yoke" (Mill 1988a, pp. 127, 129). Eventually, "their human capacities are withered and starved", so much so that they become mere husks of human beings (Ibid., p. 129). Thus, in order to reach their full potential, individuals must have intellectual vigor and manifold diversity, which come from freedom and variety of situations' union. In other words, people need originality that is agreeable to each individual and all of society (Ibid., pp. 101, 132). Moral agents, in order to realize their intellectual potential as human persons, must be given free reign to think and discuss opinions as they choose, make informed decisions, and then act upon them. The alternative is far too costly for it takes from the individual what, in part, makes him a person with intrinsic worth.

However, Mill does not promulgate a system of absolute freedom to do as one pleases whenever the desire strikes one to do so. Freedom in action, though not in opinion, may be limited if the agent makes "himself a nuisance to other people" (Mill 1988a, p. 124). Nevertheless, if the action does not entail harm to anyone other than the actor, and perhaps a more minor injury to others than that would have been suffered by the loss of autonomy to the agent, then it is permissible to do it.

It is clear that eating choices should be limited only in extreme circumstances. If people do not wish to consume certain foods because of dietary preferences, religious beliefs, or for other morally permissible reasons, then they should not be coerced by lack of choice or tricked into doing so (Fox 1999, p. 21). There is therefore, a *prima facie* obligation to provide sufficient market alternatives from which they can choose and to label all foods containing transgenics so that consumers can avoid them if they wish.

Genetic scientists and producers might respond by arguing, roughly, that by preventing genetic engineering, then not only would their freedom be limited illicitly, but that of possible consumers of the products as well. Hence, if the above claim about impermissible coercion is true, then both consumers and members of the genetic engineering industry would also be unethically coerced if the genetic alteration of foodstuffs was halted. Since this is an interesting argument which, if true, might justify genetic engineering on Kantian grounds, it is worth time briefly to examine the issue of when it is permissible for one agent to interfere with another's freedom and autonomy.[52]

What I will argue for here is a practical set of Kantian principles which classify when one person's actions may permissibly interfere with the autonomy of another. Before beginning, however, it is vital to list the intuitively appealing, though perhaps not self-evident, assumptions that I make. First, individual uses of autonomy can be classified on an ordinal, sliding scale, where the most and least valuable uses are designated as significant or trivial, respectively.[53] The trivial exercise of auton-

[52]Unmediated actions affecting only the act's agent are unproblematic, because by definition, no one's autonomy is stymied. Hence, they will not be considered.

[53]At this time, I want to make a distinction between two uses of the word "autonomy". First, "autonomy" can refer to the ontological state of an agent having autonomy or being an autonomous being. A person may have autonomy according to this definition, hence, in the same way that a courageous person has the virtue of courage as part of his ontological state. "Autonomy" can also

omy is characterized by the most inconsequential of actions, for example, choosing between different types of stick candies at a sweet shop. Autonomy's significant uses, on the other hand, deal with making major life altering decisions, such as whether or not to marry this particular person at this particular time. When comparing two different, individual uses of autonomy together in order to rank them, which is the only reasonable way to rank uses, we must exercise our common moral sense to determine which seems more important. A decision to marry is clearly a more valuable use of autonomy than picking out a sweet due both to the central role it plays in the person's identity and the impact such a decision will have on the life of the individual agent performing the action. Furthermore, since a use of autonomy is non-cardinal, we avoid problems caused by being able to add a group of non-trivial actions together to outrank an exercise of autonomy that is more significant than any one of the individual members of the group.

The second assumption is that people are always obligated to respect others in both thought and deed. What is entailed by this obligation has been discussed extensively in Chapter 2, and I will not belabor the requirements again here.

The first practical, Kantian principle (KP1) states that in situations in which there is a conflict between significant uses of autonomies of an agent and those affected by the action, the agent performing the action's exercise of autonomy trumps the autonomies of the others affected by the action. Suppose, for example, that two morally identical strangers, Mary and John, are up for the same professorship in a university, and this is the only job either candidate has a chance of receiving. Due to her better academic qualifications, Mary is selected as the department's first choice, and John, second. If Mary decides to take the job, then she will limit John's ability to use his autonomy. However, even though she reduces another person's autonomy, Mary still does something morally permissible, according to KP1.

The justification for KP1 rests upon self-interest. Of great self-interest is the preservation of one's autonomy to act in ways that maintain or increase one's flourishing. A person may permissibly perform an act even if other people's autonomy is affected significantly, as long as the agent or some other innocent person would lose something of comparable value – the significant exercise of personal autonomy – if the act were not done. Intuitively, people must see to their own vital self-interests before that of people they do not know; otherwise, what we know as supererogatory actions, such as heroism, become obligatory, rather than being beyond the call of duty. The next most important interests to pursue are those of intrinsically valuable entities, based upon their ranking of importance and intrinsic worth, which cannot do it for themselves. Of course, whether or not a moral agent is severely limiting a

---

refer to how an agent acts in particular situations. One person can, for example, limit another's autonomy by limiting the number of choices open to the individual, while not affecting the autonomous nature of the individual himself. Hence, autonomy can be understood in two different ways, by either eliminating the autonomy of the individual or limiting the autonomous actions that the agent may perform. It is this latter sense of the word that I will assume for the rest of this chapter.

person's use of autonomy, as is in all cases, the agent is obligated to respect everyone affected by the action as ends in themselves while performing her action.[54]

The second practical, Kantian principle (KP2) states that if one agent's autonomy will be trivially used, while another agent's use of autonomy will be severely limited, then, *ceteris paribus*, the insignificantly affected agent is obligated not to perform the action. Suppose Cecil takes the last handicapped parking spot, even though he is perfectly healthy. The reason he does not park in one of the spots at the end of the lot is that of laziness; he wants to avoid a long walk to the store. When Linda, who needs the space because of a medical condition, comes along a moment later, she cannot utilize it. Given that mobility is both a physically and psychologically important part of achieving one's plans and living an ordinary life, Linda's choices are severely curtailed for trivial reasons. Hence, according to KP2, Cecil is wrong for parking in the handicapped space.

The genetic engineering of some foodstuffs initially seems to be a case of KP1 rather than KP2. Clearly, scientists and corporations' decision to proceed with genetic engineering is a significant exercise of autonomy.[55] Going ahead with these plans is a life decision that will affect the agents in many important ways by altering their identity and intrinsic worth. After all, these actions will be part of their careers, competition, and how they make their livings. Simultaneously, the agents carrying out these decisions, also, severely limit the autonomy of others affected by their actions, especially those who do not want transgenics in their markets. So it is a conflict between significant uses of autonomy, which would normally support those who want to pursue genetic engineering, according to KP1.

Nevertheless, there are differences in value between the two types of autonomy exercised. When compared together, the genetic engineers and corporations' use of autonomy is more significant than that of others. Although transgenic opponents would have the autonomy conflict be analogous to the case of Mary and John – and KP1 – it is more similar to that of Cecil and Linda – and KP2. Since choices about technological advancement and competitive markets are more important than assuaging the nonrational or irrational fears of those who know little about transgenics or are merely against transgenic technology in general, it follows that the use of autonomy regarding making a living and competing, backed by scientific research showing TOs are GRAS if not safer than conventional and organic products, is more important than exercises of obstructionism based on fear or other emotion. Hence, government approved or monitored transgenics' free access to other markets inherently does not illicitly violate anyone's autonomy or disrespect anyone.

Although disrespect may be perceived, transgenic producers do not have to ensure that everyone always gets what he wants. TO products do not necessarily violate anyone's rights or devalue anyone's autonomy. In fact, they can encourage

[54]The same result is found in those situations in which the agent's use of autonomy is significant, while all those he affects would use their autonomy in a trivial way. This could be labeled principle three.

[55]Though some may disagree that the use is trivial.

greater autonomy and respect through greater diversity. If fair market practices and competition are maintained by those responsible for doing so, then TOs can provide additional alternatives from which to choose. Not only can genetic scientists, universities, and corporations decide to engage in TO research, consumers will have a wider range of choices when selecting food at stores than merely organic and conventional products.

By restricting markets, there is sufficient evidence for a reasonable person to confirm that transgenics' opponents have illicitly limited utility, rights, and a just distribution of society's benefits and burdens. Pioneer Hi-Bred International's Macy Merriman claims that European consumers do not have the freedom to choose or reject any product containing transgenic organisms "because of the activists who have convinced supermarkets to keep it out of their stores" (Brasher 2002, p. 3). If Merriman is correct, then consumer freedom is wrongfully restricted by activists' actions, i.e., convincing businesses, as well as governments, to prevent consumers from making a choice about purchasing TO products. By limiting consumers' options to basically non-transgenic products, the activists and governments have unjustifiably restricted buyers' freedom to make decisions affecting their lives. With the greater number of alternatives, consumers can make improved decisions about what is right for them. With more goods from which to select, consumers can better match products to their desires or needs to achieve maximum satisfaction of their self-interest. The market for organically grown food, for instance, has shown that diversity can work to both producers and consumers' benefits. To make up for the additional costs of information and production, organic foods cost more, but consumers are willing to pay extra to ensure they receive what they want.

Extrapolating from Merriman's contention, if transgenic products were allowed on the market, consumers could make their own choices about whether or not to purchase them. If people decided against transgenic products, then the loss would be borne by the sellers, who took a chance and failed. On the other hand, if consumers bought TO products, the sellers will be able to make a profit. In both alternatives, the freedom and autonomy of the market's consumers is respected. Hence, Pioneer believes that buyers and sellers should be able to freely enter or leave the market, and external parties should not over-regulate the quantity, or quality of any of the goods being bought and sold in it.

Furthermore, in support of their position, TO proponents strongly feel that the present dislike of transgenics is based on ignorance of the technology and products. If governments allowed consumers to have access, then the latter will like and buy transgenic organisms and the products made from them, especially with the second wave of products providing benefits to both producers and consumers as has been suggested in many studies on the subject (Anand et al. 2007, p. 4). In fact, they might even pay a small premium for them (Ibid., p. 15). Primarily, the desired approach is to permit the goods to freely enter the market, and then additional information, TOs high quality, and other market forces will create a demand for them, if buyers do not already want transgenics. For example, with the current soaring food prices and global grain shortages, many Japanese, South Korean, and European consumers, government officials and companies that were opposed to transgenics have begun adopting it (Pollack 2008). Once consumers become comfortable with transgenic products and their advantages, familiarity does not breed contempt but acceptance to the direct and indirect benefit of all.

On the other hand, by blocking transgenics, transgenic producers are not treated with the proper respect any person deserves. Creating unnecessary hardships for innocent people they would not have otherwise have had is impermissible even if the action is primarily intended to alleviate perceived but scientifically unsupported risk to oneself. Granted, it would be total freedom if each person could fulfill every desire he has, it cannot be done permissibly by imposing unnecessary harm on others without their permission. Otherwise, the autonomy of those who do the deciding will be superior to that of those for whom they make the decisions; the former's freedom is necessarily made greater at the expense of the latter's. This is a violation of Kant's tenet that people *qua* people are equal. The equality of people makes it impossible that any one person could have power to make such paternalistic decisions for others, who neither want nor deserve to have their autonomy circumscribed merely because some people are irrationally terrified of new technology.

5.4.3 Autonomy and Labeling Argument

Although TO producers do not inherently disrespect those who do not want their products marketed, some have stated that labeling goods containing transgenic organisms is morally obligatory. Labeling, it is claimed, respects consumers' right to know and autonomy by giving them the freedom of choosing to eat or not eat TOs, while still allowing producers to sell them. Given the emphasis on democracy in PMC, what the public wants should play a role in what happens in markets. From some polls, it appears that the majority of consumers in many countries want labeling (Zepeda 2004, p. 200).

Some governments have listened to this demand. The EU's current regulatory framework, for example, requires labeling in many cases. This in turn mandates that producers keep extensive records so that products which cannot be tested for each and every transgene, such as highly refined canola oil, are labeled correctly and for farm to fork tracing of the source of potentially mixed TOs and non-TOs.

The EU argues that its traceability condition is legitimate for three reasons. First, traceability allows for the control and verification of the labeling process. Since some end products do not contain easily identifiable transgenic proteins, it is necessary to be able to trace their ingredients back to their sources to verify whether transgenics were used. Second, traceability allows for the targeted monitoring of potential environmental effects. If harm occurs, then the EU can quickly act to minimize damages and hold those responsible legally liable for any injury produced. Third, if a transgenic organism causes an unforeseeable threat to human health or the environment, then the product can be withdrawn efficiently.

In order to achieve its three goals, the following traceability conditions of the EU framework need to be met.

1. Operators shall have systems and procedures in place to identify to whom and from whom products are made available.
2. For [TOs] intended for the deliberative release into the environment, operators must transmit specified information on the identity of the individual [TO(s)] a product contains.
3. For [TOs] intended for food, feed, or for processing, business operators may either transmit the specified information mentioned above or transmit a declaration that the product shall only be used as food or feed for processing, together with the identity of the [TO(s)] that the product *may* contain.
4. For food and feed produced by [TO(s)] operators shall inform the next operator in the chain that the products are produced from [TO(s)]
5. Operators shall retain the information for a period of 5 years and make it available to competent authorities on demand. (EC 2003a, pp. 6–7)

Since it has already been shown that two of the three justifications – monitoring potential environmental effects and unforeseeable risks – are not scientifically, rationally, or ethically justified, the five requirements will be interpreted in light of the need to ensure labels and the labeling system's accuracy. Although treated as separate in the EU's regulatory framework, the traceability and labeling requirements do overlap. In order to ensure labeling is accurate, it is necessary to have farm to fork traceability.

In addition to rigorous traceability regulations, the EU's labeling rules are very strict.[56] Any product containing more than 0.9% of approved TOs and 0.5% of unapproved TOs from intentional or unintentional mixing must be labeled as containing transgenic organisms (EC 2003b, p. 3). In fact, the EU requires labeling for "All foods produced from [TOs] irrespective of whether there is DNA or protein present of [TO] origin in the final product," which entails that even the highly refined soy or corn oils having no transgenic proteins in them, must be labeled as containing transgenics (EC 2003b, p. 2).[57] Given producers' control of growing and transportation, they are thought to be best suited to manage the labeling. End sellers would have a harder time because they lack the same information and could never be sure if transgenics were used in the process, especially for products no longer containing transgenic proteins. Also, producers can incorporate labeling costs into production expenditures and product pricing. Transgenic producers and companies can then internalize regulatory expenditures by pricing products with transgenic organisms at a value reflecting the actual production expense, which will then be passed on to consumers.

---

[56]Fred H. Degnan convincingly argues that the FDA is unable to mandate labeling products containing TOs due to the fact that there is no material difference between transgenics and their non-transgenic counterparts (Degnan 2007, pp. 26–7).

[57]There has already been a disagreement over soy oil. In September 2000, Thailand informed the WTO committee that Egypt restricted tuna canned with transgenic soy oil, and was baring importation of tuna from Thailand based upon the assumption it used such oil in its canning, which it does not (Wolff 2001, p. 2). The oddest part of the complaint was the failure to recognize that refined soy oil does not contain the genetic material about which Egypt was concerned.

Mandatory labeling can be a legitimate exercise of control if it is not a *de facto* trade barrier proscribed by WTO trade agreements, such as SPM or TBT, or interferes unnecessarily with free markets. The EU believes it has a duty to its citizens to inform them of something they clearly do not want, but unless the SPM or TBT legitimizes it, the EU cannot prevent the goods from coming into it markets (EC 2003a, p. 5). The legitimate objectives offered for justifying traceability and labeling rules are,

1. Education and raising awareness,
2. Environmental protection and the Precautionary Principle approach,
3. Food safety and health considerations,
4. Consumers' right to know what they are eating,
5. The ability to influence foreign and domestic precaution practices, and
6. Assorted ethical and religious concerns. (Appleton 2000, pp. 567–8)[58]

Of the six reasons, the strongest is the fourth, consumers' right to know what they are eating. Since eating is such a vital and personal issue for individuals, the argument is that they, as autonomous and intrinsically valuable moral agents, should have the information they require to make informed decisions about life quality issues.

There have been criticisms of market oriented solutions in support of transgenics and against labeling. Using the language of rights, Alan Rubel and Robert Streiffer argue that labeling does not violate anyone's right to fair completion because a right to guaranteed economic return in free markets does not exist (Rubel and Streiffer 2005). No one would disagree that transgenic producers spend large amounts of resources in developing transgenic goods, as happened in the case of Monsanto's Round-Up Ready wheat. However, the mere fact that government regulation will make the product less likely to be successful does not entail that the transgenic producers' entitlements have been compromised. In fact, Rubel and Streiffer contend that consumers have a right to know what is in the products they consume; therefore, in the absence of a different right which overrides the right to information, labeling is permissible even though it harms some sellers.[59]

However, although consumer rejection is often cited for restrictive market measures, it is not all that clear that the market as a whole is in favor of labeling, farm to fork traceability, or preventing TOs from entering their diets. Critics often cite as providing evidence of current and future consumer choices and European mar-

[58]Stilwell and van Dyke's reasons for labeling are the same as the first four of Appelton's six reasons.

[59]It is interesting to consider the effect of putting all information any consumer might want on each product in a market. If a product is controversial, then there would be many written lines for consumers to read. The questions then arise as to who would be responsible for creating and paying for additional packaging to carry the information if it exceeds what the producer currently uses, what reading comprehension level is appropriate, and so on. More importantly, we should be concerned that too much information might be less effective than desired. Consumers might ignore or take the information less seriously if it appears to be excessive or overwrought.

ket conditions the Eurobarometer survey finding 62% that Europeans believe food safety over the last 10 years has decreased (Wisner et al. 2006, p. 12). But the information is taken out of context. Those high results resulted only when participants were prompted to consider TOs and other food safety issues. Looking at the rest of the study's results causes a different picture to emerge. In general, Europeans do not think about TOs in their food all that much. First, when asked the very general question about what their attitude toward food purchasing is, only 7% were concerned about production methods (Eurobarometer 2006, p. 9). On the other hand, quality and price, the top two choices, received 42 and 40% of the responses. Second, when asked the general question about attitudes to food risk, 16% were concerned about food poisoning – the number one response – while only 8% expressed a worry about TOs – the fifth highest response (Ibid., p. 12).[60] In addition, there is a disconnect between consumers' stated preferences and what they actually do in markets. In a comprehensive study of consumer behavior in the Netherlands and China, it was found that labeling had no real effect on purchasing TO products (Kalaitzandonakes et al. 2007). If the majority of consumers are not that interested in transgenics unless they are prompted to be, then they are not disrespected if products are unlabeled. Labeling is pointless for them and costly for producers, especially since issues more important to consumers are being ignored.

Given consumers' relative apathy, why do countries impose *de facto* moratoriums and strict TO standards that make market entry and, possibly, sales difficult at best? It might be that many European and other market consumers have inaccurate views of food safety which influence their governments' representatives. In the Eurobarometer survey, respondents were less concerned about their own risky sanitary behavior and more worried about pesticide residues, new viruses, and food hygiene outside of their homes, even though the greatest risk to them stems from their own poor hygiene practices (Eurobarometer 2006, p. 15). This result means that consumers are not terribly good at assessing the risks they face, and will attack an outside threat before addressing a more pressing personal issue. The reason for overly restrictive markets also has a great deal to do with a lack of information and the fear or worry generated by it. Although extensive EC-sponsored research has shown that health concerns have been overblown, anti-TO campaigns have been successful in convincing many people that serious problems exist. The main reason for their accomplishment is that when people do not understand transgenic technology, they fear it, while they simultaneously assume that conventional and other breeding techniques are safe because they are the *status quo* and have not been exposed to any questions about the latter's safety (Beringer, http://ec.europa.eu/research/quality-of-life/gmo/general-intro.html). If these are the individuals whose concerns are supposed to justify labeling, then governments would more efficiently pursue safety if

---

[60]In American studies funded by the pro-biotechnology International Food Information Council, only 3% of American expressed concerns about food biotechnology, while microbial food borne illness and the improper handling of food were most often mentioned at 36% and 35%, respectively (IFIC, p. 2). In addition, 82% of respondents stated that they were not interested in having more information added to food labels (Ibid.).

they properly educated their citizens on transgenic technology and trained them in safe food handling.

There are another two groups of consumers who want foods labeled, neither of which is making their decisions based on unwarranted emotional responses. They are consumers with food allergies and those who, after they have performed extensive research on the issue, dislike TOs (McGarity 2007). The former group has good reason for concern given the example of Pioneer Hi-Bred and its genetically engineered soybeans modified with Brazil nut DNA that was intended to be used in animal feeds. The soybean was designed to help cattle and chickens absorb nutrients more efficiently than they could when consuming non-TO soybeans. Unfortunately, the product was also capable of causing health problems from mild skin irritation to death in those with nut allergies (Nordlee et al. 1996, pp. 688–92). If Pioneer had marketed this product which could not be kept separate from other soybeans used in human food, then the labeling could have alerted those with allergies about the ingredients. Given that labeling is justified and already done on products that might contain some peanut or other allergen residue from manufacturing, even though they are not product ingredients, someone could argue plausibly that labeling is also required to satisfy anyone concerned about the effects of TOs on his or any other valuable thing's health and safety, regardless of how unjustified the belief is. The consumer is always right about his health in this argument, and in order to respect all consumers, their decisions must be carried out by companies.

Although legitimate health issues justify labeling in some cases, requiring it for transgenics and traceability violate QCI unless TO producers and companies autonomously agree to do it. Even without the dangerous precedent being set of labeling on demand, the regulations are unnecessarily harmful. The EU framework imposes the burden of paying for "the system of traceability to identify to whom and from whom [transgenic] products are made available (EC 2003a, p. 2). Given the additional compliance expense, the regulations can keep the EU's market closed to many transgenics and products containing them. The reason why is that the overall framework's requirements may be too onerous and expensive due to costly and difficult testing, ensuring and enforcing compliance and pricey separation maintenance (USTR 2003, p. 112; Pew Initiative 2003a, p. 12; Wolff 2001, p. 3). Segregating transgenic from non-transgenic products is estimated by some to increase the processes costs by 10–30% with no measurable gain in safety (Bailey 2002, p. 40). Labeling, therefore, will be too restrictive.

Furthermore, it creates unfair burdens for the producers, especially for farmers in the developing world. In order to be able to know what to label and how, as required by EU regulations, producers must have systems in place allowing them to trace their products from farm to fork. In very poor areas of the world, establishing and maintaining the system is cost prohibitive. In one study, identity preservation has been shown to add 6–17% to farm costs (Anderson and Nielson 2000, pp. 3–4). That, in turn, will cause poor farmers in developing nations to pay higher prices for seeds than they do now, since seed companies will have to pass the costs of creating and maintaining such a complex system on to the producers. Making matters worse is that in some developing nations 75% of the population lives on less than $2 per day, which means that they already have too little income to be able to pay for any additional production expenses (Tutwiler and Straub 2005, p. 1).

The farm to fork requirement of traceability is also a form of unfair competition that only wealthy countries can win. According to the African Biotechnology Stakeholders Forum, farmers comprise 74–80% of the total population in Africa, which makes it impractical to trace the origin of any transgenic organism or product containing more than 0.9% of approved transgenic ingredients (ABSF 2003b, p. 4). Farmers in the developed world, on the other hand, make up only 2% of the population. It is much easier, as a result, to follow transgenics from farm to fork in the developed world than it is in the developing world. The result is that traceability creates a tariff like mechanism for developing world producers, and tariffs generally have a negative impact on developing world economies (Tutwiler and Straub 2005, p. 9).[61] Hence, contrary to the WTO's Agreement on Agriculture, which states that differential treatment should be to the benefit of the least developed countries, the EU's rules actually make it harder for the developing world to compete in its large markets; thereby handing a large advantage to wealthier producers in developed countries (WTO, docsonline.wto.org/ Part IX, Article 15 and Part X, Article 16).

Furthermore, although the right or freedom of consumers to choose what they eat is emphasized repeatedly in the EU's regulations,[62] those documents say nothing about the reduction of choices for producers, especially poor ones (ABSF 2003b, p. 1). Transgenics have proven advantages for African producers and citizens, such as avoiding up to 70% of crop loss, reducing dependency on expensive chemicals and preserving sensitive wild life habitats from being converted to crop production (ASBF 2003b, p. 2). If TOs are not used, even though pesticide consumption there has been on the rise, African producers will have to return to a system in which there is a 40% crop loss to diseases and pests (ASBF 2003b, p. 1). The result is greater dangerous chemical exposure for many people lacking adequate health care, and reduced yield. In addition, as was stated previously, Indian cotton farmers have benefited enormously from transgenics. Their crop yield increased significantly, while the pesticide use was cut to less than a third for conventional cotton. If these conditions can be obtained for all transgenic crops, then there is much less risk to all intrinsically valuable things in the developing world. Overly restrictive regulations would endanger those gains. By allowing transgenics into the developed world now, it will help it move into poorer countries faster; thereby providing them with a faster track to competitiveness and sustainability (Bailey 2002, p. 46).

---

[61] Restricting markets tends to make the resulting economies perform worse than more open market competition. Closed economies grow on average less than 1% per year, while open economies grow about 3.5% annually (Tutwiler and Straub 2005, p. 2).

[62] See the European Commission's "Wallstrom and Byrne welcome EP acceptance of a trustworthy and safe approach to GMOs and GM food and feed." europa.eu.int/rapid/start/cgi/guesten.ksh?reslist 2 July 2003, 1–4.

Finally, labeling a product as containing transgenic organisms has been shown to reduce consumer demand for the product. As John S. Fredland states, non-EU businesses will be forced to label products in a way repulsive to EU consumers (Fredland,http://law.vanderbilt.edu/journal/33-01/33-1-5.html, pp. 4, 12). In a study conducted by Tegen et al., the information about transgenics on the packets of three items – potatoes, vegetable oil, and corn tortilla chips – had an effect on the prices human subjects were willing to pay for the goods. If the information was anti-TOs, then the bid value decreased by 35%, If there was a combination of pro and anti-transgenic information, then the bids were 16, 24, and 29% lower than before the information was given to the human participants. Finally, even when scientific information was added to the pro and anti-transgenic information, then the bid reduction ranged from 0 to 11%. The results were consistent with the conclusion that consumers place the greatest weight on negative information (Tegen et al. 2003, pp. 1–3; Huffman et al. 2003, pp. 500–2; Anand et al. 2007, p. 4). Given these findings, it is clear that labeling products containing transgenic organisms as such is likely to make consumers less likely to buy them.

And this is what will actually happen. The United States' State Department believes that labeling will cost the US alone over $4 billion in lost sales per year (Pew Initiative 2003a, p. 12). If the cost is so great for a wealthy country such as the United States, it will be worse for poor or developing nations. Hence, even though there is insufficient scientific evidence to establish the anti-transgenic groups' claims, the marketability of transgenic products is reduced. This will, of course, negatively impact the ability to make profits from the technology regardless of the form in which it appears.

The unfairness of requiring producers and others in the transgenic market to pay for any country's labeling and traceability system is clear. First, the system is expensive to implement and maintain when it is not needed because many people do not care, and those who do are often willing to pay higher prices for non-transgenic products. Second, labeling a product as containing transgenics reduces many consumers' desire for it, even when they normally would not be interested in the information. In consequence, those legally obligated to maintain the system and label their products, regardless whether there is nothing meriting such requirements, have to pay the costs for the very system treating them unjustly. At the same time, conventional producers receive the unfair benefits of not having to go through the same stringent approval process, paying for the traceability system, and labeling non-transgenic products, while having their products preferred over transgenics merely because of uninformed or biased consumers. The EU's and other market's discrimination, on these grounds alone, violates the SPM, TBT, and Mill's prohibition on unnecessary harm.[63]

---

[63]On these grounds, PMC would not be able to classify labeling regulations as morally right. No reasonable person would reasonably believe the regulations would likely lead to the best outcome, especially given the harm done to the developing world. Moreover, the unfair treatment of moral agents would never satisfy QCI.

Besides those vehemently opposed to transgenics, there are very few who benefit from labeling. In fact, Arthur Appelton argues that labeling actually violates consumers' right to know because the negative information, unsupported by science, deceives consumers into thinking there is something wrong with the product when there is objective evidence to think that is not the case (Appleton 2000, p. 577). Hence, the labeling regulations actually defeat the goal those who want to protect consumers have set for themselves; consumers are worse off than they otherwise would have been before the mandatory labeling.

As I have argued elsewhere, labeling and farm to fork traceability required to label products accurately should not be borne by TO producers, but by those who desire the information.[64] According to the two competitive market conditions, it should be left to transgenic producers alone to decide whether to label their products. If they choose to label, then they are free to do so. If they elect against it, then they are free to do so. If Monsanto and its producers choose not to label, buyers who do not want to purchase unlabelled products do not have to do business with them. In a perfectly competitive free market, buyers are free to find producers who are willing to mark their products to the buyers' satisfaction, which they value enough to increase their probability of purchasing the product, as occurred in the case of organic milk (Kiesel and Villas-Boas 2007, p. 4). Hence, both buyers and sellers have the freedom to do as they wish without undue influence from others.

5.4.4 A Suggestion

Although the EU and other countries regulations for the authorization and labeling of transgenic organisms and products containing TOs cannot be justified on the grounds of SPM, TBT or Mill, they could achieve their goals in a legal and ethical way. Under WTO agreements, member states have no restrictions on regulating their own producers (Appleton 2000, p. 570). That is, states can impose stiffer burdens upon domestic producers without violating either SPM or TBT than they can on foreign producers. If states are concerned about labeling, traceability, and authorization, then they could make the regulatory framework under discussion mandatory for domestic producers and voluntary for foreign producers. The result would be domestic products would have the labels the country desires for its consumers, while imported products would have them only if the producers and companies chose to supply them. If consumers want to have non-transgenic products, then he or she could buy the domestic products guaranteed not to contain them. If the consumer does not care, then he or she could buy labeled or unlabeled products.

There are many benefits to this suggestion. First, countries can satisfy the demands of consumers while honoring their trade agreements and avoiding unnecessary harm to others. Second, consumers have the freedom to vote with their wal-

[64] See D.R. Cooley's "Transgenic Organisms, the European Union and the World Trade Organization"

lets. Third, businesses and nations can have access to markets that would have been closed. Fourth, the freedom of producers and companies is not limited by required labeling and traceability. Monsanto, for one, has endorsed the voluntary labeling of products containing transgenic organisms. Finally, one of the greatest benefits of this approach is that nations will not be involved in long and costly legal battles, which cause strained international relations.

5.5 Legal/Moral Questions

Although there has been a great deal of discussion on transgenic organisms and their impact on the environment and markets, relatively little attention has been paid to the legal questions being asked more frequently as the transgenic market continues to expand in size and complexity.[65] With the introduction of new transgenic organisms into the market, similar queries will be raised. In order to formulate and promulgate moral solutions to these questions, which will attract majority acceptance, it is imperative to answer them now, rather than trying to construct solutions as cases work their way through the court system and the litigants' anger and entrenchment in their positions increase.

A clear system of legal rules having widespread support of all persons affected by transgenics would be a great benefit to all concerned. The adversarial nature of our court system not only makes the process expensive it also encourages seeing others as opponents to be demonized. And opponents, once demonized, are not seen as having interests that are legitimate or arguments which deserve to be taken seriously. If we are to achieve a broad consensus on these issues we had better try to do as much work as possible before the various groups involved start thinking of each other as enemies. Indeed many problems could be avoided by taking simple reasonable steps. For example, if producers take the time to talk to their neighbors, then they can avoid having their crops pollinate at the same time, which will reduce cross pollination problems. If corn or other crops are staggered in planting, then an organic crop, for example, will pollinate before the TOs get to that state. The result is that the probability of cross pollination is significantly reduced, which will avoid some of the legal issues. Furthermore, by working together, people will create relationships to help them build consensus, or at least, allow them to garner greater understanding and tolerance for others rather than increasing acrimony between neighbors.

Finally, instead of resorting to litigation, individuals could take more responsibility for their own actions. Even though the unethical actions of others might affect us, it is our *prima facie* duty to minimize the harm done to us, if doing so does not sacrifice something of comparable moral worth.[66] For example, if a conventional

[65]For example, the questions of who should be held liable for damages caused by cross pollination and why they are liable were debated in StarLink cross pollination cases (Laidlaw,www.organicconsumers.org/gefood/surrender.cfm, pp. 1–6; Harris, pp. 1–4).

[66]Even though it has not been developed to a point of being able to defeat all attacks upon it, Peter Singer's principle of giving away one's goods until the point that a further loss would be

farmer need only to plant crops five days later on the land bordering his neighbor's transgenic crops in order to prevent a trespass, then he *prima facie* ought to do so. To do otherwise and then complain about how the neighbor damaged him is not only imprudent but unethical. All he has done is to create a much worse system overall and disrespect himself because he has decided to victimize himself rather than taking charge of his destiny in a rather simple way. In order to have a decent life, we must recognize that living with other people requires us to compromise because it is implausible, if not pathological, to believe we can always have whatever we want whenever we want it.

5.6 Conclusion

Given the animosity between the various stakeholders, questions about the permissibility of marketing transgenic organisms have only controversial answers. However, the guidelines given here should be of some help in resolving dilemmas and establishing policies and lows, or at least preventing some of the conflict and worst case scenarios. To accurately decide what we and technology businesses should do, we need to use both utilitarian and Kantian theories, as well as a practical axiology that incorporates as many people's beliefs as possible, on a case by case basis. Although not everyone will agree with every decision made with PMC, if they are reasonable people, they will at the very least respect it.

So what is the big lesson to be learned from this work? I think that it has to be that new technology, if it is too different from what preceded it, frightens or discomforts people so much that they become less likely to use their reason than is required of every moral agent. It might be an instinctual or societal response to reject seemingly sudden and not fully understood change, but whatever the background story for it, such a reaction should not serve as a basis of moral decision making in general. We are moral agents who need to act ethically not only for ourselves but for our communities as well. Thus, contentious technology should merely signal the need for greater investigation to find evidence for and against one's position, so that each person can act as a reasonable agent to pursue goals that are good for her and treat her as she should be treated, as well as being good for and respecting everyone else affected by them.

---

too great of a significant moral worth is plausible. In general, if trying to prevent further harm to the individual is impossible, without being harmed even more in some other way, then the agent cannot be required to do it.

References

Abraham, K.S. *The Forms and Functions of Tort Law*, 2nd edition (Foundation Press: New York 2002).
African Biotechnology Stakeholders Forum (ABSF). "A Statement on Draft European Legislation on the Tolerance Levels For the Unintended Presence of GM Material In Non-GM Agricultural and Food Products" http://www.absfafrica.org/state_euro.htm, 27 June 2003a: 1–4.
African Biotechnology Stakeholders Forum (ABSF). "A Statement From ABSF in Response to the Draft European Legislation On the Tolerance Levels for the Unintended Presence of GM Material in Non-GM Agricultural and Food Products" http://www.absfafrica.org/response_gm.htm, 27 June 2003b: 1–6.
Alden, E. "Greens and Free-Traders Cheer GM Crop Deal" http://www.ft.com/nbearchive/email-ftibwcq3370da.htm
Aldwinckle, H.S. "Testimony for New York Assembly Hearing on Genetically Modified Organisms" http://www.cce.cornell.edu/issues/2000/oct/testimony_HSAldwinckle.cfm, 27 September 2000: 1–5.
Allison, J.R. and R.A. Prentice. *The Legal Environment of Business* (The Dryden Press: Chicago, IL, 1990).
Altieri, M.A. "Promoting Genetically Modified Crops in Developing Countries is not Ethical", in *The Ethics of Genetic Engineering*, L. Yount (editor) (Greenhaven Press: Detroit, MI, 2002): 65–70.
Altieri, M.A. and P. Rosset. "Ten Reasons Why Biotechnology Will Not Ensure Food Security, Protect the Environment, and Reduce Poverty in the Developing World" AgBioForum, 1999a, 2(3&4), http://www.agbioforum.org/n234/v2n34a03-altieri.htm: 155–62.
Altieri, M.A. and P. Rosset. "Strengthening The Case For Why Biotechnology Will Not Help The Developing World: A Response to McGloughlin" AgBioForum, 1999b, 2(3&4), http://www.agbioforum.org/n234/v2n34a14-altieri.htm: 226–36.
American Pet Products Manufacturing Association, "2005/2006 National Pet Owners Survey" Accessed 13 June 2006. http://www.appma.org/press_industrytrends.asp
American Society for Biochemistry and Molecular Biology (ASBMB). Code of Ethics. January 1998. http://www.asbmb.org/asbmb/site.nsf/Sub/CodeofEthics?opendocument
American Society of Agronomy (ASA). Statement of Ethics. Adopted by ASA Board of Directors 1 Nov. 1992. https://www.agronomy.org/pdf/statement_of_ethics.pdf
American Society of Plastic Surgery (ASPS). "2000 Average Surgeon/Physician Fees: Cosmetic Surgery" Accessed 13 June 2006. http://www.plasticsurgery.org/mediatr
Anand, A., R.C. Mittelhammer, and J.J. McCluskey. "Consumer Response to Information and Second-Generation Genetically Modified Food in India", *Journal of Agricultural & Food Industrial Organization*, 5, 2007: 1–18.
Anderson, K. and C.P. Nielsen. "How will the GMO debate affect the WTO and farm trade reform?" Agricultural Policy Discussion Paper No. 17, Centre for Applied Economics and Policy Studies, Massey University, 8 November 2000 http://econ.massey.ac.nz/CAPS/: i–27.

D.R. Cooley, *Technology, Transgenics and a Practical Moral Code*, The International Library of Ethics, Law and Technology 4, DOI 10.1007/978-90-481-3021-4_BM2,

Anscombe, G.E.M. *Intention* (Cornell University Press: Ithaca, NY, 1963).

Appleby, M.C. "Tower of Babel: Variation in Ethical Approaches, Concepts of Welfare and Attitudes to Genetic Manipulation", *Animal Welfare*, 1999, 8: 381–90.

Appleton, A.E. "The Labeling of GMO Products Pursuant to International Trade Rules," *N.Y.U. Environmental Law Journal*, 2000, 8: 566–78.

Aquinas, T. "Creatures Have Their Own Activity" On the Power of God, q. 3, a. 7, in *An Aquinas Reader*, M.T. Clark (Editor). (Fordham University Press: New York, 1972): 329–33.

Aquinas, T. *Summa Theologia: A Concise Translation*, T. McDermott (editor) (Christian Classics: Allen, TX, 1989).

Aristotle, "De Anima", J.A. Smith (trans.) in *The Basic Works of Aristotle*, R. McKeon (editor) (Random House: New York, 1941a): 533–603.

Aristotle, "Nicomachean Ethics", in *The Basic Works of Aristotle*, R. McKeon (editor) (Random House, Inc.: New York, 1941b): 935–1126.

Aristotle, "Physics", in *The Basic Works of Aristotle*, R. McKeon (editor) (Random House, Inc.: New York, 1941c): 218–397.

Aristotle, "Politica", in *The Basic Works of Aristotle*, R. McKeon (editor), (Random House, Inc.: New York, 1941d): 1113–316.

Association for Computing Machinery (ACM). Code of Ethics. 16 October 1992. http://beta.acm.org/about/code-of-ethics

Atkinson, C. "Farmer Puzzled by Weed Killer's Surprise Crop" http://www.smh.com.au/news/0002/05/world/world13.html

Aurelius, M. *Meditations*, translated by Maxwell Staniforth (Barnes and Noble, Inc.: New York, 1996).

Babcock, B.A. and J. Beghin. "Potential Market for Non-GMO Corn and Soybeans" CARD Briefing Paper 99-BP 27, 1999, http://www.card.iastate.edu/publications/texts/99bp27.pdf: 1–4.

Bailey, R. "Genetically Modifying Food Crops is Ethical", in *The Ethics of Genetic Engineering*, L. Yount (editor) (Greenhaven Press: Detroit, MI, 2002): 34–47.

Bailey, P.D., G. Haq, and A. Gouldson. "Mind the Gap! Comparing Ex Ante and Ex Post Assessments of the Costs of Complying with Environmental Regulation", *Wiley InterScience*, 2002: 245–56.

Banerjee, S.B. "Organizational Strategies for Sustainable Development: Developing a Research Agenda for the New Millennium", *Australian Journal of Management*, Special Issue, 2002, 27:1005–18.

Banerjee, S.B. "Who sustains whose development? Sustainable development and the reinvention of nature", *Organizational Studies*, 2003, 24(1): 143–74.

Banerjee, S.B., and S. Linstead. "Globalization, Multiculturalism and Other Fictions: Colonialism for the New Millennium?" *Organization: The Interdisciplinary Journal of Organization, Theory, and Society*, 2001, 8: 683–722.

Bankowski, Z., and R.J. Levine, editors. *Ethics and Research on Human Subjects: International Guidelines* (Geneva: CIOMS, 1993): 26.

Barboza, D. "Biotech Companies Take on Critics of Gene-Altered Food" New York Times, 12 November 1999a, www.biotech-info.net/GE_critics.html

Barboza, D. "Monsanto Faces Growing Skepticism on Two Fronts" New York Times, 5 August 1999b, C1.

Bartsch, D., S. Driessen, A. Gathmann, A. Hoffman, M. Lehnen, T. Muecher, C. Saeglitz, U. Wehres, and I. Schuphan. "Monitoring the Environmental Consequences of Gene Flow from Transgenic Sugar Beet" in Proceedings: Ecological and Agronomic Consequences of Gene Flow from Transgenic Crops to Wild Relatives, Ohio State University, 5–6 March 2002: 78–89.

Baxter, W.F. "People or Penguins: The Case for Optimal Pollution", in *Philosophy and Contemporary Issues*, J.R. Burr and M. Goldinger (editors) (Prentice Hall: Upper Saddle River, NJ, 1974): 217–22.

Baxter, W.F. "A 'Good' Environment: Just One of the Set of Human Objectives", in *Applied Ethics: A Multicultural Approach*, L. May, S. Collins-Chobanian, and K. Wong (editors) (Prentice Hall: Upper Saddle River, NJ, 2002): 142–7.
Baylis, F. and J.S. Robert. "Primer on Ethics and Crossing Species Boundaries" ActionBioscience.org article, May 2006, http://www.actionbioscience.org/biotech/baylis_robert.html# Primer: 1–9.
BBC News. "Britons report 'psychic powers'" Accessed 1 June 2007. http://news.bbc.co.uk/2/hi/uk_news/5017910.stm
Beauchamp, T.L. and L. Walters. *Contemporary Issues in Bioethics* (Thomson Wadsworth: Belmont, CA, 2003).
Benatar, S. "Distributive Justice and Clinical Trials in the Third World" *Theoretical Medicine and Bioethics*, 2001; 22(3): 169–76.
Benbrook, C. *Harvest at Risk: Impacts of Roundup Ready Wheat in the Northern Great Plains* (Western Organization of Resource Councils: Billings, MT, 2005).
Beringer, J.E., "EC-sponsored Research on Safety of Genetically modified Organisms-A Review of Results: Forward" http://ec.europa.eu/research/quality-of-life/gmo/general-intro.html
Berry, W. "Religion and the Environment", in *American Environmentalism: Readings in Conservation History*, R.F. Nash (editor) (McGraw-Hill Publishing Company: New York, 1990): 40–44.
BIOTHAI (Thailand), CEDAC (Cambodia), DRCSC (India), GRAIN, MASIPAG (Philippines), PAN-Indonesia and UBINIG (Bangladesh). "Grains of Delusion: Golden rice seen from the ground" February 2001, www.grain.org/publications/delusion-en-p.htm
Bishop, J.A. and L.M. Cook, editors. *Genetic Consequences of Man Made Change* (Academic Press Inc.: London, England, 1981).
Blatz, C.V. "It is Morally Permissible to Manipulate the Genome of Domestic Hogs", *Journal of Agricultural and Environmental Ethics*, 1991, 4(2): 166–76.
Boatright, J.R. *Ethics and the Conduct of Business* (Prentice Hall: Upper Saddle River, NJ, 1997).
Bok, S. "Lying and Lies to the Sick and Dying", in *Ethical Issues in Professional Life*, J.C. Callahan, editor (Oxford University Press: New York, 1988): 141–50.
Branford, S. "Why Argentina Can't Feed Itself-How GM Soya is Destroying Livelihoods and the Environment in Argentina", *The Ecologist*, 2002, 32(8), http://OrganicConsumers.org/gefood/argentinasoya111802.cfm: 1–3.
Brasher, P. "Fear threatens US crop sales in Europe" Des Moines Register, 11 November 2002, 1–3.
Brasher, P. "Senator: Biosafety Deal Hurts U.S." http://www.dailynews.yahoo.com/h/ap/20000201/pl/biotech_crops_2.html
Brennan, A. and N. Kubasek. *The Legal Environment of Business* (Macmillan Publishing Company: New york, 1998).
Brennan, A. and Y. Lo. "Environmental Ethics" Accessed 12/23/2003. http://plato.stanford.edu/entries/ethics-environment/
Brower, D.R. "Friendship with the Earth", in *American Environmentalism: Readings in Conservation History*, R.F. Nash (editor) (McGraw-Hill Publishing Company: New York, 1990): 40–4.
Brown, L. *Eco-Economy* (W.W. Norton & Company: New York, 2001).
Bucchini, L. and L.R. Goldman. "Starlink corn: a risk analysis", *Environmental Health Perspective*, Jan. 2002, 110(1): 5–13.
Burns, G. "Europe shows little taste for US biotech crops" Chicago Tribune, 2002, http://www.chicagotribune.com/business/chi-021030029oct30,6391188.story?coll=chi%2Dbusiness%2Dhed
Byrnes, M. "Consumer Pressure for GE Food Labels Threatens US Exports to Asia & Pacific Region" Reuters North America, 1 September 1999, http://www.organicconsumers.org/ge/usgeexport.cfm
Cahen, H. "Against the Moral Considerability of Ecosystems", in *Environmental Ethics: An Anthology* (Blackwell Publishing: Malden, MA, 2003): 114–28.

Cajete, G. "Look to the Mountain: Reflections in Indigenous Ecology", in *Applied Ethics: A Multicultural Approach*, L. May, S. Collins-Chobanian, and K. Wong (editors) (Prentice Hall: Upper Saddle River, NJ, 2002): 148–57.

Callicott, B.J. *In Defense of the Land Ethic: Essays in Environmental Philosophy* (State University of New York Press: Albany, NY, 1989).

Callicott, B.J. *Beyond the Land Ethic: More Essays in Environmental Philosophy* (State University of New York Press: Albany, NY, 1999).

Campbell, B.G. "How Much in the Human condition Is without Counterpart in Other Animals?", in *Evolution and Human Values*, R. Wesson and P.A. Williams (editors) (Rodopi: Atlanta, GA, 1995): 117–37.

Carman, J. "Is GM food safe to eat?", in *Recoding Nature: Critical Perspectives on Genetic Engineering*, R. Hindmarsh and G. Lawrence (editors) (UNSW Press: Sydney, NSW, 2004): 82–93.

Carroll, J. "Judge Will Approve a Settlement on Use of StarLink Corn Products", *Wall Street Journal*, 7 March 2002: A4.

CAST "Backgrounder: FDA-CDC Blood Serum Testing Methodologies" http://www.cast-science.org/biotechnology/200106-cdc.htm, 11 June 2001: 1–4.

Charman, K. "'Biotechnology Will Feed the World' and Other Myths" accessed 21 May 2006, http://www.vegsource.com/articles/gmo_feed_myth.htm: 1–4.

Cheng, C. "On the Environmental Ethics of the Tao and the Ch'i", in *Applied Ethics: A Multicultural Approach*, L. May, S. Collins-Chobanian, and K. Wong (editors) (Prentice Hall: Upper Saddle River, NJ, 2006): 200–07.

Clark, S.R.L. "New Issues: Genetic and Other Engineering", *Journal of Applied Philosophy*, 1994, 11(2): 233–7.

Clark, E.A. "Food Safety of GM Crops in Canada: Toxicity and Allergenicity" accessed 21 May 2006, www.plant.uoguelph.ca/research/homepages/eclark/safety.htm

Cohen, H. "Speeding up the Evolutionary Process", *The Scientist*, 2001, 15(23), http://www.the-scientist.com/yr2001/nov/cohen_p12_011126.html: 12.

Colby, C. "Introduction to Evolutionary Biology" 7 January 1996, http://www.talsorigins.org/faqs/faq-intro-to-biology.html: 1–32.

Colditz, G.A., M. Samplin-Salgado, C.T. Ryan, H. Dart, L. Fisher, A. Tokuda, and B. Rockhill. "Harvard Report on Cancer Prevention, Volume 5, Fulfilling the potential for cancer prevention: policy approaches", *Cancer Causes and Control*, 2002; 13: 199–212.

Collins, H.B. and R.W. Krueger. "Potential Impact of GURTs on Smallholder Farmers, Indigenous & Local Communities and Farmers Rights: The Benefits of GURTs" Paper made available to the CBD's Ad Hoc Technical Expert Group on the Impact of GURTs on Smallholder Farmers, Indigenous People and Local Communities, 19–21 February 19–21 2003: 1–4. The paper is presented as the official position paper of the International Seed Federation.

Commission of the European Communities (CEC). COMMUNICATION FROM THE COMMISSION on the precautionary principle, COM (2000) 1, Brussels 02.02.2000.

Comstock, G.L. "Should We Genetically Engineer Hogs?" *Between the Species*, Fall 1992: 196–202.

Comstock, G.L. *On the Ethical Case Against Agricultural Biotechnology* (Kluwer Academic Publishers: Boston, MA, 2000a).

Comstock, G.L. *Vexing Nature? On the Ethical Case Against Agricultural Biotechnology* (Kluwer Academic Publishers: Norwell, MA, 2000b).

Comstock, G.L. *Life Science Ethics* (Iowa State University Press: Ames, IA, 2002).

Conservation Technology Information Center (CTIC). "Conservation Tillage and Plant Biotechnology: How New Technologies Can Improve the Environment By Reducing the Need to Plow" Accessed 23 October 2002, http://www.ctic.purdue.edu/CTIC/BiotechPaper.pdf: 1–20.

Cooley, D.R. "Distributive Justice and Clinical Trials in the Third World" *Theoretical Medicine and Bioethics*, 2001, 22(3): 151–67.

Cooley, D.R. "The CIOMS's Distributive Justice Principle: A Reply to Dr. Benatar", *Theoretical Medicine*, 2002a, 23: 11–18.

Cooley, D.R. "So Who's Afraid of Frankenstein Foods?" *Journal of Social Philosophy*, 2002b, XXXIII(3): 442–63.
Cooley, D.R. "Transgenic Organisms and the failure of a free market argument", *Business Ethics: A European Review*, 2004a, 13(4): 354–71.
Cooley, D.R. "Transgenic Organisms and Some Legal Ethics", *Public Affairs Quarterly*, 2004b, 18(2): 91–110.
Cooley, D.R. "Transgenic Organisms, the European Union, and the World Trade Organization", in *The Ethics of Genetic Commerce*. R.W. Kolb (editor) (Blackwell Publishers: Blackwood, NJ, 2007): 87–108.
Cooley, D.R. "Genetically Modified Organisms and Business Duties", in *Ethical Theory and Business*, T.L. Beauchamp, N. Bowie, and D. Arnold (editors) (Prentice Hall: Upper Saddle River, NJ, 2008): 541–545.
Cooley, D., G. Goreham, and G. Youngs. "Practical Moral Codes in the Transgenic Organism Debate", *Journal of Agricultural and Environmental Ethics*, 17(4): 1–8.
Council for International Organizations of Medical Sciences (CIOMS). International Ethical Guidelines for Biomedical Research Involving Human Subjects. Geneva: CIOMS, 1993.
Crawford, M.A. "Genetically Modified Foods", *The Lancet*, 1999, 353: 1531.
Crouch, R.A. and J.D. Arras. "AZT Trials and Tribulations", *Hastings Center Report*, 1998, 28(6): 26–34.
Cummins, J. "Homologous Recombination in Plants", 2002, lists.ifas.ufl.edu/cgi-bin/wa.exe?A2=ind0204&L=sanet-mg&P=4689
Curtin, D. *Environmental Ethics for a Postcolonial World* (Rowman & Littlefield Publishers, Inc.: Lanham, MD, 2005).
Curtis, H. *Invitation to Biology* (Worth Publishers, Inc.: New York, 1972).
Davies, P. "Gene flow and genetically engineered crops", in *Recoding Nature: Critical Perspectives on Genetic Engineering*, R. Hindmarsh and G. Lawrence (editors) (UNSW Press: Sydney, NSW, 2004): 71–81.
de Tocqueville, A. *Democracy in America*, J.P. Mayer and M. Lerner (editors), G. Lawrence (translator) (Harper & Row: New York, 1966).
Deane-Drummond, C.E. *Genetically modified organisms For the Environment: Ethical Implications of the Biotechnology Revolution* (Blackwell Publishers, Ltd.: Oxford, England, 1995).
Deckers, J. "Are Scientists Right and Non-Scientists Wrong? Reflections on Discussion of GM", *Journal of Agricultural and Environmental Ethics*, 18(5), September 2005: 451–78.
Degnan, F.H. "Biotechnology and the Food Label", in *Labeling Genetically Modified Food*, P. Weirich (editor) (Oxford University Press: New York, 2007): 17–32.
Detmer, D. "Response: Of Pigs and Primitive Notions", *Between the Species*, Fall 1992: 203–08.
Devall, B. and G. Sessions. *Deep Ecology* (Peregrine Smith: Layton, UT, 1985).
Dixon, B. "Engineering Chimeras for Noah's Ark", *The Hastings Center Report*, 1984, 10: 10–12.
Dobson, A. "Genetic Engineering and Environmental Ethics", *Cambridge Quarterly of Healthcare Ethics*, 1997, 6: 205–21.
Donaldson, T. *The Ethics of International Business* (Oxford University Press: New York, 1989).
Edwards, F. "Genetic Engineering Can Be Ethical", in *The Ethics of Genetic Engineering*, L. Yount (editor) (Greenhaven Press: Detroit, MI, 2002): 10–23.
Elliot, R. and A. Gare (editors). *Environmental Philosophy: A Collection of Readings* (University of Queensland Press: St Lucia, Queensland, 1983).
Elliott, R. "Environmental Ethics", in *Contemporary Moral Issues*. L.M. Hinman (editor) (Prentice Hall: Upper Saddle River, NJ, 2000): 567–76.
Environmental Media Services (EMS). "Liability" http://www.ems.org/biotech/liability_insurance.html: 1–3.
ETC Group. "New Enclosures: Alternative Mechanisms to Enhance Corporate Monopoly and BioSerfdom in the 21st Century" Communique, November–December 2001, Issue no. 73: 1–20.
Eurobarometer. "Risk Issues" European Commission, February 2006, http:/www.mdrgf/org/-uipp/Pdf/com_summary_eurobarometer_en1.pdf

European Commission (EC). "Question and Answers on the regulation of GMOs in the EU" EU Institutions press releases, Accessed 4 March 2003a http://europa.eu.int/rapid/start/cgi/guesten.ksh?reslist: 1–23.

European Commission (EC). "Wallstrom and Byrne welcome EP acceptance of a trustworthy and safe approach to GMOs and GM food and feed." Accessed 2 July 2003b, http://europa.eu.int/rapid/start/cgi/guesten.ksh?reslist: 1–4.

European Court of Justice, Judgment of the Court in Case C-236/01, Monsanto Italia SpA and Others v Presidenza del Consiglio dei Ministri, Accessed 9 September 2003b http://curia.eu.int/en/actu/communiques/cp03/aff/cp0367en.htm: 1–2.

European Food Safety Authority (EFSA). GMO Opinions, http://www/efsa.europa.eu/en/science/gmo/gmo_opinions.html

European Parliament (EP). "Plenary vote on two GMO Commission proposals (traceability and labeling-food and feed)" 2000, http://europa.eu.int/comm/trade/speeches_articles/spla114_en.htm

Evangelical Lutheran Church in America (ELCA). "Social Policy Resolution on Genetically Modified Organisms in the Food Supply" Church Council Action, November 2004, www.elca.org/socialpolicyresolutions: 1–9.

Faden, R.R. and T.L. Beauchamp. *A History and Theory of Informed Consent* (Oxford University Press: New York, 1986).

Feder, B.J. "Monsanto Bars Selling of Seeds Made Infertile" New York Times on the Web, 19 April 1999a, http://www.nytimes.com/library/national/science/100599sci-gm-seeds.html

Feder, B.J. "Plant Sterility Research Inflames Debate on Biotechnology's Role in Farming" New York Times on the Web, 19 April 1999b, http://www.nytimes.com/library/national/science/041999sci-agri-biotech.html

Feinberg, J. "The Rights of Animals and Unborn Generations", in Today's Moral Problems, R.A. Wasserstrom (editor) (Macmillan Publishing Co.: New York, 1979): 581–601.

Feldman, F. *Doing the Best We Can* (D. Reidel Publishing Company: Boston, MA, 1986).

Felsot, A. "AgBioView Special: Some Corny Ideas About Gene Flow And Biodiversity" Agrichemical & Environmental News, 2002, 193, http://aenews.wsu.edu: 1–12.

Fiester, A. "Justifying a Presumption of Restraint in Animal Biotechnology Research", *The American Journal of Bioethics*, June 2008, 8(6): 36–44.

Filteau, S., A. Tomkins et al., "Promoting vitamin A status in low-income countries (Commentary)", *The Lancet*, 1999, 353(9163): 1458.

Fincham, J.R.S, Ravetz, J.R. *Genetically Engineered Organisms: Benefits and Risks* (University of Toronto Press: Toronto, Ontario 1991)

Finkelstein, E.A., I.C. Fiebelkorn, and G. Wang. "State-Level Estimates of Annual Medical Expenditures Attributable to Obesity", *Obesity Research*, 2004, 12: 18–24.

Food Standards Agency (FSA). "A Proposal for a Regulation of the European Parliament and of the Council on genetically modified food and feed" 2002, www.food.gov.uk/foodindustry/regulation/ria/ria2002/gmfoodfeedria: 1.

Fox, M.W. *Killer Foods* (The Lyons Press: Guilford, CN, 1999).

Frankena, W. *Ethics* (Prentice Hall, Inc: Engelwood Cliffs, NJ, 1973).

Fredland, J.S. "Unlabel Their Frankenstein Foods!: Evaluating a U.S. Challenge to the European Commission's Labeling Requirements for Food Products Containing Genetically-Modified Organisms" http://law.vanderbilt.edu/journal/33-01/33-1-5.html: 1–22.

Frewer, L.J., C. Howard, and R. Shepherd. "The influence of realistic product exposure on attitudes toward genetic engineering of food," *Food Quality and Preference*, 1996, 7: 61–7.

Frey, R.G. "Rights, Interests, Desires, and Beliefs", in *Living Well: Introductory Readings in Ethics*, S. Luper (editor) (Harcourt Brace College Publishers: Fort Worth, TX, 2000): 436–42.

Friedlander, B.P., Jr. "Toxic Pollen from Widely Planted, Genetically Modified Corn Can Kill Monarch Butterflies, Cornell Study Shows", Accessed May 1999, http://www.news.cornell.edu/releases/May99/Butterflies.bpf.html

Friends of the Earth Europe (FoEE). "US Biotech Industry out of control: Colonic Cancer Genes Used to Engineer New Crop Plants by Speeding up Evolution" 25 November 2002, http://www.foeeurope.org/press/GR_25.11.02_colonic.htm: 1–2.

Friends of the Earth Europe (FoEE). "FoEE Biotechnology Programme and European GMO Campaign" 21 May 2006, http://www.foeeurope.org/GMOs/Index.htm

GM Science Review Panel "GM Science Review First Report" July 2003, http://www.gmsciencedebate.org.uk/report/pdf/gmsci-report1-full.pdf

GM Science Review Panel "GM Science Review Second Report" January 2004, http://www.gmsciencedebate.org.uk/report/pdf/gmsci-report2-full.pdf

Gert, B. *Morality: Its Nature and Justification* (Oxford University Press: New York, 1998).

Glaeser, B. *Environment, development, agriculture: Integrated policy through human ecology* (M.E. Sharpe: Armonk, NY, 1995).

Glantz, L.H., G.J. Annas, M.A. Grodin, and W.K. Mariner. "Research in Developing Countries: Taking 'Benefit' Seriously" Hastings Center Report, November–December 1998, 28(6): 38–42.

Glaser, B.G. and A.L. Strauss. *The Discovery of Grounded Theory: Strategies for Qualitative Research* (Aldine: Chicago, IL, 1967).

Glenn, L.M. "When Pigs Fly?", *The Physiologist*, 46(5), October 2003, http://www.the-aps.org/publications/tphys/2003html/oct03/randall.htm

Glenn, L.M. "Ethical Issues in Genetic Engineering and Transgenics" ActionBioscience.org. June 2004, http://www.actionbioscience.org/biotech/glenn.html

Global Knowledge Center on Crop Biotechnology (GKCCB). "Developing Countries should Adopt Biotech, Says UN Panel" 27 September 2002, http://www.biotechknowledge.com/BIOTECH/knowcenter.nsf/0/5aa5452798cfdf6e86256c4100478562/$FILE/Crop%20Biotech%20Update%2027%20September%202002.htm: 1–2.

Goad, M. "Mainers favor labels on engineered food" Portland Press Herald, 12 November 2002: 1B.

Golkany, I.M. *The Precautionary Principle* (Cato Institute: Washington, DC, 2001).

Goode, E. "Among the Inept, Researchers Discover, Ignorance is Bliss" 18 January 2000. http://www/nytimes.com/library/national/s. . .alth/011800hth-behavior-incompetents.html

Goreham, G. "Ethical Perspectives on the Transgenic Organism Debate" unpublished PowerPoint presentation of research data, North Dakota State University, 2002: 1–10.

Grady, C. "Science in the Service of Healing" Hastings Center Report, November–December 1998, 28(6): 34–8.

Grady, D. "Study Says Children with Cancer Suffer More Than Necessary" New York Times, 22 May 2007, http://query.nytimes.com/gst/fullpage.html?sec=health&res=9E0DEEDA123FF930A35751C0A9669C8B63

Graff, J. "Read Any Good Fruit?" *Time*, 2002, http://www.time.com/time/europe/eu/printout/0,9869,169369,00.html

Graham, G. *Genes: A Philosophical Inquiry* (Routledge: New York, 2002).

Gray, J. "Genetic Engineering Will Be Used for Destructive Purposes", in *The Ethics of Genetic Engineering*, M. Siedler (editor) (Greenhaven Press: Detroit, MI, 2005): 26–32.

Greenpeace International "The Precautionary Principle" accessed 22 May 2007, http://www.greenpeace.org/international/campaigns/trade-and-the-environment/the-precautionary-principle

Greenpeace International Summary of the Cartagena Protocol on Biosafety" http://archive.greenpeace.org/~geneng/highlights/bio/daily-updatem9.htm: 1–4.

Greenpeace. "No GMO releases through seed contamination" accessed 22 May 2007 http://archive.greenpeace.org/~geneng/highlights/gmo/EUseedcon.pdf: 1–4.

Gremmen, B. "Methodological Pragmatism in Bioethics: A Modest Proposal?" in *Pragmatist Ethics for a Technological Culture*, J. Keulartz, M. Korthals, M. Schermer, and T. Swierstra (editors) (Kluwer Academic Publishers: Boston, MA, 2002): 99–103.

Grossman, M.R. "European Community Legislation for Traceability and Labeling of Genetically Modified Crops, Food, and Feed", in *Labeling Genetically Modified Food*, P. Weirich (editor) (Oxford University Press: New York, 2007): 32–62.

Guha, R. "Radical American Environmentalism and Wilderness Preservation: a Third World Critique", in *Contemporary Moral Issues*, L.M. Hinman (editor) (Prentice Hall: Upper Saddle River, NJ): 605–15.

Gura, T. "New Genes Boost Rice Nutrients", *Science*, 13 August 1999, 285: 98–99.

Hallman, W.K., W.C. Hebden, H.L. Aquino, C.L., Cuite, J.T. Lang. Public Perceptions of Genetically Modified Foods: A National Study of American Knowledge and Opinion. (Publication number RR-1003-004). (New Brunswick, NJ: Food Policy Institute, Cook College, Rutgers-The State University of New Jersey, 2003).

Harder, B. "Deprived of Darkness, The unnatural ecology of artificial light at night," Science News. 161(16), April 2002, http://findarticles.com/p/articles/mi_m1200/is_16_161/ai_85465122/, accessed 6 July 2009.

Hardin, G. "On Not Feeding the Starving", in *The Right Thing to Do*, J. Rachels (editor) (Random House: New York, 1989): 184–90.

Hardin, G. "Carrying Capacity as an Ethical Concept," in *Applied Ethics: A Multicultural Approach*, L. May, S. Collins-Chobanian, and K. Wong (editors) (Prentice Hall: Upper Saddle River, NJ, 2002): 194–202.

Hare, R.M. *Moral Thinking: Its Levels, Method and Point* (Oxford University Press: New York, 1992).

Hargrove, E.C. *Foundations of Environmental Ethics* (Environmental Ethics Books: Denton, TX, 1989).

Hargrove, E.C. "Weak Anthropocentric Intrinsic Value", in *Environmental Ethics: An Anthology*, A. Light and H. Rolston III (editors) (Blackwell Publishers: Malden, MA, 2003): 175–90.

Harlander, S.K. "The Evolution of Modern Agriculture and its Future with Biotechnology", *Journal of the American College of Nutrition*, 2002, 21(3): 161S–165S.

Harremoes, P., D. Gee, M. MacGarvin, A. Stirling, J. Keys, B. Wynne, S. Guedes Vaz (editors). *The Precautionary Principle in the 20th Century* (Earthscan Publications Ltd: Sterling, VA, 2002).

Harris, A. "GE Corn Pollution Spawns 30 Lawsuits" http://www.organicconsumers.org/gefood/starlinklawsuits092102.cfm: 1–4.

Harrison, R.K. (editor) *The Encyclopedia of Biblical Ethics* (Testament Books: New York, 1992).

Hart, J. *Sacramental Commons: Christian Ecological Ethics* (Rowman & Littlefield Publishers, Inc.: Lanham, MD, 2006).

Hill, T.E. Jr. "Servility and Self-Respect", in Vice and Virtue in Everyday Life, C. Sommers and F. Sommers (editors) (Harcourt Brace College Publishers, 1997): 536–45.

Hindmarsh, R. and K. Hulsman. "Beyond the Lab! Precautionary Weakness", in *Recoding Nature: Critical Perspectives on Genetic Engineering*, R. Hindmarsh and G. Lawrence (editors) (UNSW Press: Sydney, NSW, 2004): 53–68.

Hindmarsh, R. and G. Lawrence. "Recoding Nature: Deciphering the script", in *Recoding Nature: Critical Perspectives on Genetic Engineering*, R. Hindmarsh and G. Lawrence (editors) (UNSW Press: Sydney, NSW, 2004): 23–40.

Hinman, L.M. "An Introduction to the Ethical Issues in Our Relationships with the Environment", in *Contemporary Moral Issues*, L.M. Hinman (editor) (Prentice Hall: Upper Saddle River, NJ): 559–66.

Ho, M.-Wan "Report on horizontal gene transfer-Department of Public Prosecution versus Gavin Harte and others, New Ross, Ireland" Institute of Science in Society, 22 March 1999, http://www.i-sis.org.uk/ireaff99.php: 1–7.

Ho, M.-Wan "Science, Ethics and Nature" Forward to Recoding Nature: critical perspectives on genetic engineering Richard Hindmarsh and Geoffrey Lawrence (editors) (UNSW Press: Sidney, NSW, 2004): 15–20.

Ho, M.-Wan "Perils Amid Promises of Genetically Modified Foods", in *Applied Ethics: A Multicultural Approach*, L. May, S. Collins-Chobanian, and K. Wong (editors) (Prentice Hall: Upper Saddle River, NJ, 2006): 282–92.

Holmes, R.L. *Basic Moral Philosophy* (Thomson Wadsworth: Belmont, CA, 2003).

Holmes, R.L. "Richard Taylor, 1919-2003: memorial minutes" in Proceedings and Addresses of The American Philosophical Association, 2004, 77(5): 170–1.

Holt, J. "Prevalence and Management of Herbicide-Resistant Weeds" in Proceedings: Ecological and Agronomic Consequences of Gene Flow from Transgenic Crops to Wild Relatives, Ohio State University, 5–6 March 2002: 47–57.

House of Commons Select Committee on Environment, Food, and Rural Affairs. "Memorandum submitted by the Federal Ministry of Consumer Protection, Food and Agriculture, Germany: GM GOVERNMENT DECISION" April 2004, http://www.publications.parliament.uk/pa/cm200304/cmselect/cmenvfru/607/607we12.htm

Howard, T. and J. Rifkin. *Who Should Play God?: The Artificial Creation of Life and What it Means for the Future of the Human Race* (Delacorte Press: New York, 1977).

Huffman, W.E. "Production, Identity Preservation and Labeling in a Market Place with Genetically Modified (GM) and Non-GM Foods" Plant Physiology, January 2004: 3–10.

Huffman, W.E., J. Shogren, M. Rousu, A. Tegene. "Consumer Willingness to Pay for Genetically Modified Food Labels in a Market with Diverse Information: Evidence from Experimental Auctions," *Journal of Agricultural and Resource Economics*, December 2003, 28: 481–502.

Hume, D. "An Enquiry Concerning The Principles of Morals", in Hume: Moral and Political Philosophy, H.D. Aiken (editor) (Hafner Press: New York, 1948a): 173–291.

Hume, D. *A Treatise of Human Nature*, in *Moral and Political Philosophy*, H.D. Aiken (editor) (Hafner Press: New York, 1948b): 3–169.

Hund, A. "Monsanto: Visionary or Architect of Bioserfdom? A Global Socio-Economic Examination of Genetically Modified Organisms" http://www.organicconsumers.org/Monsanto/bioserf.cfm: 1–34.

Hutchings, J. "Tradition and test tubes: Maori and GM", in *Recoding Nature: Critical Perspectives on Genetic Engineering*, R. Hindmarsh and G. Lawrence (editors) (UNSW Press: Sydney, NSW, 2004): 179–91.

Ingelfinger, F.J. "Informed (but Uneducated) Consent," in Ethical Issues in Professional Life, J.C. Callahan (editor) (Oxford University Press, 1988): 180–2.

International Assessment of Agricultural Knowledge, Science and Technology for Development. *Global Summary for Decision Makers.* April 2008, http://www.agassessment.org/docs/Global_SDM_050508_FINAL.pdf

International Food Information Council (IFIC). Food Biotechnology: A Study of U.S. Consumer Attitudinal Trends 2006 Report.

Ismeal, Y., R. Bennett, and S. Morse. "Benefits of Bt Cotton Use by Smallholder Farmers in South Africa", *AgBioForum*, 2002 5(1): 1–5.

Johnson, R.U. "Aesthetics and Conservation", in *American Environmentalism: Readings in Conservation History*, R.F. Nash (editor) (McGraw-Hill Publishing Company: New York, 1990): 90–3.

Johnson, V.R. and A. Gunn. *Studies in American Tort Law* (Carolina Academic Press: Durham, NC, 1994).

Joly, P. and S. Lemarie. "Industry Consolidation, Public Attitude, And The Future Of Plant Biotechnology In Europe", AgBioForum, 1998, 1(2), http://www.agbioforum.org/v1n2a10-joly.htm: 85–90.

Jonas, H. "Technology and Responsibility: The Ethics of an Endangered Future", in *Responsibilities to Future Generations*, E. Partridge (editor) (Prometheus Books: Buffalo, NY, 1981) 23–36.

Jones, J. "Transgenic Plant Technology: Threat or Promise?" Accessed 22 May 2007, www.ifgene.org/jones.htm

Kalaitzandonakes, N.L.A. Marks, and S.S. Vickner. "Consumer Response to Mandated Labeling of Genetically Modified Foods", in *Labeling Genetically Modified Food*, P. Weirich (editor) (Oxford University Press: New York, 2007): 106–27.

Kant, I. *Groundwork of the Metaphysic of Morals*, translated by H.J. Paton (Harper & Row Publishers: New York, 1956).

Kant, I. *The Metaphysics of Morals*, translated by M.J. Gregor (Harper & Row Publishers: New York, 1964).

Kant, I. "Crimina Carnis", in *Philosophical Perspectives on Sex and Love*, R.M. Stewart (editor) (Oxford University Press: New York, 1995): 144–5.

Kant, I. "Why We Have No Obligations to Animals", in *The Right Thing to Do: Basic Readings in Moral Philosophy*, J. Rachels (editor) (Random House: New York, 1999): 209–11.

Kass, L.F. "The Wisdom of Repugnance", *New Republic*, 1997, 2: 17–26.

Keeton, W.P., D. Dobbs, R. Keeton, and D. Owen. *Prosser and Keeton on Torts* (West Publishing Co: St. Paul, MN, 1984).

Kelemu, S., G. Mahuku, M. Fregene, D. Pachio, N. Johnson, L. Clavert, I. Rao, R. Buruchara, T. Amede, P. Kimani, R. Kirkby, S. Kaaria, K. Ampfo. "Harmonizing the agricultural biotechnology debate for the benefit of African farmers", *African Journal of Bioechnology*, November 2003, 2(11): 394–416.

Keulartz, J., M. Korthals, M. Schermer, and T. Swierstra. "Ethics in a Technological Culture", in *Pragmatist Ethics for a Technological Culture*, J. Keulartz, M. Korthals, M. Schermer, and T. Swierstra (editors) (Kluwer Academic Publishers: Boston, MA, 2002): 3–21.

Khatchadourian, H. "Terrorism and Morality", in *Applied Ethics: A Multicultural Approach*, L. May, S. Collins-Chobanian, K. Wong (editors) (Prentice Hall: Upper Saddle River, NJ, 2001): 291–300.

Kheel, M. *Nature Ethics: An Ecofeminist Perspective* (Rowman & Littlefield: Lanham, MD, 2008).

Kiesel, K and S.B. Villas-Boas. "Got Organic Milk? Consumer Valuations of Milk Labels after the Implementation of the USDA Organic Seal", *Journal of Agricultural & Food Industrial Organization*, 5, 2007: 1–38.

Klee, K. "Frankenstein Foods" Newsweek, 13 September 1999: 33–5.

Keulartz, J., M. Korthals, M. Schermer, and T. Swierstra (editors). *Pragmatist Ethics for a Technological Culture* (Kluwer Academic Publishers: Boston, MA, 2002).

Korthals, M. "A Multi-Practice Ethics of Domesticated and "Wild" Animals" in *Pragmatist Ethics for a Technological Culture*, Jozef Keulartz, Michiel Korthals, Maartje Schermer, and Tsalling Swierstra (editors) (Kluwer Academic Publishers: Dordrecht, The Netherlands 2002).

Korthals, M. *Before Dinner* (Kluwer Academic Publishers: Dordrecht, The Netherlands 2004)

Krebs, A.V. "Corporate Agriculture Continues to Dominate Farm Policies in Washington" http://www.organicconsumers.org/corp/farmbill12601.cfm: 1–15.

Kripke, S.A. *Naming and Necessity* (Harvard University Press: Cambridge, MA, 1980).

Kumar, S. "Sterilization by Quinacrine Comes Under Fire in India" The Lancet, London, 17 May 1997.

Kunkel, H.O. *Human Issues in Animal Agriculture* (Texas A & M University Press: College Station, TX, 2000).

Kuzma, J. "Food Biotechnology: An Ethical Perspective Commentary on Prof. Paul Thompson's Book" APA Central Division Meeting, 19 April 2007: 1–10.

Kyne, P. "Official Says Many Chinese Fear Genetically Modified Food" Dow Jones International News, 8 November 2002: 1–2.

Laidlaw, S. "StarLink Fallout Could Cost Billions" www.organicconsumers.org/gefood/surrender.cfm: 1–6.

Lawson, M.E., R.G. Stover, D.J. Hess, and G.A. Gorham. *Agricultural Biotechnology and the Midwestern Agricultural Producer* (Brookings, SD: SDSU Rural Life & Census Data Center, May 2003): 1–31.

Lee, K. "The Source and Locus of Intrinsic Value: A Reexamination", in Environmental Ethics: An Anthology (Blackwell Publishers: Malden, MA, 2003): 154–62.

Lehman, H. *Rationality and Ethics in Agriculture* (The University of Idaho Press: Moscow, ID, 1995).

Leiser, B. "Is Homosexuality Unnatural?" in *The Right Thing to Do: Basic Readings in Moral Philosophy*, J. Rachels (editor) (McGraw-Hill Education: Boston, MA, 2007): 126–34.

Lenski, R.E. "Evaluating the fate of genetically modified microorganisms in the environment: are they inherently less fit?", *Experentia*, 1993, 49(3): 201–9.

Leonard, W.R. "Food for Thought" Scientific American.com, 13 November 2003, http://www.sciam.com/article.cfm?articleID=0007B7DC-6738-1DC9-AF71809EC588EEDF: 1–5.

Leopold, A. "The Land Ethic", in *Applied Ethics: A Multicultural Approach*, L. May, S. Collins-Chobanian, and K. Wong (editors) (Prentice Hall: Upper Saddle River, NJ, 2001): 129–41.

Levine, C. "Placebos and HIV: Lessons Learned" Hastings Center Report, November–December 1998, 28(6): 43–8.

Lheureux, K., M. Libeau-Dulos, H. Nilsagard, E. Rodriguez Cerezo, K. Menrad, M. Menrad, D. Vorgrimler. "Review of GMOs Under Research and Development and In the Pipeline in Europe" European Science and Technology Observatory, Joint Research Centre: European Commission, Institute for Prospective Technological Studies, http://www.jrc.es/gmoreview.pdf, March 2003: 1–127.

Light, A. "Compatabilism in Political Ecology", in *Environmental Pragmatism*, A. Light and E. Katz (editors) (Routledge: New York, 1996): 161–84.

Light, A. "A Modest Proposal: Methodological Pragmatism for Bioethics", in *Pragmatist Ethics for a Technological Culture*, J. Keulartz, M. Korthals, M. Schermer, and T. Swierstra (editors) (Kluwer Academic Publishers: Boston, MA, 2002): 79–97.

Light, A. and E. Katz (editors). *Environmental Pragmatism* (Routledge: New York, 1996).

Light, A. and H. Rolston, III (editors). *Environmental Ethics: An Anthology.* (Blackwell Publishers: Great Britain, 2003).

Liu, Y., B.E. Tabashnik, T.J. Dennehy, A.L. Patin, and A.C. Bartlett. "Development time and resistance to *Bt* Crops", Nature, 5 August 1999, 400: 519.

Livingston, A. "Kantian Duties Towards the Nonrational: On the Possibility of a Kantian Ethics of Environment. Accessed 12/23/2003. http://artsandscience.concordia.ca/philosophy/Gnosis/vol_v/Kant_Enviro.pdf

Locke, J. Second Treatise of Government (Harlan Davidson, Inc: Wheeling IL, 1982) London Radio Service (LRS) "Advertising authority upholds GM ads complaints" 1999 http://www.netlondon.com/news/1999-32/3AA45F92303120C3802.html: 1–3.

Lucassen, E. "The Ethics of Genetic Engineering", *Journal of Applied Philosophy*, 1996, 13(1): 51–61.

Mack, S.K. "Consumers worry about pesticide use UM survey" Bangor Daily News, 13 November 2002: 1.

Mackey, M.A. and C.R. Santerre. "Agricultural Biotechnology Provides Many Benefits Worldwide", in *The Ethics of Genetic Engineering*, M. Siedler (editor) (Greenhaven Press: Detroit, MI, 2005): 42–57.

Manning, R.C. *Speaking from the Heart: A Feminist Perspective on Ethics* (Rowman & Littlefield Publishers, Inc: Lanham, MD, 1992).

Marsh, G.P. "Human Responsibility for the Land", in *American Environmentalism: Readings in Conservation History*, R.F. Nash (editor) (McGraw-Hill Publishing Company: New York, 1990): 40–4.

Marshall, R. "Wilderness", in *American Environmentalism: Readings in Conservation History*, R.F. Nash (editor) (McGraw-Hill Publishing Company: New York, 1990): 160–5.

Marx, K. "Capital, Volume One," The Marx-Engels Reader (W. W. Norton & Company: New York, 1978a).

Marx, K. "Capital, Volume Three," The Marx-Engels Reader (W. W. Norton & Company: New York, 1978b).

Mather, R. *A Garden of Unearthly Delights: Bioengineering and the Future of Food* (Dutton, Penguin Group: New York, 1995).

Maulik, S. and S.D. Patel. *Molecular Biotechnology: Therapeutic Applications and Strategies* (Wiley-Liss, Inc.: New York, 1997).

McGarity, T.O. "Frankenfood Free: Consumer Sovereignty, Federal Regulation, and Industry Control in Marketing and Choosing Food in the United States", in *Labeling Genetically Modified Food*, P. Weirich (editor) (Oxford University Press: New York, 2007): 128–50.

McGloughlin, M. "Ten Reasons Why Biotechnology Will Be Important To The Developing World" AgBioForum, 1999, 2(3&4), http://www.agbioforum.org/n234/v2n34a04-mcgloughlin.htm: 163–74.

Meadows, D., D.L. Meadows, and J. Randers. *Beyond the Limits* (Chelsea Green Publishing Company: Post Mills, Vermont, 1992).

Mepham, B. "'Wurde Der Kreatur' and the Common Morality", *Journal of Agricultural and Environmental Ethics*, 2000, 13: 65–78.

Merriman, W.G. "Innate Morality." Accessed 12/23/2003. http://www.iserv.net/~merriman/natmoral.htm

Mill, J.S. *On Liberty*, in *Utilitarianism, on Liberty and Considerations on Representative Government* (J.M. Dent and Sons Ltd.: New York, 1988a): 69–185.

Mill, J.S. *Utilitarianism*, in *Utilitarianism, on Liberty and Considerations on Representative Government* (J.M. Dent and Sons Ltd.: New York, 1988b): 1–67.

Miller, R.B. "How the Belmont Report Fails" Essays in Philosophy, 4(2), accessed 1 July 2003, http://www.humboldt.edu/~essays/miller.html

Miller, R.B. "Rethinking Pragmatism as A Philosophical Method", Presented at North Dakota State University's Department of History, Philosophy, and Religious Studies' colloquia series, October 2007, revised edition July 2009: 1–39.

Moeller, D. "GMO Liability Threats for Farmers" http://www.mindfully.org/GE/GE3/GMO-Liability-For-FarmersNov01.htm: 1–9.

MolecularFarming.com "Contamination of food crops we can help avoid" http://www.molecularfarming.com/contamination.htm: 1–4.

Monsanto. Monsanto, Inc. Technology/Stewardship Agreements (2002 and 2003).

Monsanto. Pledge, Accessed 14 June 2004, http://www.monsanto.com/monsanto/layout/our_pledge/letter_grant.asp

Monsanto. "Monsanto to Realign Research Portfolio, Development of Roundup Ready Wheat Deferred: Decision Follows Portfolio Review, Consultation with Growers" Accessed 14 June 2006, http://www.monsanto.com/monsanto/layout/media/04/05-10-04.asp

Monsanto. "Monsanto Reaffirms Commitment to Product Stewardship" 2000, Accessed 28 June 2007, http://www.monsanto.co.uk/news/2000/december2000/041200monsanto.html: 1.

Moore, G.E. *Principia Ethica* (Prometheus Books: Buffalo, New York, 1988).

Moran, M. *Rethinking the Reasonable Person* (Oxford University Press: New York, 2007).

Muir, J. "A Voice for Wilderness", in *American Environmentalism: Readings in Conservation History*, R.F. Nash (editor) (McGraw-Hill Publishing Company: New York, 1990): 94–7.

Murrell, J.C. and L.M. Roberts (editors). *Understanding Genetic Engineering* (Ellis Horwood Limited: Chichester, England, 1989).

Myerson, A.R. "Monsanto Paying Delta Farmers to Settle Genetic Seed Complaints", *The New York Times*, 22 May 2007, http://www.nytimes.com/library/national/science/022498monsanto-seed.html

Nader, R. "Genetically Modifying Food Crops is not Ethical", in *The Ethics of Genetic Engineering*, L. Yount (editor) (Greenhaven Press: Detroit, MI, 2002): 48–51.

Nash, R.F. "Wilderness Advocacy", in *American Environmentalism: Readings in Conservation History*, R.F. Nash (editor) (McGraw-Hill Publishing Company, New York, 1990): 94–7.

National Commission for the Protection of Human Subjects of Biomedical and Behavioral Research (NCPHSBBR). The Belmont Report: Ethical Principles and Guidelines for the Protection of Human Subjects of Research 18 April 1979, http://206.102.88.10/ohsrsite/guidelines/belmont.html

New Zealand Herald. "Lobbyist says seed companies must be liable for GE contamination" http://www.nzherald.co.nz/storydisplay.cfm?thesection=news&thesubsection=&storyID=2997567&reportID=53009, 4 October 2002: 1–2.

Nicholl, D.S.T. "An Overview of the Ethical Debates in Genetic Engineering", in *The Ethics of Genetic Engineering*, M. Siedler (editor) (Greenhaven Press: Detroit, MI, 2005): 11–15.

Nishiguchi, M. and F. Motoyoshi. "Resistance Mechanisms of Tobacco Mosaic Virus Strains in Tomato and Tobacco", in *Plant Resistance to Viruses* (John Wiley & Sons Ltd.: Chichester, England, 1987).

Non-StarLink Farmer Litigation. "Frequently Asked Questions" Accessed 22 June 2007, http://www.non-starlinkfarmerssettlement.com/faq.php3

Nordlee, J.A., S.L. Taylor, J.A. Townsend, L.A. Thomas, R.K. Bush. "Identification of Brazil-Nut Allergen in Transgenic Soybeans", *New England Journal of Medicine*, 1996, 334: 688–92, http://www.nejm.org/content/1996/0334/0011/0688.asp

Norris, F. "Public Misinformed on Genetically Modified Foods" The New York Times, 17 December 1999, http://www.nytimes.com/library/financial/columns/121799norris-column.html

Nozick, R. *Anarchy, State and Utopia* (Basil Blackwell: Oxford, UK, 1974).

Nuffield Council on Bioethics. *Genetically Modified Foods: The Ethical and Social Issues* (Nuffield Council on Bioethics: London, GB, 1999).

Nuffield Council on Bioethics. *The Use of Genetically Modified Crops in Developing Countries* (Nuffield Council on Bioethics: London, GB, 2003).

Office of Technology Assessment. "Impacts of Applied Genetics: Micro-Organisms, Plants, and Animals", Congress of the United States, 1981: 197–265.

Olmsted, F.L. "The Value of the Care of Parks", in *American Environmentalism: Readings in Conservation History*, R.F. Nash (editor) (McGraw-Hill Publishing Company: New York, 1990): 45–51.

Oneworld. "ENGINEERING SOLUTIONS TO MALNUTRITION: GRAIN" Accessed 22 May 2007, http://www.biotech-info.net/engineering.html

Organic Valley "G.M.O. Alert" Accessed 1 July 2003, http://www.organicvalley.com/member/forum-gmo1.html

O'Neill, J. "The Varieties of Intrinsic Value", in *Environmental Ethics: An Anthology*, A. Light and H. Rolston III (editors) (Blackwell Publishers: Malden, MA, 2003): 131–42.

Paarlburg, R. "Promoting Genetically Modified Crops in Developing Countries is Ethical", in *The Ethics of Genetic Engineering*, L. Yount (editor) (Greenhaven Press: Detroit, MI, 2002): 52–64.

Parfit, D. *Reasons and Persons* (Oxford University Press: New York, 1992).

Pariza, M.W. "A Scientific Perspective on Labeling Genetically Modified Food", in *Labeling Genetically Modified Food*, P. Weirich (editor) (Oxford University Press: New York, 2007): 3–9.

Pascal, B. *Pensees*. Translated by A.J. Krailsheimer (Penguin Books: New York, 1966).

Pennisi, E. "Gene evolution: Cannibalism and prion disease might have been rampant in ancient humans", *Science*, 11 April 2003, 300(5617): 227–28.

Perez, J. "The Future of G.M.OS (Genetically Manipulated Organisms): The G.M. Homo-Neandertal (Genetically Manipulated Homo-Neandertal's Organism)" September 2000, http://www.genum.com/gmo_evolution/gmo_evolution.htm: 1–9.

Peters, P. *Biotechnology: A Guide to Genetic Engineering* (WCB McGraw-Hill: New York, 1993).

Peterson, G., S. Cunningham, L. Deutsch, J. Erikson, A. Quinlan, E. Raez-Luna, R. Tinch, M. Troell, P. Woodbury, and S. Zens. "The Risks and Benefits of Genetically Modified Crops: A Multidisciplinary Perspective", *Ecology and Society*, 4(1), 27 March 2000, http://www.ecologyandsociety.org/vol4/iss1/art13/

Pew Charitable Trusts "Biotech and Organic: Reaching Détente" AgBiotech Buzz: Peaceful Coexistence, 2000, 6(2), http://pewagbiotech.org/buzz/printfull.php3?IssueID=13

Pew Initiative on Food and Biotechnology (Pew Initiative). "Have Transgenes, Will Travel: Issues Raised by Gene Flow from Genetically Engineered Crops" http://pewagbiotech.org/resources/issuebriefs/geneflow.pdf, August 2003a: 1–15.

Pew Initiative on Food and Biotechnology (Pew Initiative). "U.S. vs. EU: An Examination of the Trade Issues Surrounding Genetically Modified Food" http://pewagbiotech.org/resources/issuebriefs/europe.pdf, August 2003b: 1–35.

Pew Initiative on Food and Biotechnology (Pew Initiative). "Public Sentiment About Genetically Modified Food" http://pewagbiotech.org/research/2006update/, 2001–6.

Pierce, C. *Immovable Laws, Irresistible Rights: Natural Law, Moral Rights, and Feminist Ethics* (University Press of Kansas: Lawrence, KA, 2001).

Pollack, A. "Talks on Biotech Food Day in Montreal Will See U.S. Isolated" The New York Times, 24 January 2000a, http://nytimes.com/library/national/science/012400sci-gm-regulate.html

Pollack, A. "Talks on Biotech Food Turn on a Safety Principle" The New York Times, 28 January 2000b, http://nytimes.com/yr/mo/day/news/world/biotech-food-gm.html

Pollack, A. "We Can Engineer Nature. But Should We?" The New York Times, 6 February 2000c, http://www.nytimes.com/library/review/020600genetic-food-review.html

Pollack, A. "In lean times, biotech grains are less taboo" The New York Times, 21 April 2008.

Pound, R. *An Introduction to the Philosophy of Law* (Yale University Press: New Haven, CN, 1959).

Powell, J. "US Farmers Union Calls for Moratorium on GMO Patents" http://www.organicconsumers.org/patent/farmersunion031102.cfm: 1–3.

Powledge, T.M. "GM animals could threaten environment" The Scientist, 28 May 2003, http://www.biomedcentral.com/news/20020822/06/: 1–2.

Prakash, C.S. "Academic and Science Community Applauds WTO Ruling" AgBioWorld, 7 Feb. 2006, http?/www.agbioworld.org/biotech-info/pr/wtoruling.html

Pray, C., J. Huang, R. Hu, and S. Rozelle. "Five Years of Bt Cotton in China – the Benefits Continue", *The Plant Journal*, 31(4), 2002: 423–30.

Preston, C. "GM soy affects posterity, or misuse of science?" AgBioView, 31 October 2005, http://www.agbios.com/main.php?action=ShowNewsItem&id=6980

Pusztai, A. "Genetically Modified Foods: Are They a Risk to Human/Animal Health?" ActionBioscience.org, June 2001, http://www.actionbioscience.org/biotech/pusztai.html

Qaim, M. and D. Zilberman. "Yield effects of genetically modified crops in developing countries", *Science*, 2003, 299: 900–2.

Rachels, J. *The Right Thing to Do: Basic Readings in Moral Philosophy* (Random House: New York, 1989).

Rachels, J. *The Elements of Moral Philosophy* (McGraw Hill: Boston, 2003).

Rampton, R. "Canada Wheat Board cheers Monsanto GMO decision" PlanetArk world Environment News, Accessed 22 May 2007, www.planetark.com/dailynewsstory.cfm/newsid/25064/story.htm

Ranii, D. "Drug Pairs Don't Always Turn Out Peachy", The News & Observer, 4 February 2000, 1D.

Rawls, J. *A Theory o f Justice* (Harvard University Press, 1971).

Regan, T. *All That Dwell Therein: Animal Rights and Environmental Ethics* (University of California Press: Berkeley, CA, 1982): 165–83.

Regan, T. "The Case For Animal Rights", in *The Right Thing to Do*. J. Rachels (editor) (Random House: New York, 1989): 211–25.

Reiss, M.J. and R. Straughan. *Improving Nature? The science and ethics of genetic engineering* (Cambridge University Press: Cambridge, UK, 2001).

Rescher, N. *Risk* (University of America Press: Lanham, MD, 1983).

Reuters. "U.S. seizes tainted toothpaste from China" 1 June 2007, http://www.reuters.com/article/topNews/idUSWAT00764420070601

Richards, A.I. "Characteristics of Ethical Systems in Primitive Human Society", in Biology & Ethics:Institute of Biology Symposia: Number 18 F.J. Ebling (editor) (Academic Press: New York, 1969).

Rifkin, J. "The Biotech Century: Playing Ecological Roulette with Mother Nature's Designs", Accessed 22 May 2007, http://www.emagazine.com/view/?632

Ripstein, A. *Equality, Responsibility, and the Law* (Cambridge University Press: New York, 2001).

Rivermouth Action Group, Inc (RAG). "The Worldwide Debate Over GMO Foods: Three Key Disagreements" 2003, http://www.rag.org.au/baa/gmofoods.htm: 1–4.

Robert, J.S. and F. Baylis. "Crossing Species Boundaries", *American Journal of Bioethics*, 2003, 3(3): 1–13.

Roberts, R.M. "Genetically Modified Organisms for Agricultural Food Production: The Extent of the Art and the State of the Science", in *Labeling Genetically Modified Food*, P. Weirich (editor) (Oxford University Press: New York, 2007): 10–16.

Robinson, S.W. III, "Opinion in Canterbury v. Spence" U.S. Court of Appeals, District of Columbia Circuit, 464 Federal Reporter, 2nd Series (West Publishing Company: Saint Paul, MN, 1972).

Rollin, B.E. "The 'Frankenstein Thing': The Moral Impact of Genetically modified organisms of Agricultural Animals on Society and Future Science" Agricultural Bioethics: Implications of Agricultural Biotechnology (Iowa State University Press: Ames, Iowa 1990): 292–308.

Rollin, B.E. "Environmental Ethics and International Justice", in *Applied Ethics: A Multicultural Approach*, L. May and S.C. Sharratt (editors) (Prentice Hall: Englewood Cliffs, NJ, 1994): 75–84.

Rollin, B.E. *The Frankenstein Syndrome*. (Cambridge University Press: Cambridge, UK, 1996)

Rollin, B.E. "The Moral Status of Nonhuman Things", in *Living Well: Introduction Readings in Ethics*, S. Luper (editor) (Harcourt Brace College Publishers: Fort Worth, TX, 2000): 430–36.

Rollin, B.E. *Science and Ethics* (Cambridge University Press: New York, 2006).

Rolston, H. III, "Are Values in Nature Subjective or Objective?", in *Environmental Philosophy: A Collection of Readings*, R. Elliot and A. Gare (editors) (Pennsylvania State University Press: University Park, PA, 1983): 135–65.

Rolston, H. III, *Environmental Ethics* (Temple University Press: Philadelphia, PA, 1988).

Rolston, H. III, *Conserving Natural Value* (Columbia University Press: New York, 1994).

Rolston, H. III, "Challenges in Environmental Ethics", in *Contemporary Moral Issues*, L.M. Hinman (editor) (Prentice Hall: Upper Saddle River, NJ, 2000): 587–604.

Rolston, H. III, "Value in Nature and the Nature of Value", in *Environmental Ethics: An Anthology*, A. Light and H. Rolston III (editors) (Blackwell Publishers: Malden, MA, 2003): 143–53.

Rolston, H. III, *Science and Religion* (Templeton Foundation Press: Philadelphia, PA, 2006).

Ross, W.D. *The Right and the Good* (Hackett Publishing Company: Indianapolis, IN, 1988).

Royal Society of London, the U.S. National Academy of Sciences, the Brazilian Academy of Sciences, the Chinese Academy of Sciences, the Indian National Academy of Sciences, the Mexican Academy of Sciences, and the Third World Academy of Sciences Transgenic Plants and World Agriculture (National Academy Press: Washington, D.C., July 2000).

Royal Society of London. "Genetically modified plants for food use and human health—an update" Policy Document 4/02, February 2002, www.royalsoc.ac.uk: 1–22.

Rubel, A and R. Streiffer. "Respecting the Autonomy of European and American Consumers: Defending positive labels on GM Foods", *Journal of Agriculture and Environmental Ethics*, 2005, 18: 75–84.

Russow, L.M. "Chapter 5: Environment", in *Life Science Ethics*, G.L. Comstock (editor) (Iowa State Press: Ames, IA, 2002): 93–123.

Sagoff, M. "On Preserving the Natural Environment", in *Today's Moral Problems*, R.A. Wasserstrom (editor) (Macmillan Publishing Co.: New York, 1979): 613–23.

Sagoff, M. *Price, Principle, and The Environment* (Cambridge University Press: New York, 2004).

Sagoff, M. *The Economy of the Earth* (Cambridge University Press: New York, 2008).

Sandler, R. "A Virtue Ethics Perspective on Genetically Modified Crops" in *Environmental Virtue Ethics*, R. Sandler and P. Carfaro (editors) (Rowman Littlefield Publishers, Inc: Lanham, MD, 2005): 215–231.

Sandrasagra, M. "Western Demand Shifts Away From Transgenic Produce" http://www.twnside.org.sg/title/shifts.htm: 1–3.

Sartre, J.P. *Being and Nothingness*, translated by Hazel E. Barnes (Simon & Schuster, Inc.: New York, 1956).

Saxena, D., S. Florest, and G. Stotzky. "Insecticidal Toxin in Root Exudates from *Bt* Corn", *Nature*, 2 December 1999, 402: 480.

Schmickle, S. "Resistance to Genetically Altered Crops Grows" Anderson Independent Mail, January 6, 2000.

Schmit, J. "U.S. food imports outrun FDA resources" USA TODAY, Accessed 8 June 2007. http://www.usatoday.com/money/industries/food/2007-03-18-food-safety-usat_N.htm

Schmitz, S.A. "Agricultural Biotechnology Companies Sacrifice the Environment for Economic Gain", in *The Ethics of Genetic Engineering*, M. Siedler (editor) (Greenhaven Press: Detroit, MI, 2005): 58–67.

Schubert, R. "Nebraska organic farmer bears costs of GM testing" 11 April 2001, http://www.cropchoice.com/leadstry.asp?recid=289

Schubert, R. "Colorado wheat grower sees transgenic threats to family farmers" http://www.cropchoice.com/leadstry.asp?recid=255: 1–7.

Sessions, G. and B. Devall. "Deep ecology", in *American Environmentalism:Readings in Conservation History*, R.F. Nash (editor) (McGraw-Hill Publishing Company: New York, 1990): 36–9.

Shadid, A. "Genetic Drift Threatens US Organic Farmers" http://www.organicconsumers.org/Organic/geneticdrift.cfm: 1–5.

Shamoo, A.E. and D.B. Resnik. *Responsible Conduct in Research* (Oxford University Press: Oxford, UK, 2002).

Shelton, A.M., and R.T. Roush. "False Reports and the Ears of Men", *Nature Biotechnology*, September 1999, 17: 832.

Sherlock, R. "Three Concepts of Genetic Trespassing", in *Ethical Issues in Biotechnology*, R. Sherlock and J.D. Morrey (editors) (Rowman & Littlefield Publishers, Inc.: Lanham, MD, 2002): 149–59.

Sherlock, R. and J.D. Morrey (editors). *Ethical Issues in Biotechnology* (Rowman & Littlefield Publishers, Inc.: Lanham, MD, 2002).

Shiva, V. "Development, Ecology and Women", in *Applied Ethics: A Multicultural Approach*, L. May, S. Collins-Chobanian, and K. Wong (editors) (Prentice Hall: Upper Saddle River, NJ, 2002): 177–86.

Siegrist, M. 2003. "Perception of gene technology, and food risks: results of a survey in Switzerland", *Journal of Risk Research*, 2003, 45–60.

Singer, P. "Famine, Affluence, and Morality", in *Applied Ethics: A Multicultural Approach*, L. May, S. Collins-Chobanian, and K. Wong (editors) (Prentice Hall: Englewood Cliffs, NJ, 2006): 253–62.

Smith, A. *The Wealth of Nations* (Alfred A. Knopf: New York, 1991).

Smith, B.H. "Belief and Resistance: A Symmetrical Account", in *Questions of Evidence*, J. Chandler, A.I. Davidson, and H. Harootunian (editors) (University of Chicago Press: Chicago, IL, 1994): 139–53.

Snyder, G. "Respect For Nature", in American Environmentalism: Readings in Conservation History, R.F. Nash (editor) (McGraw-Hill Publishing Company: New York, 1990): 166–70.

Stearns, J.B. "Ecology and the Indefinite Unborn", in *Today's Moral Problems*, R.A. Wasserstrom (editor) (Macmillan Publishing Co.: New York, 1979): 602–13.

Stein, J. "Biotech crops may be safer" Wisconsin State Journal, 9 July 2003 http://www.madison.com/wisconsinstatejournal/special/52505.php

Stevenson, C.L. *Facts and Values* (Yale University Press: New Haven, 1963).

Stilwell, M. and B. Van Dyke. "An Activist's Handbook On Genetically Modified Organisms and the WTO" Center for International Environmental Law, July, 1999, http://www/consumerscouncil.org/policy/handbk799.htm: 1–16.

Story, N. and L.M. Ward. *American Business Law and The Regulatory Environment* (South-Western Publishing Co.: Cincinnati, OH, 1989).

Straughan, R. "Ethics, Morality and Animal Biotechnology" Biotechnology and Biological Science Research Council, 29 September 2000 http://www.bbsrc.ac.uk/tools/download/ethics_animal_biotech/ethics_animal_biotech.pdf: 1–28.

Strieffer, R. "In Defense of the Moral Relevance of Species Boundaries", *American Journal of Bioethics*, 2003, 3(3): 37–8.

Streiffer, R. and S.G. Ortiz, "Ethical Issues Arising from Enviropigs: A Cooperative Learning Ethics Case Study" University of Wisconsin – Madison Extension, 19–20 November 2002: 1–82.

Sunstein, C.R. *Risk and Reason* (Cambridge University Press: New York, 2002).

Sunstein, C.R. *Laws of Fear* (Cambridge University Press: New York, 2005).

Syvanen, M. "Conserved Regions in Mammalian ß-globins: Could They Arise by Cross Species Gene Exchange?" Journal of Theoretical Biology, 1984, 107: 685–96.

Syvanen, M. "Cross-species Gene Transfer: Implications for a New Theory of Evolution", *Journal of Theoretical Biology*, 1985, 112: 333–43.

Syvanen, M. "On the Occurrence of Horizontal Gene Transfer Among an ArbitrarilyChosen Group of 26 Genes" *Journal of Molecular Evolution*, 2002a, 54: 258–66.

Syvanen, M. "Recent emergence of the modern genetic code: a proposal" *TRENDS in Genetics*, 2002b, 18(5): 245–48.

Tabashnik, B.E., R.T. Roush, E.D. Earle, and A.M. Shelton, "Resistance to Bt Toxins", *Science*, 7 January 2000, 287: 42.

Tabor, J.M., editor, *Genetically modified organisms Technology in Industrial Pharmacy: Principles and Applications* (Marcel Dekker, Inc.: New York, 1989).

Taylor, R. Superhumans: Like it or not, in a few short years we'll have the power to control our own evolution. New Scientist, 1998, 160, 24. Retrieved November 26, 2007, from http://geneticsandsociety.org/article.php?id=129

Taylor, P.W. *Respect For Nature: A Theory of Environmental Ethics* (Princeton University Press: Princeton, NJ, 1986):

Team Canada. "Canada's International Business Strategy: 2001–2002" http://atn-riae.agr.ca/info/can/e3204.htm

Tegen, A., W.E. Huffman, M. Rousu, and J.F. Shogren. "The Effects of information on Consumer Demand for Biotech foods: Evidence from Experimental Auctions" United States Department of Agriculture, Electronic Report form the Economic Research Service, March 2003, www.ers.usda.gov: i–28.

Teitel, M. "Genetic Engineering is not Ethical", in *The Ethics of Genetic Engineering*, L. Yount (editor) (Greenhaven Press: Detroit, MI, 2002): 24–33.

The Holy Bible: Revised Standard Version (Penguin Group: New York, 1974).

Third World Network Biosafety Information Service (TWNBIS). "GE Wheat Dumping in Africa" Accessed 14 June 2004. http://www.twnside.org.sg/title/service92.htm

Thiroux, J.P. *Ethics: Theory and Practice* (Prentice Hall: Upper Saddle River, NJ, 2004): 428–53.

Thompson, J.L. "Preservation of Wilderness and the Good Life", in *Environmental Philosophy: A Collection of Readings*, R. Elliot and A. Gare (editors) (Pennsylvania State University Press: University Park, PA, 1983): 85–105.

Thompson, P.B. *The Spirit of the Soil* (Routledge: New York, 1995).

Thompson, P.B. *Food Biotechnology in Ethical Perspective* (Springer: AA Dordrecht, 2007).

Thompson, J.L. "A Refutation of Environmental Ethics", in Contemporary Moral Issues, L.M. Hinman (editor) (Prentice Hall: Upper Saddle River, NJ): 576–87.

Thoreau, H.D. "The Value of Wildness", in *American Environmentalism: Readings in Conservation History*, R.F. Nash (editor) (McGraw-Hill Publishing Company: New York, 1990): 36–9.

Tutwiler, M.A. and M. Straub. "Making Agricultural Trade Reform Work for the Poor" International Food & Agriculture Policy Council Position Paper, June 2005: 1–20.

UNAIDS/WHO – 2002, Joint United Nations Programme on HIV/AIDS (UNAIDS) and World Health Organization (WHO), *AIDS Epidemic Update December 2002* UNAIDS/02.46E, http://www.unaids.org/worldaidsday/2002/press/Epiupdate.html: 1–42.

United States Department of Agriculture (USDA). "Biotechnology Permits – 7 Code of Federal Regulations part 340" May 1997, http://www.aphis.usda.gov/biotech/7cfr340.html: 340.0-340.9

United States Department of Agriculture (USDA). "Japan Biotechnology: Update on Japan's Biotechnology Safety Approval and Labeling Policies" 29 August 2001, http://www.fas.usda.gov/gainfiles/200108/125681754.pdf

United States Department of Agriculture (USDA). "Tobacco: World Markets and Trade" Circular Series FT-09-05, September 2005, http://www.fas.usda.gov/tobacco/circular/2005/092005/TBL8.pdf

United States Department of Agriculture (USDA). "Biotechnology Risk Assessment Research", June 21, 2006, http://www.csrees.usda.gov/nea/biotech/in_focus/biotechnology_if_risk.html

United States Department of Agriculture (USDA). "NASA Principles for the Ethical Care and Use of Animals" Accessed 22 May 2007, http://warp.nal.usda.gov/awic/legislat/nasa.htm

United States Department of Agriculture, Animal and Plant Health Inspection Services (APHIS). Monsanto Company and Forage Genetics International Petition 04-110-01p for Determination

of Nonregulated Status for Roundup Ready Alfalfa Events J101 and J163: USDA/APHIS Environmental Assessment and Finding of No Significant Impact, May 2005: 1–29.
United States Department of Agriculture, National Agricultural Statistics Service (NASS). Acreage 30 June 2006, http://usda.mannlib.cornell.edu/usda/nass/Acre//2000s/2006/Acre-06-30-2006.pdf: 1–43.
United States Department of Commerce, Russian Chamber of Commerce and Industry, and U.S.-Russia Business Development Committee. Basic Guidelines for Codes of Business Conduct Accessed 30 May 2007. http://www.trade.gov/goodgovernance/adobe/basicguideng2.pdf
United States Office of Science and Technology Policy, Coordinated Framework for Regulation of Bioetchnology, 51 FR 23302, 26 June 1986.
United States Environmental Protection Agency (EPA) Public Law 104-170 – Aug. 3, 1996 111 STAT. 1489.
United States Environmental Protection Agency (EPA) "The Food Quality Protection Act (FQPA) Background" http://www.epa.gov/oppfead1/fqpa/backgrnd.htm
United States Food and Drug Administration (FDA) "Statement of Policy: Foods Derived From New Plant Varieties" Federal Register, Vol. 57, No. 104, 29 May 1992: 22984–3001.
United States Food and Drug Administration (FDA) "Guidance for Industry: Voluntary Labeling Indicating Whether Foods Have or Have Not Been Developed Using Bioengineering" Center for Food Safety and Applied Nutrition, January 2001, http://vm.cfsan.fda.gov/~dms/biolabgu.html: 1–8.
United States Geological Survey (USGS) "Dreissena polymorpha" http://nas.er.usgs.gov/zebra.mussel/docs/sp_account.html#HDR7
United States International Trade Commission. *Harmonized Tariff Schedule of the United States*, 2007, http://www.usitc.gov/tata/hts/bychapter/index.htm
United States Office of the United States Trade Representative (USTR) 2003 National Trade Estimate Report on Foreign Trade Barriers http://www.ustr.gov/reports/nte/2003/eu.pdf: 107–135.
United States Office of the United States Trade Representative (USTR) 2006 National Trade Estimate Report on Foreign Trade Barriers http://www.ustr.gov/assets/Document_Library/Reports_Publications/2006/2006_NTE_R eport/asset_upload_file929_9220.pdf
United States Supreme Court. J.E.M. AgSupply, Inc. V. Pioneer HiBredInternational, Inc (99-1996) 200 F.3d 1374. Thomas, C http://supct.law.cornell.edu/supct/html/99-1996.ZO.html: 1–11. Breyer, J., dissenting, http://supct.law.cornell.edu/supct/html/99-1996.ZD.html: 1–6.
Van den Eede, G., S. Kay, E. Anklam, and H. Schimmel. "Analytical challenges: bridging the gap from regulation to enforcement", *Journal of AOAC International*, 2002, 85(3): 757–61.
Velasquez, M.G. *Business Ethics: Concepts and Cases* (Prentice Hall: Upper Saddle River, NJ, 1998).
Verhoog, H. "The concept of intrinsic value and genetic animals", *Journal of Agriculture and Environmental Ethics*, 5, 1992: 147–60.
Verhoog, H. "Genetic Modification of Animals: Should Science and Ethics Be Integrated?" Accessed 22 May 2007, http://www.anth.org/ifgene/verhoog1.htm
Vidal, J. "GM genes found in human gut" The Guardian, 17 Wednesday 2002, http://www.guardian.co.uk/gmdebate/Story/0,2763,756666,00.html
Vogt, W. "The Global Perspective", in *American Environmentalism: Readings in Conservation History*, R.F. Nash (editor) (McGraw-Hill Publishing Company: New York, 1990): 166–70.
Volk, T. Gaia's Body (MIT Press: Cambridge, MA, 2003).
Wade, N. "Eek!; The Hidden Traps in Fooling Mother Nature" The New York Times, 5 September 1999, http://www.nytimes.com/library/review/090599sci-ge-review.html
Walgate, R. "French refusal to pull up GMO 'contaminated' maize splits government" The Scientist, 18 July 2000, http://www.biomedcentral.com/news/20000718/03/
Waller, H. "What the Competition Title in the Farm Bill Would Have Done and Not Have Done for the Nation's Farmers and Ranchers" AGRIBUSINESS EXAMINER, 18 November 2001, http://www.organicconsumers.org/corp/farmbill112601.cfm: 4–6.

Walliman, T. "Bt Toxin: Assessing GM Strategies", *Science*, 2000, 287: 41.

Warren, M.A. "The Rights of the Nonhuman World", in *Environmental Philosophy: A Collection of Readings*, R. Elliot and A. Gare (editors) (Pennsylvania State University Press: University Park, PA, 1983): 109–33.

Watanabe, M.E. "Studying SIV to Understand HIV: HIV hyperactivates, not suppresses, the immune system, causing CD4+T-cell depletion", The Scientist, 17(11): http://www.the-scientist.com/yr2003/jun/research1_030603.html: 1–3.

Wenegrat, B. "Belief in Unseen Beings: Its Evolutionary Basis and Its Effect on Morality", in *Evolution and Human Values*, R. Wesson and P.A. Williams (editors) (Rodopi: Atlanta, GA, 1995): 139–68.

Wenz, P. "Synergistic Environmental Virtues: Consumerism and Human Flourishing", in *Environmental Virtue Ethics*, R. Sandler and P. Cafaro (editors) (Rowman & Littlefield Publishers, Inc.: Boulder, CO, 2005): 197–213.

Westra, L. "A 'Transgenic Dinner'? Social and Ethical Issues in Biotechnology and Agriculture", *Journal of Social Philosophy*, 1993, 24(3): 215–32.

Whiteside, K.H. *Precautionary Politics* (The MIT Press: Cambridge, MA, 2006).

Williams, B. "Food Fight" The Raliegh News and Observer, 7 May 2000: 23A & 26A.

Winston, M.L. *Travels in the Genetically Modified Zone* (Harvard University Press: Cambridge, MA, 2002).

Wisner, R. and Western Organization of Resource Councils, "Potential Market Impacts for Commercializing Round-Up Ready Wheat: September 2006 Update, http://www.wroc.org/pdfs/Market%20Risks%20Update%20Final%208-06.pdf

Wolff, C. "Resolving Possible Conflicts Between Domestic and International Biotechnology Regulations and WTO Rules" ICTSD Workshop on Biotechnology, Biosafety and Trade: Issues for Developing Countries, Bellevue, Switzerland, International Centre for Trade and Sustainable Development, www.ictsd.org, 18-20 July 2001: 1–10.

Wood, H.W. Jr. "Modern Pantheism as an Approach to Environmental Ethics." Accessed 12/23/2003. http://www.pantheist.net/society/modern_panthesim_approach.html

World Medical Association, Declaration of Helsinki, October 2000, http://www.wma.net/e/policy/b3.htm

World Trade Organization "Agreement on the Application of Sanitary and Phytosanitary Measures." http://docsonline.wto.org/GEN_searchResult.asp?RN=0&searchtype=browse&q1=%28@meta_Symbol+LT%FCUR%FCA1-A%FC2%29+%26+%28@meta_Types+Legal+text%29, 15 April 1994a.

World Trade Organization "Agreement on Technical Barriers to Trade" http://docsonline.wto.org/GEN_searchResult.asp?RN=0&searchtype=browse&q1=%28@meta_Symbol+LT%FCUR%FCA-1A%FC10%29+%26+%28@meta_Types+Legal+text%29, 15 April 1994b.

World Trade Organization, EC-Approval and Marketing of Biotech Products, DS291, DS292, DS293, September 29, 2006a.

World Trade Organization "European Communities – Measures Affecting the Approval and Marketing of Biotech Products" WT/DS291/R, WT/DS292/R, WT/DS293/R, Documents Number 06-4318, 29 September 2006b.

World Trade Organization "Agreement on Agriculture" http://docsonline.wto.org/GEN_searchResult.asp?RN=0&searchtype=browse&q1=%28

World Trade Organization "General Agreement on Tariffs and Trade 1994" http://docsonline.wto.org/GEN_searchResult.asp?RN=0&searchtype=browse&q1=%28@meta_Symbol+LT%FCUR%FCA-1A%FC1%FCGATT%FC1++%29+%26+%28@meta_Types+Legal+text%29.

Yoon, C.K. "Few Federal Checks Exist on the Growing of Crops Whose Genes Are Altered" The New York Times, 3 November 1999, http://www.nytimes.com/library/national/science/110399sci-ge-squash.html

Yoon, C.K. "E.P.A. Announces New Rules on Genetically Altered Corn" The New York Times, 17 June 2000, http://www.nytimes.com/library/national/science/011700sci-ge-crop.html

Zemetra, R.S., C.A. Mallory-Smith, J. Hansen, Z. Wang, J. Snyder, A. Hang, L. Kroiss, O. Riera-Lizarazu, and I. Vales. "The Evolution of Biological Risk Program: Gene flow between Wheat

(Titicum aestivum L.) and Jointed Goatgrass (Aegilops cylindrica Host)" Gene Flow Workshop Proceedings, The Ohio State University, 5 and 6 March 2002: 169–78.

Zepeda, L. "Food Sovereignty or Trade Obstruction? Labelling of Genetically Engineered Foods", *Journal of International Biotechnology Law*, September 2004, 1(5): 200–4.

Zimdahl, R.L. *Agriculture's Ethical Horizon* (Academic Press: Boston, MA, 2006).

Zimmerman, A. "Summary of Current Lawsuits on the Frankenfood Front: Summary of Lawsuits & Litigation on GE Food Issues in the US and Canada" http://www.organicconsumers.org/ge/gelawsuits.cfm: 1–4.

Anonymous

"Allergies to Transgenic Foods – Questions of Policy" editorial, *New England Journal of Medicine*, 1996, 334(11), http://www.nejm.org/content/1996/0334/0011/0726.asp: 688–92.

"Eat Your Genes: How Safe are GM Crops?" *Nature*, 22 April 1999, http://www.nature.com/cgi-bin. . ...:36fd5973.def:36f84c90.def:36352b7b.def:3

Index

MIX
Papier aus verantwortungsvollen Quellen
Paper from responsible sources
FSC® C105338

If you have any concerns about our products,
you can contact us on
ProductSafety@springernature.com

In case Publisher is established outside the EU,
the EU authorized representative is:
Springer Nature Customer Service Center GmbH
Europaplatz 3, 69115 Heidelberg, Germany

Printed by Libri Plureos GmbH
in Hamburg, Germany